O. Borůvka

Foundations of the Theory of Groupoids and Groups

O. Borůvka

Foundations of the Theory of Groupoids and Groups

1975 Birkhäuser Verlag Basel and Stuttgart

O. Borůvka
Grundlagen der Gruppoid- und Gruppentheorie
VEB Deutscher Verlag der Wissenschaften
Berlin 1960
Translated from the German by Milada Borůvkova

This title is published by VEB Deutscher Verlag der Wissenschaften
in their series „Hochschulbücher für Mathematik“ volume 46

Licenced edition of
Birkhäuser Verlag Basel, 1975

ISBN 3-7643-0780-3

PREFACE TO THE ENGLISH EDITION

In the thirties the theory of sets started to exert considerable influence upon the majority of mathematical branches. This happened especially in case of algebraic disciplines where it finally led to the study of general algebras. At that time the author of this book consequently applied the set-theory in laying the foundation of the theory of groups. He started from the congruences on groups and the decompositions of the carrier of a group, which describe these congruences. Thus an extensive theory of decompositions of sets and factoroids on groupoids was originated. In outline it was developed during World-War II, at a time of scientific isolation. A natural consequence of this isolation is a remarkable originality of conception which remained devoted to employing decomposition of a set as the fundamental concept, whereas other authors treated similar problems by means of relations.

This book is not intended to be an introduction into the study of general algebras, not even of binary systems. Nevertheless it shows one of the ways to gain an idea of the elementary notions of binary systems. It is a clear way even if, owing to the nature of the subject, not always simple.

The object of the present book is, as it was already stated, the theory of decompositions of sets and its application to binary systems. It has the chief characteristics of originality and is unique in the literature. Thus it suggests topics which might yield a new approach to and aspect of the theory of general algebras.

It is in these properties that the actual value of Borůvka's book consists. We therefore believe that it is most helpful and worthy of merit on the part of the Publishing House VEB Deutscher Verlag der Wissenschaften to publish this new edition.

F. Šik and M. Sekanina
J. E. Purkyně University, Brno

AUTHOR'S REMARK

The present English edition of the German original: Grundlagen der Gruppoid- und Gruppentheorie (1960) differs from the latter only by some insignificant details in which it conforms to the Czech edition (1962).

Since the original work contains a number of new concepts for which adequate English terms had be found, the translation was not easy and I wish to thank Dr. M. Borůvková for the care with which she treated it. I also take this opportunity to thank Mrs. O. Fialová for invaluable technical help in preparing the manuscript and, last but not least, to the VEB Deutscher Verlag der Wissenschaften who have offered to publish an English translation of my work. It is the friendly help due to their indefatigable endeavour to foster further development of mathematical sciences which led to the final result.

Brno, April 1972 O. Borůvka

PREFACE

This book contains the foundations of the theory of groupoids and groups. A groupoid is a nonempty set on which there is defined a binary operation, called multiplication, associating with every two-membered sequence of elements of the set again an element of the latter. Generally there are no postulates as regards the multiplication. The concept of a groupoid forms the basis of an extensive theory of groupoids which, though rather general, considerably approximates the properties of groups. The theory of groupoids is founded on the theory of the decompositions in sets, on the one hand, and on the concept of homomorphic mapping, on the other hand.

The theory of decompositions in sets was founded by the author in about 1939, independently of the theory of equivalence relations developed at about the same time by P. Dubreil and M.-L. Dubreil-Jacotin (1937) and O. Ore (1942). Between the two theories there is no essential difference; the theory of decompositions in sets can, however, in certain cases be more conveniently applied because its basis, the concept of a decomposition in a set is purely of set-theoretical character and so less complicated than the concept of an equivalence relation. Since their origin, both the theories have been considerably developed and often applied to problems of various branches of mathematics. Employing the concepts and methods relative to sets and lattices, the theory of decompositions in sets describes situations occurring in connection with its basic concept. The decompositions in sets implied in the theory of groupoids are mostly of algebraic character, that is to say, are bound by certain relations with the multiplication, as—for example—the decompositions corresponding to homomorphic mappings. The theory of such decompositions is, in fact, the essence of the theory of groupoids. This is, naturally, true even for groups, which are groupoids with special properties of the multiplication.

The present book is based on two editions of my text-book "Introduction into the theory of groups" which met with most favourable criticism in the literature. It has, however, been largely extended and contains a number of genuine results due to the mentioned concept of the subject; the latter are, for the most part, closely connected with the classical theorems of the theory of groups. That applies, in particular, to the theory of series of decompositions in sets and their application to scientific classifications as well as to the corresponding algebraic theories of the series of factoroids and factor groups.

The book consists of three chapters of about the same length: I. Sets, II. Groupoids, III. Groups. The chapters are, so to say, simply mapped onto one another, since to the single situations concerning sets and dealt with in Chapter I there correspond, in Chapters II and III, analogous algebraic situations concerning groupoids and groups, respectively. This method of exposition seems particularly useful from the didactic point of view because the simple notions relative to sets take on more complicated forms in case of groupoids and groups; that leads to a better understanding of the structure of the concepts and methods of the algebraic theories in question and helps to formulate the most satisfactory proofs. The book also suggests many new ways of developing the mentioned theory and leads the reader to independent scientific work.

On this occasion I wish to thank my collaborators for invaluable help and advice in preparing the book, in particular, to Dr. M. Sekanina for carefully revising the manuscript, Dr. F. Šik for writing the Bibliography and Dr. M. Kolibiar for helping me to correct the proofs. I am also much obliged to the VEB Deutscher Verlag der Wissenschaften zu Berlin for their kind and correct cooperation.

Brno, August 1959 O. Borůvka

CONTENTS

I. SETS

1. Basic concepts

We shall first introduce some basic concepts from the theory of sets and found the following considerations on these.

1.1. The notion of a set

A *set* is a number of partikular things which are called the *elements* or *points* of the set. *Every set in uniquely determined by its elements.* Two sets consisting of the same elements are called *equal.*

All about us we can see examples of sets such as:

[1] the set consisting of the symbol a;

[2] the set consisting of all the words in this book;

[3] the set of all natural numbers.

In this book we shall often deal with sets of sets, that is to say, sets whose elements are again sets; for convenience we shall call them *systems of sets.*

In a system of sets there are elements of the system, namely sets, on the one hand, and elements of these sets, on the other hand. In such cases we generally use the terms elements of the systems and points of these elements.

A system of sets is, for example,

[4] the set whose elements are sets of natural numbers; one of these sets consists of all prime numbers 2, 3, 5, 7, 11, ..., another of all the products of two prime numbers, another of all the products of three prime numbers, etc.

1.2. Notation of sets

Sets will generally be denoted by Latin capitals, e.g. A, and the elements of sets by small Latin letters, e.g. a. But in case of systems of sets, both the systems and their elements would, by this rule, be denoted by Latin capitals; we shall therefore use the notation $\bar{A}, \bar{a}$; $\bar{B}, \bar{b}$, etc. for systems and their elements.

If a and b denote the same object, we say that a, b *are equal* and write $a = b$ or $b = a$. The opposite case, i.e., the inequality of a, b, is expressed by the formula $a \neq b$ or $b \neq a$. If the sets A, B consist of the same elements, then $A = B$ and, in the opposite case, $A \neq B$. If a is an element of A, we write $a \in A$.

If a set A consists of elements denoted $a, b, c, \ldots$, then we write $A = \{a, b, c, \ldots\}$. Thus $\{a\}$ and $\{1, 2, 3, \ldots\}$ are symbols of the above sets [1] and [3], respectively. Nevertheless, we shall not always stick to the chosen terminology literally but, if convenient, change it a little, if there is no danger of misunderstanding, of course. Instead of "the set A is the collection of the elements $a, b, c, \ldots$" we can say "the set A consists of the elements $a, b, c, \ldots$" or "the set A contains the elements $a, b, c, \ldots$ and no others"; instead of "a is an element of (in) the set A" we may say "a belongs to the set A", and similarly.

1.3. Further notions

As a set we also introduce the so-called *empty set*, characterized by the property that it has no elements. Since every set is uniquely determined by its elements, there exists only one empty set. We shall denote it by the symbol $\emptyset$. Later we shall see that the introduction of the empty set is useful as regards the formulation of our considerations in special cases.

Every set whose elements are certain symbols, e.g. letters, the meaning of which is not precisely determined, is called *abstract*; the above set [1], for instance, is abstract.

Every set consisting of a finite number of elements is called *finite*; in the opposite case it is *infinite*; e.g., the sets [1], [2] are finite, whereas [3], [4] are infinite.

By the *order* of a finite nonempty set we mean the number of its elements. The set [1], for example, has the order 1. It will also be useful to assign, to every infinite set, the order 0. The empty set has no order.

1.4. Subsets and supersets

Suppose A, B are sets. If each element of A isi simultaneously an element og B, then we say that A is a *subset of* B or that B is a *superset of* A. We can also say that A is a *part of* B or that B *contains* A. Then we write $A \subset B$ or $B \supset A$, respectively. The empty set is considered to be a part of any set; in particular, $\emptyset \subset \emptyset$.

If $A \subset B$, then B may (but need not) contain elements that do not belong to A. If B includes at least one element that does not belong to A, then A is said to be a *proper subset* of B and B a *proper superset* of A. In the opposite case, A (B) is a *non-proper* subset (superset) of B (A) and we see that it equals B (A): $A = B$ ($B = A$).[1])

The set of all prime numbers, for example, is a proper subset of the set [3], for every prime number is an element of the set [3] and the latter also contains numbers that are not prime, e.g. 4. If A is a non-proper subset of B, then each element of A is an element of B and, at the same time, each element of B is also an element of A; that is to say, there simultaneously holds $A \subset B$ and $B \subset A$. It is clear that both these relations together express the equality $A = B$.

It is easy to see that each subset of B is either proper or equal to B. Note that the equality $A = B$ is equivalent to the relations $A \subset B, B \subset A$ in the sense that, if $A = B$, then $A \subset B$, $B \subset A$ and vice versa. Generally we can tell whether two sets are equal just by verifying that either of them is a subset of the other.

1.5. The sum (union) of sets

By the *sum* or *union of the set A and the set B* we understand the set of all elements that belong to A or to B.

Since this definition determines all the elements that belong to the sum of A and B and every set is uniquely determined by its elements, there exists only one sum of A and B, denoted $A \cup B$. From the above definition there follows: $A \cup B = B \cup A$. Therefore we generally speak about the *sum of A and B* regardless of whether we mean the sum of A and B or B and A. We observe that the sum of A and B is the set of all the elements that belong to, at least, one of them. Either of the set A, B is a subset of $A \cup B$, for each element of, e.g., A belongs, at least, to one of the sets A, B, namely to A; so we can write: $A \subset A \cup B$, $B \subset A \cup B$. The sum of the set of all positive even numbers and the set of all positive odd numbers, for example, is the set [3] because there holds

$$\{2, 4, 6, \ldots\} \cup \{1, 3, 5, \ldots\} = \{1, 2, 3, \ldots\}.$$

The sum of the set consisting of a single word *and* and the set [2] is again the set [2].

The notion of the sum of two sets can easily be extended to the sum of systems of sets: by the *sum* or *union of anys ystem of sets*, $\bar{A}$, we mean the set of all the points belonging to, at least, one of the sets that are elements of $\bar{A}$.

---

1) In this form we express: In the opposite case A is a non-proper subset of B an dwe see that it equals the set B: $A = B$; simultaneously, B is a non-proper superset of A and we see that it equals the set A: $B = A$.

A similar abbreviated form of expression will often be used throughout the book.

There holds, again, that the system $\bar{A}$ has exactly one sum and that every set which is an element of $\bar{A}$ is a subset of the sum of $\bar{A}$. The sum of $\bar{A}$ is generally denoted by $\mathbf{s}\bar{A}$; if the elements of $\bar{A}$ are denoted by $\bar{a}_1, \bar{a}_2, \ldots$, then the sum of $\bar{A}$ is denoted by $\bar{a}_1 \cup \bar{a}_2 \cup \ldots$, briefly $\bigcup \bar{a}$ or similarly, which is clear from the context.

1.6. The intersection of sets. Incident and disjoint sets

The *intersection of the set A and the set B* is the set of all the elements that belong to A as well as to B.

In a similar way as in the case of the sum, we can verify that there exists only one intersection of A and B; let us denote it by $A \cap B$. Moreover, we observe that $A \cap B = B \cap A$. So we generally speak about the *intersection of A and B* without paying any attention to whether we mean the intersection of A and B or that of B and A. It is obvious that the intersection of A and B is the set of all the elements that belong to both A and B. The intersection $A \cap B$ is a part of either A and B, for each element of $A \cap B$ belongs, e.g., to A. Note that, even if A and B have no common elements, the definition of the intersection of A and B applies because, in that case, $A \cap B$ is the empty set. And we realize that the notion of the empty set is of advantage; without it we could only speak about intersection in case of certain sets. Nevertheless, it is convenient to have special terms for sets that have common elements and for those that have not.

If A and B have common elements, they are called *incident* and A (B) is said to be *incident with* B (A). In the opposite case A and B are called *disjoint*. In the first case there holds: $A \cap B \neq \emptyset$, in the second: $A \cap B = \emptyset$.

Examples: The set consisting of the single word *and* and the set [2] are incident; their intersection is the former set. The set of all even natural numbers and the set of all odd natural numbers are disjoint; their intersection is obviously $\emptyset$.

The notion of the intersection of two sets can be extended to the intersection of a system of sets: The *intersection of any system of sets*, $\bar{A}$, is the set of all the points belonging to each of the sets that are elements of $\bar{A}$.

There again holds that $\bar{A}$ has exactly one intersection which is a subset of each element of $\bar{A}$. The intersection of $\bar{A}$ is denoted by $\boldsymbol{p}\bar{A}$; if the elements of $\bar{A}$ are denoted by $\bar{a}_1, \bar{a}_2, \ldots$, we write $\bar{a}_1 \cap \bar{a}_2 \cap \ldots$, briefly $\bigcap \bar{a}$, or similarly.

1.7. Sequences

By a *sequence on a* (non-empty) *set A*, briefly: a sequence, we mean the set A whose elements are numbered. Exactly one element is marked as the first, exactly one as the second, etc., each element of A being marked at least once. The element marked by the (natural) number γ is called the *γ-th member of* the sequence or the *member with index* γ or the *member of the rank* γ. The rank of a member is generally expressed by the adequate index; e.g., $a_1, a_2, \ldots$. Two different members of a sequence, for instance, a_1, a_2, may be the same element of A numbered once by 1 and another time by 2.

If the last member of a sequence is $\bar{a}_\alpha$, then the sequence is called *finite* or, more precisely, *α-membered* and α is its *length*. In that case there corresponds, to each number $\gamma = 1, 2, \ldots, \alpha$ exactly one member a_γ of the rank γ but the sequence does not comprise any members of a rank higher than α. Accordingly, such a sequence is denoted $(a_\gamma)_{\gamma=1}^{\alpha}$ or $(a_1, \ldots, a_\alpha)$ or in a similar way. If a sequence has no last member, we say that it is *infinite* or that its length is infinite. In an infinite sequence there corresponds, to each positive integer γ, precisely one member of the rank γ; notation $(a_\gamma)_{\gamma=1}^{\infty}$, $(a_1, a_2, \ldots)$, and similarly. If a sequence contains a finite number of different elements, then it is either finite or infinite; in the opposite case it is infinite.

Let the sequence $(a) = (a_1, a_1, \ldots)$ be either finite or infinite. Every sequence $(a_1', a_2', \ldots)$ generated from (a) by omitting some members a_γ is called a *partial sequence* or a *part of* (a). The sequence (a) is considered to be a part of itself. A partial sequence $(a_1, \ldots, a_\gamma)$ consisting of the first γ members of (a) is called the *γ-th main partial sequence* or the *γ-th main part of* (a); γ denotes a positive integer which, of course, in an α-membered sequence is not higher than α. If (a) consists of α members, then its main part $(a_1, \ldots, a_\gamma)$ has, for $\gamma = 1, \ldots, \alpha - 1$, exactly one *successor*, namely $(a_1, \ldots, a_\gamma, a_{\gamma+1})$; then the α-th main part of (a), naturally, coincides with (a). If (a) is infinite, then each of its main parts has exactly one successor.

The sequences $(a) = (a_1, a_2, \ldots)$, $(b) = (b_1, b_2, \ldots)$ are considered equal if and only if they have the same length and their members with the same indices are equal elements: $(a) = (b)$ means $a_1 = b_1$, $a_2 = b_2, \ldots$.

Now let us consider the above notions in case of sets of sequences.

Suppose $\mathcal{A}$ is a nonempty set consisting of finite, e.g., α-membered sequences. The main parts of the elements of $\mathcal{A}$, of length γ, where $1 \leqq \gamma \leqq \alpha$, form a nonempty set called the *γ-th set of the main parts belonging to $\mathcal{A}$*; notation: $\mathcal{A}_\gamma$. To the set $\mathcal{A}$ therefore belong the sets $\mathcal{A}_1, \ldots, \mathcal{A}_\alpha$; $\mathcal{A}_\alpha$, naturally, coincides with $\mathcal{A}$, i.e., $\mathcal{A}_\alpha = \mathcal{A}$. Furthermore, in case of $\gamma < \alpha$, there corresponds to each element $a^{(\gamma)} \in \mathcal{A}_\gamma$ a nonempty set of the successors of the element $a^{(\gamma)}$. This set consists of the main parts of all the elements of $\mathcal{A}$, of length $\gamma + 1$, beginning with $a^{(\gamma)}$; notation, e.g., $N(a^{(\gamma)})$. There evidently holds: $N(a^{(\gamma)}) \subset \mathcal{A}_{\gamma+1}$.

A remarkable example of a set consisting of α-membered sequences is an arbitrary nonempty set of points in an α-dimensional coordinate space. In that case, any point a is identical with a certain α-membered sequence $(a_1, \ldots, a_\alpha)$, the coordinates $a_1, \ldots, a_\alpha$ being real or complex numbers. The γ-th main part $(a_1, \ldots, a_\gamma)$, where $1 \leqq \gamma < \alpha$, is the "projection" of the point a into the γ-dimensional space given by the equations $a_{\gamma+1} = \cdots = a_\alpha = 0$.

1.8. The Cartesian product of sets. Cartesian powers

The *Cartesian product of the set A and the set B* is the set of all ordered pairs (a, b) such that a and b are elements of A and B, respectively. If either of the sets A, B is empty, then the Cartesian product of A and B is defined as the empty set.

This definition determines exactly one Cartesian product of the sets A and B, denoted by $A \times B$. A (B) is the first (second) *factor* of the Cartesian product $A \times B$ and a (b) is the first (second) *coordinate* of its element (a, b). From the above definition we see that, generally, $A \times B \neq B \times A$.

The *Cartesian second power* or the *Cartesian square of the set A* is the Cartesian product $A \times A$. The latter is, therefore, the set of all ordered pairs (a, b) such that a, b are elements of A. If the set A is empty, then the same applies to the Cartesian square $A \times A$. For example, the Cartesian square of the set [3] is the set of all ordered pairs formed by two equal or different positive integers; consequently, this Cartesian square is a set of points in the plane, both coordinates of which are positive integers.

The extension of the notion of the Cartesian product to more than two factors as well as the notion of the Cartesian second power of a set to higher Cartesian powers is easy and will be left to the reader. These, more general, notions will be omitted here, since we shall not make any use of them in the following considerations.

Note that the Cartesian products belong to sets consisting of finite sequences of elements.

1.9. α-grade structures

In the following study we shall come across even more complicated figures based on the concept of a set, in particular, the so-called α-grade structures.

Let α $(\geqq 1)$ be a positive integer and $(A) = (A_1, \ldots, A_\alpha)$ a sequence of nonempty sets.

An *α-grade set structure with regard to the sequence* (A), briefly, an *α-grade structure* is a nonempty set $\tilde{A}$ of the following form: Each element $\bar{\bar{a}} \in \tilde{A}$ is an α-membered sequence $(\bar{a}) = (\bar{a}_1, \ldots, \bar{a}_\alpha)$ such that each of its members $\bar{a}_\gamma$ is a nonempty part of the set A_γ; $\gamma = 1, 2, \ldots, \alpha$.

We shall, in particular, meet with the case when $A_1, \ldots, A_\alpha$ consist of nonempty sets, so that every set A_γ is a system $\bar{A}_\gamma$ of nonempty sets; $1 \leqq \gamma \leqq \alpha$. Such α-grade structures $\tilde{A}$ are therefore of the following form: Every element $\bar{\bar{a}} \in \tilde{A}$ is an α-membered sequence, $(\bar{a}) = (\bar{a}_1, \ldots, \bar{a}_\alpha)$, each member $\bar{a}_\gamma$ of the latter being a nonempty sub-system of $\bar{A}_\gamma$; $\gamma = 1, 2, \ldots, \alpha$.

1.10. Exercises

1. $A \cup \emptyset = A$; $A \cup A = A$; $A \cap \emptyset = \emptyset$; $A \cap A = A$.
2. $A \cup (A \cap B) = A$; $A \cap (A \cup B) = A$.
3. If $A \subset B$, then $A \cup B = B$, $A \cap B = A$; conversely, if either of these equalities is correct, then $A \subset B$.
4. $(A \cup B) \cup C = A \cup (B \cup C)$; $(A \cap B) \cap C = A \cap (B \cap C)$.
5. $(A \cup B) \cap C = (A \cap C) \cup (B \cap C)$; $(A \cap B) \cup C = (A \cup C) \cap (B \cup C)$.
6. A set of a finite number n ($\geqq 0$) of elements has 2^n subsets.
7. The Cartesian product of a set of m ($\geqq 0$) elements and a set of n ($\geqq 0$) elements consists of $m \,.\, n$ elements.
8. A part of the Cartesian product $A \times B$ is not necessarily the Cartesian product of a subset of A and a subset of B.

2. Decompositions (partitions) in sets

2.1. Decompositions in a set

Let G stand, throughout the book, for an arbitrary nonempty set.

A *decomposition* (*partition*) *in* G is a nonempty system of nonempty and mutually disjoint subsets of G.

This notion is one of the most important in this book and is, in fact, essential to the theory of groupoids and groups we intend to develop in the following chapters.

Every decomposition in G has therefore at least one element, each of its elements is a nonempty subset of G and, of course, the intersection of any two of its elements is empty.

A simple example of a decomposition in, let us say, the set of all positive integers is the system consisting of one single element, namely of the set of all even positive integers. More generally: the system consisting of a single element which is a nonempty subset of G is a decomposition in G. The system of sets [4] in part 1.1 is an example of a decomposition in the set of all positive integers $\geqq 2$.

2.2. Decompositions on a set

Let $\bar{A}$ be an arbitary decomposition in G. Any point of G may lie at most in one element of $\bar{A}$, since every two elements of $\bar{A}$ are disjoint; it may, of course, happen that it does not lie in any element of $\bar{A}$.

If the decomposition $\bar{A}$ is such that each point of G lies in some element of $\bar{A}$ so that $\bar{A}$ covers G, then $\bar{A}$ is called a decomposition *on* G or a decomposition *of* G. The last of the above examples is a decomposition on the set of all positive integers $\geqq 2$; in fact, every natural number $\geqq 2$ is either a prime number or a product of several prime numbers and therefore lies in some element of this decomposition.

Important examples of decompositions on G are the two so-called *extreme decompositions of* G, namely the *greatest* and the *least decomposition of* G. The greatest decomposition of G, denoted by $\bar{G}_{\max}$, consists of a single element, G. The least decomposition of G, $\bar{G}_{\min}$, is the system of all one-point sets $\{a\}$, $a \in G$.

For example, the set whose only element consists of all positive integers is the greatest decomposition of the set of all positive integers, whereas the system of all one-point sets each of which consists of a single positive integer is its least decomposition.

Note that an arbitrary decomposition $\bar{A}$ *in* G is a decomposition *on* the set $\mathbf{s}\bar{A}$. Consequently, the results concerning the properties of decompositions on sets can often be employed to describe the characteristics of decompositions in sets.

2.3. Closures and intersections

Let $\bar{A}$ stand for an arbitrary decomposition in G and B for a subset of G.

The *closure of the subset B in the decomposition $\bar{A}$* is the set of all the elements of $\bar{A}$ that are incident with B. Notation: $B \sqsubset \bar{A}$ or $\bar{A} \sqsupset B$. Since every element $\bar{a} \in \bar{A}$ that is incident with B simultaneously cuts the subset $B \cap \mathbf{s}\bar{A}$ and conversely, there holds $B \sqsubset \bar{A} = (B \cap \mathbf{s}\bar{A}) \sqsubset \bar{A}$, We observe that the closure $B \sqsubset \bar{A}$ is a part of the decomposition $\bar{A}$ which may coincide with $\bar{A}$ or may even be empty. The first case, $B \sqsubset \bar{A} = \bar{A}$, occurs if and only if each element of $\bar{A}$ is incident with B. The second case, $B \sqsubset \bar{A} = \emptyset$, occurs if and only if no element of $\bar{A}$ is incident with B; it is characterized by the equality $B \cap \mathbf{s}\bar{A} = \emptyset$. When the closure $B \sqsubset \bar{A}$ is not empty, it is a decomposition in G.

The *intersection of the decomposition $\bar{A}$ and the subset B* or *of the subset B and the decomposition $\bar{A}$* is the set of all nonempty intersections of the elements of $\bar{A}$ and the subset B. Notation: $\bar{A} \sqcap B$ or $B \sqcap \bar{A}$. It is obvious that even the intersection $\bar{A} \sqcap B$ may be empty. That occurs if and only if no element of $\bar{A}$ is incident with B; this case is, as we have mentioned above, characterized by the equality $B \cap \mathbf{s}\bar{A} = \emptyset$. If the intersection $\bar{A} \sqcap B$ is not empty, it is a decomposition in the set G and even in the set B. Note that $A \sqcap \bar{B}$ is, at the same time, a decomposition on the set $B \cap \mathbf{s}\bar{A}$. There obviously holds: $\mathbf{s}(\bar{A} \sqcap B) = B \cap \mathbf{s}\bar{A}$.

To sum up: The closure $B \sqsubset \bar{A}$ and the intersection $B \sqcap \bar{A}$ are simultaneously either nonempty or empty systems of sets according as there holds $B \cap \mathbf{s}\bar{A} \neq \emptyset$ or $B \cap \mathbf{s}\bar{A} = \emptyset$. If $B \neq \emptyset$ and $\bar{A}$ covers G, then $B \sqsubset \bar{A}$ and $B \sqcap \bar{A}$ are nonempty systems, the former being a part of $\bar{A}$ and the latter a decomposition on B. Every decomposition

$\bar{A}$ on G and a nonempty set B in G therefore determine: 1) a certain nonempty subset of $\bar{A}$, namely, the closure $B \sqsubset \bar{A}$, 2) a certain decomposition on B, namely, the intersection $A \sqcap \bar{B}$.

The notions of a closure and an intersection described above will now be extended to the case when the subset $B \subset G$ is replaced by a decomposition in G. Thus it will be a question of the closure of a decomposition in a decomposition and of the intersection of two decompositions.

Let $\bar{A}$, $\bar{B}$ denote decompositions in G.

The *closure of the decomposition* $\bar{B}$ *in* $\bar{A}$ is the set of all elements $\bar{a} \in \bar{A}$ such that $\bar{a}$ is incident with some element of $\bar{B}$. Notation: $\bar{B} \sqsubset \bar{A}$ or $\bar{A} \sqsupset \bar{B}$. Since every element $\bar{a} \in \bar{A}$ incident with an element of $\bar{B}$ is simultaneously incident with $\mathbf{s}\bar{B}$, and vice versa, there holds: $\bar{B} \sqsubset \bar{A} = \mathbf{s}\bar{B} \sqsubset \bar{A}$. This relation reduces the new concept of a closure to the notion of the closure of a subset in a decomposition. If the decomposition $\bar{B}$ consists of one single element B, then of course: $\bar{B} \sqsubset \bar{A} = B \sqsubset \bar{A}$.

The *intersection of the decomposition* $\bar{A}$ *and the decomposition* $\bar{B}$ is the set of all nonempty intersections of the individual elements of $\bar{A}$ and the elements of $\bar{B}$. Notation: $\bar{A} \sqcap \bar{B}$. It is obvious that $\bar{A} \sqcap \bar{B} = \bar{B} \sqcap \bar{A}$; with respect to this symmetry, we also speak about the intersection of the decompositions $\bar{A}$ and $\bar{B}$. Since, for every two elements $\bar{a} \in \bar{A}, \bar{b} \in \bar{B}$, there holds $\bar{a} \cap \bar{b} = (\bar{a} \cap \mathbf{s}\bar{B}) \cap (\bar{b} \cap \mathbf{s}\bar{A})$, we have: $\bar{A} \sqcap \bar{B} = (\bar{A} \sqcap \mathbf{s}\bar{B}) \sqcap (\bar{B} \sqcap \mathbf{s}\bar{A})$. Either system $\bar{A} \sqcap \mathbf{s}\bar{B}$, $\bar{B} \sqcap \mathbf{s}\bar{A}$ is a decomposition on $\mathbf{s}\bar{A} \cap \mathbf{s}\bar{B}$ or is empty according as there holds $\mathbf{s}\bar{A} \cap \mathbf{s}\bar{B} \neq \emptyset$ or $= \emptyset$. We see that, in case of $\mathbf{s}\bar{A} \cap \mathbf{s}\bar{B} \neq \emptyset$, the intersection of the decompositions $\bar{A}$ and $\bar{B}$ coincides with the intersection of the decompositions $\bar{A} \sqcap \mathbf{s}\bar{B}$ and $\bar{B} \sqcap \mathbf{s}\bar{A}$ which are on $\mathbf{s}\bar{A} \cap \mathbf{s}\bar{B}$, whereas, in case of $\mathbf{s}\bar{A} \cap \mathbf{s}\bar{B} = \emptyset$, it is empty. Note that the intersection of any two decompositions lying on the same set is always a decomposition of the latter. If $\bar{B}$ consists of one single element B, then of course, $\bar{A} \sqcap \bar{B} = \bar{A} \sqcap B$.

2.4. Coverings and refinements of a decomposition

Let $\bar{A}$, $\bar{B}$, $\bar{C}$ be decompositions in G.

The decomposition $\bar{A}$ ($\bar{B}$) is called a *covering* (*refinement*) *of the decomposition* $\bar{B}$ ($\bar{A}$) if every element of $\bar{B}$ is a part of some element of $\bar{A}$. Then we write $\bar{A} \geqq \bar{B}$ or $\bar{B} \leqq \bar{A}$. For example, the greatest decomposition of G is a covering of $\bar{A}$ and the least decomposition of $\mathbf{s}\bar{A}$ is a refinement of $\bar{A}$: $\bar{G}_{\max} \geqq \bar{A}$, $\mathbf{s}\bar{A}_{\min} \leqq \bar{A}$. In particular, ($A = B$), the decomposition $\bar{A}$ is its own covering and refinement. If $\bar{A} \geqq \bar{B}$ and, at the same time, $\bar{A} \neq \bar{B}$, then $\bar{A}$ ($\bar{B}$) is said to be a *proper covering* (*refinement*) *of* $\bar{B}$ ($\bar{A}$). Notation: $\bar{A} > \bar{B}$ or $\bar{B} < \bar{A}$.

From the definition of the meaning of the symbol $\geqq$ *there follows*:

a) $\bar{A} \geqq \bar{A}$;
b) *if* $\bar{A} \geqq \bar{B}$, $\bar{B} \geqq \bar{C}$, *then* $\bar{A} \geqq \bar{C}$;
c) *if* $\bar{A} \geqq \bar{B}$, $\bar{B} \geqq \bar{A}$, *then* $A = \bar{B}$.

The statements a) and b) are evidently true. To prove c), we proceed as follows: Suppose there holds $\bar{A} \geqq \bar{B}$, $\bar{B} \geqq \bar{A}$. Let $\bar{b} \in \bar{B}$ stand for an arbitrary element. Then there exist elements $\bar{a} \in \bar{A}$, $\bar{b}' \in \bar{B}$ such that $\bar{b}' \supset \bar{a} \supset \bar{b}$. Consequently, the sets $\bar{b}'$, $\bar{b}$ are incident and therefore, as they are elements of the same decomposition $\bar{B}$, identical. So we have $\bar{b}' = \bar{a} = \bar{b}$ and, in fact, $\bar{b} \in \bar{A}$. Accordingly, $\bar{B}$ is a part of $\bar{A}$, i.e., $\bar{B} \subset \bar{A}$ and, analogously, $\bar{A}$ is a part of $\bar{B}$, $\bar{A} \subset \bar{B}$. Hence $\bar{A} = \bar{B}$.

If $\bar{A} \geqq \bar{B}$, then every element of $\bar{B}$ is a subset of some element of $\bar{A}$; but $\bar{A}$ may contain elements in which there does not lie any element of $\bar{B}$. If such elements of $\bar{A}$ do not exist, i.e., if every element of $\bar{A}$ contains, as a subset, some element of $\bar{B}$, then the decomposition $\bar{A}$ ($\bar{B}$) is said to be a *normal covering* (*refinement*) *of* $\bar{B}$ ($\bar{A}$); if, moreover, every element of $\bar{A}$ is the sum of some elements of $\bar{B}$, then $\bar{A}$ ($\bar{B}$) is called a *pure covering* (*refinement*) *of* $\bar{B}$ ($\bar{A}$).

The decomposition $\bar{A}$ ($\bar{B}$) is a pure covering (pure refinement) of $\bar{B}$ ($\bar{A}$) only if both the decompositions $\bar{A}$, $\bar{B}$ lie on the same set $\mathbf{s}\bar{A} = \mathbf{s}\bar{B}$; conversely, the relations $\bar{A} \geqq \bar{B}$, $\mathbf{s}\bar{A} = \mathbf{s}\bar{B}$ express that $\bar{A}$ ($\bar{B}$) is a pure covering (pure refinement) of $\bar{B}$ ($\bar{A}$). Note that *if* $\bar{A}$, $\bar{B}$ *lie on* G *and* $\bar{A} \geqq \bar{B}$, *then* $\bar{A}$ ($\bar{B}$) *is a pure covering* (*pure refinement*) *of* $\bar{B}$ ($\bar{A}$).

Let us now assume the decomposition $\bar{A}$ ($\bar{B}$) to be a pure covering (pure refinement) of $\bar{B}$ ($\bar{A}$), hence $\bar{A} \geqq \bar{B}$, $\mathbf{s}\bar{A} = \mathbf{s}\bar{B}$. Then every element $\bar{a} \in \bar{A}$ is the sum of some elements of $\bar{B}$ and, evidently, the system of these element is a decomposition of $\bar{a}$. It is also clear that the system of all subsets of $\bar{B}$, each of which consists of all the elements of $\bar{B}$ that are parts of the same element of $\bar{A}$, is a certain decomposition $\bar{\bar{B}}$ of $\bar{B}$. The decomposition $\bar{\bar{B}}$, on the other hand, determines the decomposition $\bar{A}$; the latter is formed by summing all the elements of $\bar{B}$ that lie in the same element of $\bar{\bar{B}}$. The decomposition $\bar{\bar{B}}$ is said to *enforce* the covering $\bar{A}$. So we can say that $\bar{B}$ is obtained from $\bar{A}$ if every element of $\bar{A}$ is replaced by its own convenient decomposition; conversely, $\bar{A}$ is obtained from $\bar{B}$ by choosing a suitable decomposition of $\bar{B}$, i.e. $\bar{\bar{B}}$, and summing all the elements of $\bar{B}$ that lie in the same element of $\bar{\bar{B}}$.

2.5. Chains of decompositions

Let $A \supset B$ be nonempty subsets of G.

A *chain of decompositions from* A *to* B *in* G, briefly: a *chain from* A *to* B, is an α-membered ($\alpha \geqq 1$) sequence of decompositions $\bar{K}_1, \ldots, \bar{K}_\alpha$ in G with the following properties: 1) $\bar{K}_1$ lies on A; 2) $\bar{K}_{\gamma+1}$ lies on an element of $\bar{K}_\gamma$ for $1 \leqq \gamma \leqq \alpha - 1$; 3) $B \in \bar{K}_\alpha$. Notation:

$$\bar{K}_1 \to \ldots \to \bar{K}_\alpha, \text{ briefly, } [\bar{K}].$$

The set $\mathbf{s}\bar{K}_\gamma$ is, for $1 \leqq \gamma \leqq \alpha$, denoted by $\bar{a}_\gamma$; so we have, in particular, $\bar{a}_1 = A$. Moreover, we put $\bar{a}_{\alpha+1} = B$. From the definition of a chain, we have $\bar{a}_2 \in \bar{K}_1, \ldots,$

$\bar{a}_{\alpha+1} \in \bar{K}_\alpha$ and, furthermore: $A = \bar{a}_1 \supset \cdots \supset \bar{a}_{\alpha+1} = B$. The sets A, B are called the *ends of the chain* $[\bar{K}]$. We see that any element of the decomposition $\bar{K}_\alpha$ may be an end of $[\bar{K}]$. $\bar{K}_1, \ldots \bar{K}_\alpha$ are called *members of the chain* $[\bar{K}]$; $\bar{K}_1$ and $\bar{K}_\alpha$ are the *initial* and the *final member of* $[\bar{K}]$, respectively. By the *length of* $[\bar{K}]$ we understand the number α of the members of $[\bar{K}]$.

An important type of chains are the so-called elementary chains over a decomposition.

Let $\bar{A}$ be a decomposition in G and B an element of $\bar{A}$ so that $B \in \bar{A}$. Denote $\bar{A} = \mathbf{s}\bar{A}$.

An *elementary chain of decompositions from A to B over $\bar{A}$*, briefly: an *elementary chain over $\bar{A}$*, is a chain of decompositions from A to B,

$$([\mathring{K}] =)\quad \mathring{K}_1 \to \cdots \to \mathring{K}_\alpha$$

such that, for $1 \leqq \gamma \leqq \alpha$, the member $\mathring{K}_\gamma$ is a covering of the decomposition $\bar{A}_\gamma = \bar{A} \sqcap \bar{a}_\gamma$, where $\bar{a}_\gamma = \mathbf{s}\mathring{K}_\gamma$ $(\bar{a}_1 = A)$.

In such a chain, first of all, the member $\mathring{K}_1$ is a covering of the decomposition $\bar{A}_1$ $(= \bar{A})$. The set $\bar{a}_2$, determined by $B \subset \bar{a}_2 \in \mathring{K}_1$, is a subset of $\bar{a}_1$ and, on $\bar{a}_2$, there is the decomposition $(\bar{A}_2 =)\ \bar{A} \sqcap \bar{a}_2$. There holds: $B \in \bar{A}_2 \subset \bar{A}_1$. Furthermore, $\mathring{K}_2$ is a covering of $\bar{A}_2$. The set $\bar{a}_3$, determined by $B \subset \bar{a}_3 \in \mathring{K}_2$, is a subset of $\bar{a}_2$ and, on $\bar{a}_3$, there is the decomposition $(\bar{A}_3 =)\ \bar{A} \sqcap \bar{a}_3$. There holds: $B \in \bar{A}_3 \subset \bar{A}_2$. Next, $\mathring{K}_3$ is a covering of $\bar{A}_3$, etc. Finally, the member $\mathring{K}_\alpha$ is a covering of $(\bar{A}_\alpha =)\ \bar{A} \sqcap \bar{a}_\alpha$ and there holds $B \in \mathring{K}_\alpha$.

For example, the chain consisting of a single decomposition $\bar{A}$ is an elementary chain from A to B over $\bar{A}$, of length 1. Putting, before or after $\bar{A}$, an arbitrary finite number of greatest decompositions of A or of B, respectively, we again obtain an elementary chain of decompositions from A to B over $\bar{A}$.

Let us now consider a chain $([\bar{K}] =)\ \bar{K}_1 \to \cdots \to \bar{K}_\alpha$ from A to B using the above notation.

If $\bar{K}_\gamma$ $(\gamma = 1, 2, \ldots, \alpha)$ is not the greatest decomposition on $\bar{a}_\gamma$, i.e., if $\bar{a}_{\gamma+1}$ is a proper subset of $\bar{a}_\gamma$, then $\bar{K}_\gamma$ is called an *essential member of the chain* $[\bar{K}]$. In the opposite case, $\bar{K}_\gamma$ is an *inessential member of* $[\bar{K}]$. If the chain $[\bar{K}]$ contains at least one inessential member $\bar{K}_\gamma$, then $[\bar{K}]$ is called a *chain with iteration*, since $\bar{a}_{\gamma+1} = \bar{a}_\gamma$. If all the members of $[\bar{K}]$ are essential, then $[\bar{K}]$ is said to be a *chain without iteration*. The number α' of essential members of $[K]$ is the so-called *reduced length of* $[\bar{K}]$. There evidently holds $0 \leqq \alpha' \leqq \alpha$; the equality $\alpha' = \alpha$ is characteristic of a chain without iteration. If $A = B$, then all the members of $[\bar{K}]$ are inessential so that $\alpha' = 0$ and conversely. If $A \neq B$, then $[\bar{K}]$ may, by omitting all its inessential members, be *reduced*, i.e., shortened to a certain chain $[\bar{K}']$ without iteration. The length of the reduced chain $[\bar{K}']$ equals the reduced length α' of $[\bar{K}]$. The chain $[\bar{K}]$ may, on the other hand, be *lengthened* by way of inserting, between arbitrary members $\bar{K}_\gamma, \bar{K}_{\gamma+1}$ or before the initial member $\bar{K}_1$ (after the final member $\bar{K}_\alpha$) of $[\bar{K}]$, the greatest decomposition of the set $\bar{a}_{\gamma+1}$, or $\bar{a}_1$ $(\bar{a}_{\alpha+1})$ or an arbitrary finite member of such decompositions. Every shortening or lengthening of the chain $[\bar{K}]$ may be realized by gradually omitting or adding, respectively, one greatest decomposition lying on some of the subsets $\bar{a}_1, \ldots, \bar{a}_{\alpha+1}$. It is clear that every chain formed by shortening or lengthening of $[\bar{K}]$ has the same reduced length as $[\bar{K}]$.

A *refinement* $[\mathring{\bar{K}}]$ *of* $[\bar{K}]$ is a chain of decompositions in G, with arbitrary ends, A_0, B_0, satisfying the relations $A_0 \supset A \supset B \supset B_0$, i.e., a chain of the following type:

$$\begin{aligned} \bar{K}_{0,0} &\to \bar{K}_{1,1} \to \cdots \to \bar{K}_{1,\beta_1-1} \to \bar{K}_{1,\beta_1} \to \bar{K}_{2.1} \\ &\to \cdots \to \bar{K}_{2,\beta_2-1} \to \bar{K}_{2,\beta_2} \to \cdots \to \bar{K}_{\alpha,\beta_\alpha} \to \bar{K}_{\alpha+1,1} \\ &\to \cdots \to \bar{K}_{\alpha+1,\beta_{\alpha+1}-1} \to \bar{K}_{\alpha+1,\beta_{\alpha+1}} \to \bar{K}_{\alpha+2,1} \\ &\to \cdots \to \bar{K}_{\alpha+2,\beta_{\alpha+2}-1}. \end{aligned}$$

In this formula, first, $\beta_1, \beta_2, \ldots, \beta_{\alpha+2}$ stand for positive integers. Furthermore:

$$([\bar{K}'] =)\ \bar{K}_{\delta,\beta_\delta} \to \ldots \to \bar{K}_{\delta+1,\beta_{\delta+1}-1}$$
$$(\delta = 0, \ldots, \alpha + 1;\ \beta_0 = 0)$$

is a chain of decompositions in G (if $\beta_{\delta+1} = 1$, read only the initial member $\bar{K}_{\delta,\beta_\delta}$) that varies according to the value of δ: For $\delta = 0$, it is a chain from $\bar{a}_0$ $(= A_0)$ to $\bar{a}_1$ $(= A)$ which need not occur if $A_0 = A$; for $\delta = 1, \ldots, \alpha$, an elementary chain from $\bar{a}_\delta$ to $\bar{a}_{\delta+1}$ over the decomposition $\bar{K}_\delta$; for $\delta = \alpha + 1$, a chain from $\bar{a}_{\alpha+1}(= B)$ to $\bar{a}_{\alpha+2}$ $(= B_0)$ which need not occur if $B = B_0$.

We observe that any refinement of the chain $[\bar{K}]$ is obtained by replacing every member $\bar{K}_\gamma$ $(\gamma = 1, \ldots, \alpha)$ of $[\bar{K}]$ by an elementary chain from $\bar{a}_\gamma$ to $\bar{a}_{\gamma+1}$ over $\bar{K}_\gamma$ and, if convenient, adding before the initial member $\bar{K}_1$ or after the final member $\bar{K}_\alpha$ of $[\bar{K}]$ an arbitrary chain with the last member A or the first member B, respectively.

In particular, if every member of the chain $[\bar{K}]$ is replaced by the elementary chain formed by this member only, we again obtain the chain $[\bar{K}]$. Consequently, the chain $[\bar{K}]$ is its own refinement.

The reduced length of every refinement of $[\bar{K}]$ is the sum of the reduced lengths of the single elementary chains and the mentioned added chains; it therefore equals, at least, the reduced length of $[\bar{K}]$.

2.6. Exercises

1. $\mathbf{s}A \sqsubset \bar{A} = \bar{A} = \mathbf{s}A \sqcap \bar{A}$.
2. $\mathbf{s}(B \sqsubset \bar{A}) \sqsubset \bar{A} = B \sqsubset \bar{A}$;
   $\mathbf{s}(B \sqcap \bar{A}) \sqcap \bar{A} = B \sqcap \bar{A}$;
   $\mathbf{s}(B \sqsubset \bar{A}) \sqcap \bar{A} = B \sqsubset \bar{A} = \mathbf{s}(B \sqcap \bar{A}) \sqsubset \bar{A}$.

3. If $B \sqsubset \bar{A} = B \sqcap \bar{A}$, then for every element $\bar{a} \in \bar{A}$ there holds either $\bar{a} \subset B$ or $\bar{a} \cap B = \emptyset$; and conversely.
4. $\mathbf{s}(\mathbf{s}\bar{A} \sqsubset \bar{C}) \sqcap \bar{A} = \mathbf{s}\bar{C} \sqcap \bar{A}$.
5. If $B \supset C$, then there holds: a) $(C \sqsubset \bar{A}) \sqcap B = C \sqsubset (\bar{A} \sqcap B)$. With regard to this equality, the set on either side of the latter may be denoted by $C \sqsubset \bar{A} \sqcap B$. In particular, for $C = B$, we have $(B \sqsubset \bar{A}) \sqcap B = \bar{A} \sqcap B$; b) $(B \sqsubset \bar{A}) \sqcap C = \bar{A} \sqcap C$.
6. If one of the following three statements is true, then the remaining two are true as well: a) Every element of the decomposition $\bar{A}$ is incident with at least one element of the decomposition $\bar{C}$; b) $\bar{A} = \bar{C} \sqsubset \bar{A}$; c) $\mathbf{s}\bar{A} = \mathbf{s}(\mathbf{s}\bar{C} \sqsubset \bar{A})$.
7. Every lengthening of an arbitrary chain of decompositions is simultaneously its refinement.
8. The number p_{n+1} of decompositions of every finite set of order $n + 1$ ($\geqq 1$) is finite. The numbers p_{n+1} are given by the formula:

$$p_{n+1} = \sum_{\nu=1}^{n} \binom{n}{\nu} p_\nu \qquad (p_0 = 1).$$

So we have, in particular:

$$p_1 = 1,\ p_2 = 2,\ p_3 = 5,\ p_4 = 15,\ p_5 = 52,\ p_6 = 203,\ \ldots$$

3. Decompositions on sets

In this chapter we shall deal with decompositions on sets. The results are often useful (see: 2.2) when we are to describe the properties of decompositions in sets; in fact: a decomposition $\bar{A}$ in the set G is, simultaneously, a decomposition on the set $\mathbf{s}\bar{A}$.

3.1. Bindings in decompositions

Let $\bar{A}$, $\bar{B}$ stand for decompositions of G.

Consider two arbitrary elements $\bar{a}, \bar{p} \in \bar{A}$.

A *binding from $\bar{a}$ to $\bar{p}$ in $\bar{A}$ with regard to $\bar{B}$* is a finite sequence of elements of $\bar{A}$:

$$\bar{a}_1, \ldots, \bar{a}_\alpha \quad (\alpha \geqq 2)$$

such that $\bar{a}_1 = \bar{a}$, $\bar{a}_\alpha = \bar{p}$ and that every two neighbouring members $\bar{a}_\beta, \bar{a}_{\beta+1}$ ($\beta = 1, \ldots, \alpha - 1$) are incident with the same element $\bar{b}_\beta \in \bar{B}$. Such a binding is said to be *generated* by the decomposition $\bar{B}$; we speak, briefly, about the *binding* $\{\bar{A}, \bar{B}\}$ *from $\bar{a}$ to $\bar{p}$*.

Note that the individual members of the binding need not be different from one another.

If there exists a binding $\{\bar{A}, \bar{B}\}$ from $\bar{a}$ to $\bar{p}$, then we say that the element $\bar{p}$ can be *connected*, in $\bar{B}$, with the element $\bar{a}$ or, briefly, that $\bar{p}$ can be connected with $\bar{a}$.

Let us now consider the properties of bindings.

First, it is easy to see (the proof may be left to the reader) that, for arbitrary elements $\bar{a}, \bar{b}, \bar{c} \in \bar{A}$, there holds:

a) *The element $\bar{a}$ can be connected with $\bar{a}$.*
b) *If $\bar{b}$ can be connected with $\bar{a}$ and $\bar{c}$ with $\bar{b}$, then $\bar{c}$ can be connected with $\bar{a}$.*
c) *If $\bar{b}$ can be connected with $\bar{a}$, then $\bar{a}$ can be connected with $\bar{b}$.*

Taking account of the statement c), we generally speak about the binding between two elements, or say that two elements can be connected, without stressing which can be connected with which.

If the elements $\bar{a}, \bar{b} \in \bar{A}$ can be connected with an element $\bar{c} \in \bar{A}$, then they can also be connected with each other.

In fact, if the assumption is satisfied, then $\bar{a}$ can be connected with $\bar{c}$, $\bar{c}$ with $\bar{b}$ and, therefore, even $\bar{a}$ with $\bar{b}$.

Let $\bar{a}, \bar{p} \in \bar{A}$; $\bar{b}, \bar{q} \in \bar{B}$ be arbitrary elements and suppose that the elements $\bar{a}, \bar{b}$ as well as $\bar{p}, \bar{q}$ are incident.

If $\bar{a}, \bar{p}$ can be connected in $\bar{B}$, then $\bar{b}, \bar{q}$ can be connected in $\bar{A}$.

Indeed, if there exists a binding $\{\bar{A}, \bar{B}\}$ from $\bar{a}$ to $\bar{p}$ of the form:

$$\bar{a}_1, \ldots, \bar{a}_\alpha \quad (\bar{a}_1 = \bar{a}, \bar{a}_\alpha = \bar{p}),$$

then every two neighbouring elements $\bar{a}_\beta, \bar{a}_{\beta+1}$ are incident with a certain element $\bar{b}_\beta \in \bar{B}$; consequently, $\bar{b}_\beta, \bar{b}_{\beta+1}$ are incident with $\bar{a}_{\beta+1}$. Moreover, $\bar{b}$ is incident with $\bar{a}_1$ and $\bar{q}$ with $\bar{a}_\beta$. Consequently,

$$\bar{b}_0, \ldots, \bar{b}_\alpha \quad (\bar{b}_0 = \bar{b}, \bar{b}_\alpha = \bar{q})$$

is a binding $\{\bar{B}, \bar{A}\}$ from $\bar{b}$ to $\bar{q}$.

3.2. Coverings and refinements of decompositions in sets

Let us, first, introduce once more the notions of a covering and a refinement of a decomposition lying *on* the set G. These notions have already been mentioned in 2.4 and play an important part in the following deliberations.

Let $\bar{A}$, $\bar{B}$ denote arbitrary decompositions on G.

$\bar{A}$ and B are called a covering and a refinement of $\bar{B}$ and $\bar{A}$, respectively, if every element of $\bar{A}$ is the sum of some elements of $\bar{B}$. This relation between $\bar{A}$ and $\bar{B}$ is expressed by $\bar{A} \geqq \bar{B}$ or $\bar{B} \leqq \bar{A}$. In particular $(\bar{A} = \bar{B})$, every decomposition on G is both its own covering and refinement. If $A \geqq \bar{B}$, then the decomposition $\bar{B}$ is obtained, as we have seen, by replacing each element of $\bar{A}$ by its suitable decomposition; we have also noticed that the covering $\bar{A}$ is enforced by a certain decomposition lying on $\bar{B}$.

Let us now proceed to a more detailed study of the above concepts.

Let $\bar{A}$, $\bar{B}$, $\bar{C}$ be arbitrary decompositions on G.

First, we shall show that $\bar{A} \geqq \bar{B}$ *is true if and only if, for any two incident elements* $\bar{a} \in \bar{A}$, $\bar{b} \in \bar{B}$, *there holds* $\bar{a} \supset \bar{b}$.

Suppose $\bar{A} \geqq \bar{B}$. Let $\bar{a} \in \bar{A}$, $\bar{b} \in \bar{B}$ be arbitrary incident elements, hence $\bar{a} \cap \bar{b} \neq \emptyset$. Then $\bar{a}$ is the sum of certain subsets of G that are elements of $\bar{B}$. One of them is $\bar{b}$ because $\bar{a} \cap \bar{b} \neq \emptyset$ and the elements of $\bar{B}$ are disjoint. So we have $\bar{a} \supset \bar{b}$.

Let, conversely, for any two incident elements $\bar{a} \in \bar{A}$, $\bar{b} \in B$, the relation $\bar{a} \supset \bar{b}$ be true. Then $\bar{a}$ is the sum of those subsets of G that are elements of $\bar{B}$ and are incident with $\bar{a}$ and, consequently, the relation $\bar{A} \geqq \bar{B}$ applies.

Note, furthermore, that the statements set below are correct:

a) $\bar{A} \geqq \bar{A}$.
b) *From* $\bar{A} \geqq \bar{B}$, $\bar{B} \geqq \bar{C}$ *there follows* $\bar{A} \geqq \bar{C}$.
c) *From* $A \geqq B$, $B \geqq \bar{A}$ *there follows* $A = \bar{B}$.

3.3. Common covering and common refinement of two decompositions

Let $\bar{A}$, $\bar{B}$ denote arbitrary decompositions on G.

A *common covering of the decompositions* $\bar{A}$ *and* $\bar{B}$, briefly, a *covering of* $\bar{A}$, $\bar{B}$ is a decomposition on G that is a covering of $\bar{A}$ as well as of $\bar{B}$.

Analogously, by a *common refinement of the decompositions* $\bar{A}$ *and* $\bar{B}$, briefly, a *refinement of* $\bar{A}$, $\bar{B}$ we mean a decomposition on G that is a refinement of $\bar{A}$ as well as of $\bar{B}$.

For example, the greatest decomposition $\bar{G}_{\max}$ is a common covering and the least decomposition $\bar{G}_{\min}$ is a common refinement of $\bar{A}$, $\bar{B}$.

It is easy to understand that *every covering of any common covering of* $\bar{A}$, $\bar{B}$ *is again a covering of* $\bar{A}$, $\bar{B}$; similarly, *every refinement of any common refinement of* $\bar{A}$, $\bar{B}$ *is again a refinement of* $\bar{A}$, $\bar{B}$.

A remarkable progress in the trend of the above considerations leading to a number of important results is due to the notions of the least common covering and the greatest common refinement of two decompositions. We shall deal with them in 3.4; 3.5; 3.6.

3.4. The least common covering of two decompositions

In 3.3 we saw that every covering of any common covering of two decompositions $\overline{A}, \overline{B}$ is again a covering of $\overline{A}, \overline{B}$. It is important to note that among all the common coverings of two decompositions $\overline{A}, \overline{B}$ there is one least covering, $\overline{X}$; least in the sense that every common covering of $\overline{A}$ and $\overline{B}$ is a covering of $\overline{X}$. This particular covering is called the *least common covering of the decompositions* $\overline{A}$ *and* $\overline{B}$ or, briefly, the *least covering of* $\overline{A}, \overline{B}$.

Let $\overline{A}, \overline{B}, \overline{C}$ be decompositions on G.

We shall now construct a decomposition on G, denoted by $[\overline{A}, \overline{B}]$, and verify that it is the least common covering of $\overline{A}$ and $\overline{B}$.

Let $\overline{\overline{A}}$ be the system of all subsets of $\overline{A}$, characterized by the following property: Every subset $\overline{\overline{a}} \in \overline{\overline{A}}$ consists of all the elements of $\overline{A}$ that can be connected, in the decomposition $\overline{B}$, with some element $\bar{a} \in \overline{A}$.

First of all, we shall show that $\overline{\overline{A}}$ is a decomposition on $\overline{A}$.

In fact, every element $\bar{a} \in \overline{A}$ lies in some subset $\overline{\overline{a}} \in \overline{\overline{A}}$ because $\bar{a}$ can be connected with itself and therefore lies in the subset $\overline{\overline{a}} \in \overline{\overline{A}}$ consisting of all the elements of $\overline{A}$ that can be connected with $\bar{a}$.

Moreover, every two elements of the system $\overline{\overline{A}}$ are either disjoint or identical. To prove this, let us consider two arbitrary elements $\overline{\overline{a}}, \overline{\overline{b}} \in \overline{\overline{A}}$, $\overline{\overline{a}}$ consisting of all the elements of $\overline{A}$ that can be connected with an element $\bar{a} \in \overline{A}$ and $\overline{\overline{b}}$ of all the elements of $\overline{A}$ that can be connected with an element $\bar{b} \in \overline{A}$. Suppose the elements $\overline{\overline{a}}$ and $\overline{\overline{b}}$ are incident so that they have a common element $\bar{c} \in \overline{A}$. The latter lies in $\overline{\overline{a}}$ and can therefore be connected with $\bar{a}$; it also lies in $\overline{\overline{b}}$ so that it can be connected with $\bar{b}$. Hence $\bar{a}$ and $\bar{b}$ can be connected with $\bar{c}$ and, consequently, they can be connected with each other (3.1). We observe that any element $\bar{x} \in \overline{\overline{a}}$ can be connected with $\bar{a}$ and the latter again with $\bar{b}$. Thus the element $\bar{x}$ can be connected with $\bar{b}$, and we have $\overline{\overline{a}} \subset \overline{\overline{b}}$. In a similar way we verify that $\overline{\overline{b}} \subset \overline{\overline{a}}$ and we have $\overline{\overline{a}} = \overline{\overline{b}}$.

Consequently, $\overline{\overline{A}}$ is a decomposition on $\overline{A}$.

Note that any two elements of $\overline{A}$ that are in the same element of $\overline{\overline{A}}$ can be connected with each other, whereas two elements that do not lie in the same element of $\overline{\overline{A}}$ cannot be connected.

The decomposition $\overline{\overline{A}}$ enforces a certain covering of the decomposition $\overline{A}$, denoted by $[\overline{A}, \overline{B}]$. So we have

$$[\overline{A}, \overline{B}] \geqq \overline{A}.$$

Let us remark that every element $\bar{u} \in [\overline{A}, \overline{B}]$ is the sum of all the elements of $\overline{A}$ that lie in some element of $\overline{\overline{A}}$. In other words, $\bar{u}$ is the sum of all the elements of A that can be connected, in $\overline{B}$, with some element $\bar{a} \in \overline{A}$ lying in $\bar{u}$: $\bar{a} \subset \bar{u}$.

Now we shall consider the properties of the decomposition $[\overline{A}, \overline{B}]$.

First, we shall show that the *equality* $[\bar{A}, \bar{B}] = \bar{A}$ *and the relation* $\bar{A} \geqq \bar{B}$ *are simultaneously valid.*

Proof. a) Suppose $[\bar{A}, \bar{B}] = \bar{A}$. Let $\bar{a} \in \bar{A}, \bar{b} \in \bar{B}$ be arbitrary incident elements. If $\bar{a} \supset \bar{b}$ does not apply, then there exists an element $\bar{p} \in \bar{A}$ incident with $\bar{b}$ and different from $\bar{a}$. The elements $\bar{a}$, $\bar{p}$, arranged in this order, form a binding $\{\bar{A}, \bar{B}\}$ from $\bar{a}$ to $\bar{p}$, so that the set $\bar{a} \cup \bar{p}$ is a part of a certain element $\bar{u} \in [\bar{A}, \bar{B}]$. Consequently, $\bar{u}$ is the sum of at least two different elements of $\bar{A}$, therefore it is not an element of $\bar{A}$, which contradicts the assumption. Thus $\bar{a} \supset \bar{b}$ and we have $\bar{A} \geqq \bar{B}$.

b) Suppose $\bar{A} \geqq \bar{B}$. In that case every element of $\bar{B}$ is a part of an element of A. Consequently, no two different elements of $\bar{A}$ can be connected in $\bar{B}$. We observe that the above decomposition $\bar{\bar{A}}$ is the least decomposition on $\bar{A}$, hence $[\bar{A}, \bar{B}] = \bar{A}$.

Furthermore, we are going to prove that *there holds*:

a) $[\bar{A}, \bar{B}] = [\bar{B}, \bar{A}]$;
b) $[\bar{A}, \bar{A}] = \bar{A}$;
c) $\left[A, [B, \bar{C}]\right] = \left[[\bar{A}, \bar{B}], \bar{C}\right]$.

Proof. a) Let $\bar{u} \in [\bar{A}, \bar{B}]$, $\bar{v} \in [\bar{B}, \bar{A}]$ be arbitrary incident elements. Since $\bar{u}$ and $\bar{v}$ are the sums of certain elements of $\bar{A}$ and $\bar{B}$, respectively, there exist elements $\bar{a} \in \bar{A}$, $\bar{b} \in \bar{B}$ such that $\bar{a} \subset \bar{u}, \bar{b} \subset \bar{v}$ and $\bar{a} \cap \bar{b} \neq \emptyset$. Because $\bar{A}$ covers G, every point $p \in \bar{u}$ lies in an element $\bar{p} \in \bar{A}$. We see that $\bar{u} \supset \bar{p}$ and, as the elements $\bar{a}, \bar{p} \in \bar{A}$ lie in $\bar{u} \in [\bar{A}, \bar{B}]$, $\bar{p}$ can be connected, in $\bar{B}$, with $\bar{a}$. Since $\bar{B}$ covers G, p lies in an element $\bar{q} \in B$ which is, of course, incident with $\bar{p}$. Moreover, in accordance with 3.2, the element $\bar{q}$ can be connected, in $\bar{A}$, with $\bar{b}$. Hence $\bar{v} \supset \bar{q}$ and thus even $\bar{v} \supset \bar{u}$. So we have $[\bar{B}, \bar{A}] \geqq [\bar{A}, \bar{B}]$. Simultaneously, for analogous reasons, there holds the relation $\leqq$ and we have $[\bar{A}, \bar{B}] = [\bar{B}, \bar{A}]$.

b) Since there holds $\bar{A} \geqq \bar{A}$, there also holds $[\bar{A}, \bar{A}] = \bar{A}$.

c) If any two elements $\bar{a}_1, \bar{a}_2 \in \bar{A}$ are incident with some element $\bar{z} \in [\bar{B}, \bar{C}]$, then they lie in the same element of the decomposition $\left[[\bar{A}, \bar{B}], \bar{C}\right]$. In fact, in that case there exist elements $\bar{b}, \bar{q} \in \bar{B}$ such that $\bar{b}$, $\bar{q} \subset \bar{z}$, $\bar{b} \cap \bar{a}_1 \neq \emptyset \neq \bar{q} \cap \bar{a}_2$, and a binding $\{\bar{B}, \bar{C}\}$ from $\bar{b}$ to $\bar{q}$:

$$\bar{b}_1, \ldots, \bar{b}_\gamma \quad (\bar{b}_1 = \bar{b}, \quad \bar{b}_\gamma = \bar{q}).$$

Every element $\bar{b}_\delta$ of the latter is a part of a certain element $\bar{u}_\delta \in [\bar{B}, \bar{A}] = [\bar{A}, \bar{B}]$, $\delta = 1, \ldots, \gamma$. Since $\bar{a}_1$ ($\bar{a}_2$) is incident with $\bar{b}_1$ ($\bar{b}_\gamma$) and $\bar{b}_1$ ($\bar{b}_\gamma$) is a part of $\bar{u}_1$ ($\bar{u}_\gamma$), the element $\bar{a}_1$ ($\bar{a}_2$) is incident with $\bar{u}_1$ ($\bar{u}_\gamma$) and therefore $\bar{a}_1 \subset \bar{u}_1$, $\bar{a}_2 \subset \bar{u}_\gamma$. As every two neighbouring elements $\bar{b}_\delta, \bar{b}_{\delta+1}$ are incident with some element $\bar{c}_\delta \in \bar{C}$, the same holds for every two elements $\bar{u}_\delta, \bar{u}_{\delta+1}$ so that $\bar{u}_1, \ldots, \bar{u}_\gamma$ is a binding $\{[\bar{A}, \bar{B}], \bar{C}\}$ from $\bar{u}_1$ to $\bar{u}_\gamma$. Consequently, the elements $\bar{u}_1$, $\bar{u}_\gamma$, and therefore even $\bar{a}_1$, $\bar{a}_2$, lie in the same element of the decomposition $\left[[\bar{A}, \bar{B}], \bar{C}\right]$.

Now, let $\bar{u} \in \left[\bar{A}, [\bar{B}, \bar{C}]\right], \bar{v} \in \left[[\bar{A}, \bar{B}], \bar{C}\right]$ be arbitrary incident elements. Then there exists an element $\bar{a} \in \bar{A}$, $\bar{a} \subset \bar{u} \cap \bar{v}$; the element $\bar{u}$ is the sum of all the elements $\overline{p} \in \overline{A}$ such that there exists a binding $\{\overline{A}, [\overline{B}, \overline{C}]\}$ from $\bar{a}$ to $\overline{p}$:

$$\bar{a}_1, \ldots, \bar{a}_\alpha \; (\bar{a}_1 = \bar{a}, \bar{a}_\alpha = \overline{p}).$$

Every two neighbouring elements $\bar{a}_\beta$, $\bar{a}_{\beta+1}$ are incident with some element of the decomposition $[\overline{B}, \overline{C}]$ so that they lie, as we have just verified, in the same element of the decomposition $[[\overline{A}, \overline{B}], \overline{C}]$. From this and from $\overline{v} \supset \bar{a}_1$ there follows $\overline{v} \supset \bar{a}_\alpha$ so that $\overline{v} \supset \overline{p}$ and we have $\overline{v} \supset \bar{u}$. Hence, $[[\overline{A}, \overline{B}], \overline{C}] \geqq [\overline{A}, [\overline{B}, \overline{C}]]$. With regard to a), there follows:

$$\begin{aligned}[\overline{A}, [\overline{B}, \overline{C}]] &= [[\overline{B}, \overline{C}], \overline{A}] \geqq [\overline{B}, [\overline{C}, \overline{A}]] = [[\overline{C}, \overline{A}], \overline{B}] \\ &\geqq [\overline{C}, [\overline{A}, \overline{B}]] = [[\overline{A}, \overline{B}], \overline{C}] \geqq [\overline{A}, [\overline{B}, \overline{C}]],\end{aligned}$$

and then, by 3.2c),

$$[\overline{A}, [\overline{B}, \overline{C}]] = [[\overline{A}, \overline{B}], \overline{C}],$$

which completes the proof.

Now we can show that *the decomposition* $[\overline{A}, \overline{B}]$ *is the least common covering of the decompositions* $\overline{A}, \overline{B}$.

Indeed, the decomposition $[\overline{A}, \overline{B}]$ is, by its construction, a covering of $\overline{A}$ and, by a), also a covering of $\overline{B}$. Therefore it is a common covering of $\overline{A}$ and $\overline{B}$. Let, moreover, $\overline{X}$ be an arbitrary common covering of $\overline{A}$ and $\overline{B}$. Then there holds

$$[\overline{X}, \overline{A}] = \overline{X}, \quad [\overline{X}, \overline{B}] = \overline{X},$$

and, by c),

$$[\overline{X}, [\overline{A}, \overline{B}]] = [[\overline{X}, \overline{A}], \overline{B}] = [\overline{X}, \overline{B}] = \overline{X},$$

which proves that $\overline{X}$ is a covering of $[\overline{A}, \overline{B}]$.

Every common covering of $\overline{A}$ and $\overline{B}$ is, therefore, a covering of $[\overline{A}, \overline{B}]$ so that $\overline{A}, \overline{B}]$ is the least common covering of $\overline{A}$ and $\overline{B}$.

3.5. The greatest common refinement of two decompositions

In 3.3 we saw that every refinement of any common refinement of two decompositions $\overline{A}$, $\overline{B}$ on G is again their refinement. It is important to note that among all the common refinements of two decompositions $\overline{A}$, $\overline{B}$ there is one greatest refinement $\overline{Y}$; greatest in the sense that every common refinement of $\overline{A}$ and $\overline{B}$ is a refinement of $\overline{Y}$. This particular refinement is called the *greatest common refinement of the decompositions* $\overline{A}$ *and* $\overline{B}$ or, briefly, the greatest refinement of $\overline{A}, \overline{B}$.

Suppose $\overline{A}, \overline{B}, \overline{C}$ are decompositions on G.

We shall now construct a decomposition on G, denoted by $(\overline{A}, \overline{B})$, and show that it is the greatest common refinement of $\overline{A}$ and $\overline{B}$.

The construction: If every element $\bar{a} \in \overline{A}$ is replaced by its decomposition $\bar{a} \sqcap \overline{B}$, we obtain a certain decomposition on G, $(\overline{A}, \overline{B})$.

The decomposition $(\bar{A}, \bar{B})$ is therefore the system of all nonempty intersections of the elements $\bar{a} \in \bar{A}$ and the elements $\bar{b} \in \bar{B}$.

$(\bar{A}, \bar{B})$ is evidently a refinement of $\bar{A}$, i.e.,

$$(\bar{A}, \bar{B}) \leqq \bar{A}.$$

Let us now consider the properties of the decomposition $(\bar{A}, \bar{B})$.

First, *the following relations are simultaneously valid*:

$$(\bar{A}, \bar{B}) = \bar{A}, \quad \bar{A} \leqq \bar{B}.$$

Proof. a) Suppose $(\bar{A}, \bar{B}) = \bar{A}$. Let $\bar{a} \in \bar{A}$, $\bar{b} \in \bar{B}$ stand for arbitrary incident elements. Then there holds:

$$\bar{a} \cap \bar{b} \in \bar{a} \sqcap \bar{B} \subset (\bar{A}, \bar{B}) = \bar{A}$$

and, consequently, $\bar{a} \cap \bar{b} = \bar{a}$. Hence $\bar{a} \subset \bar{b}$ and, moreover, $\bar{A} \leqq \bar{B}$, by 3.2.

b) Suppose $\bar{A} \leqq \bar{B}$. Then every element $\bar{a} \in \bar{A}$ is a part of an element of $\bar{B}$ so that $\bar{a} \sqcap \bar{B}$ consists of a single element $\bar{a}$. Hence $(\bar{A}, \bar{B}) \geqq \bar{A}$. Since there simultaneously holds the relation $\leqq$ (as we have seen above), the equality $(\bar{A}, \bar{B}) = \bar{A}$ is correct (3.2c).

Furthermore, *there holds*:

a) $(\bar{A}, \bar{B}) = (\bar{B}, \bar{A})$;
b) $(\bar{A}, \bar{A}) = A$;
c) $\big(\bar{A}, (B, \bar{C})\big) = \big((\bar{A}, \bar{B}), \bar{C}\big)$.

Proof. a) Every element $\bar{v} \in (\bar{A}, \bar{B})$ is an element of the decomposition $\bar{a} \sqcap \bar{B}$ where $\bar{a}$ stands for a convenient element of $\bar{A}$. So we have $\bar{v} = \bar{a} \cap \bar{b}$ where $\bar{b} \in B$ is a convenient element. Hence: $\bar{v} \in \bar{b} \sqcap \bar{A} \subset (\bar{B}, \bar{A})$. There follows $(\bar{A}, \bar{B}) \subset (\bar{B}, \bar{A})$ and, for analogous reasons, there holds the relation $\supset$ and the proof of the equality a) is complete.

Note that the equality in question also follows from the relations $(\bar{A}, \bar{B}) = \bar{A} \sqcap \bar{B}$ and $\bar{A} \sqcap \bar{B} = \bar{B} \sqcap \bar{A}$ (valid by 2.3).

b) Since $\bar{A} \leqq \bar{A}$, there holds $(\bar{A}, \bar{A}) = \bar{A}$.

c) Let $\bar{v} \in \big(\bar{A}, (B, \bar{C})\big)$, so that $\bar{v} = \bar{a} \cap (\bar{b} \cap \bar{c})$ where $\bar{a} \in \bar{A}$, $\bar{b} \in \bar{B}$, $\bar{c} \in \bar{C}$ are convenient elements. Since $\bar{a} \cap (\bar{b} \cap \bar{c}) = (\bar{a} \cap \bar{b}) \cap \bar{c}$ and, moreover, $(\bar{a} \cap \bar{b}) \cap \bar{c} \in \in \big((\bar{A}, \bar{B}), \bar{C}\big)$, we have $\big(\bar{A}, (\bar{B}, \bar{C})\big) \subset \big((\bar{A}, \bar{B}), \bar{C}\big)$. From this and a) it follows that there also holds the relation $\supset$, which completes the proof of c).

By means of these results we can show that *the decomposition* $(\bar{A}, \bar{B})$ *is the greatest common refinement of the decompositions* $\bar{A}$ *and* $\bar{B}$.

Indeed, by its construction, $(\bar{A}, \bar{B})$ is a refinement of $\bar{A}$ and, by the relation a), it is also a refinement of $\bar{B}$. Let, furthermore, $\bar{Y}$ stand for an arbitrary common refinement of $\bar{A}$ and $\bar{B}$. Then we have:

$$(\bar{Y}, \bar{A}) = \bar{Y}, \quad (\bar{Y}, \bar{B}) = \bar{Y}.$$

Hence, by the relation c), there holds:

$$(\bar{Y}, (\bar{A}, \bar{B})) = ((\bar{Y}, \bar{A}), \bar{B}) = (\bar{Y}, \bar{B}) = \bar{Y}.$$

We see that $\bar{Y}$ is a refinement of the decomposition $(\bar{A}, \bar{B})$.

Every common refinement of the decompositions A and $\bar{B}$ is, therefore, a refinement of their common refinement $(\bar{A}, \bar{B})$. Thus $(\bar{A}, B)$ is the greatest common refinement of $\bar{A}$ and $\bar{B}$.

3.6. Relations between the least common covering and the greatest common refinement of two decompositions

Let $\bar{A}$ and $\bar{B}$ stand for arbitrary decompositions on the set G.

It is easy to show that *between the least common covering* $[\bar{A}, \bar{B}]$ *and the greatest common refinement* $(\bar{A}, \bar{B})$ *of* $\bar{A}, \bar{B}$ *there hold the following equalities*:

$$[\bar{A}, (\bar{A}, \bar{B})] = \bar{A}, \quad (\bar{A}, [\bar{A}, \bar{B}]) = \bar{A}.$$

In fact, these equalities express the relations $\bar{A} \geqq (\bar{A}, \bar{B})$ and $[\bar{A}, \bar{B}] \geqq \bar{A}$ (3.4; 3.5).

3.7. Exercises

1. Deduce, for arbitrary decompositions $\bar{A}, \bar{B}$ of the set G, on the ground of $\bar{a} \in \bar{A}, \bar{b} \in \bar{B}$, $\mathbf{s}(\bar{a} \sqsubset \bar{B}) = \mathbf{s}(\bar{b} \sqsubset \bar{A}) = \bar{u}$, the relation $\bar{u} \in [\bar{A}, \bar{B}]$.
2. For any decompositions $\bar{A}, \bar{B}, \bar{X}$ on G, where $\bar{X} \geqq \bar{A}$, there holds a) $[\bar{X}, \bar{B}] \geqq [\bar{A}, \bar{B}]$, $(\bar{X}, \bar{B}) \geqq (\bar{A}, \bar{B})$; b) $(\bar{X}, [A, \bar{B}]) \geqq [\bar{A}, (\bar{X}, \bar{B})]$.
3. Find an example to show that, under the assumptions of the previous exercise, the equality in formula b) need not be valid.
4. Two decompositions *in* G always have the least common covering but need not have the greatest common refinement. For the least common coverings of the decompositions $\bar{A}, \bar{B}, \bar{C}$ in G there hold the formulae 3.4 a) b) c).

4. Special decompositions

In this chapter we shall deal with particular kinds of relations between decompositions in or on the set G.

4.1. Semi-coupled (loosely coupled) and coupled decompositions

Let $\bar{A}$ and $\bar{C}$ be decompositions in G.

The decompositions $\bar{A}, \bar{C}$ are called *semi-coupled* or *loosely coupled* if every element $\bar{a} \in \bar{A}$ is incident with, at most, one element of $\bar{C}$ and every element $\bar{c} \in \bar{C}$ with, at most, one element of $\bar{A}$ and if, moreover, at least for one pair of the elements $\bar{a} \in \bar{A}, \bar{c} \in \bar{C}$ the incidence really occurs. If $\bar{A}$ and $\bar{C}$ are semi-coupled, then the decomposition $\bar{A}$ ($\bar{C}$) is called semi-coupled or loosely coupled with the decomposition $\bar{C}$ ($\bar{A}$).

The decompositions $\bar{A}$, $\bar{C}$ are *coupled* if every element $\bar{a} \in \bar{A}$ is incident with exactly one element of $\bar{C}$ and every element $\bar{c} \in \bar{C}$ with exactly one element of $\bar{A}$. If $\bar{A}$ and $\bar{C}$ are coupled, then the decomposition $\bar{A}$ ($\bar{C}$) is said to be coupled with the decomposition $\bar{C}$ ($\bar{A}$).

We observe that two coupled decompositions in G are always semi-coupled.

Example of coupled decompositions: If there holds, for the subset $X \subset G$ and the decomposition $\bar{Y}$ in G, the relation $X \cap \mathbf{s}\bar{Y} \neq \emptyset$, then the decompositions $X \sqsubset \bar{Y}$ and $\bar{Y} \sqcap X$ are coupled.

Let us now proceed to describe the properties of semi-coupled and coupled decompositions.

First, note that if the decompositions $\bar{A}$ and $\bar{C}$ are semi-coupled, then $\mathbf{s}\bar{A} \cap \mathbf{s}\bar{C} \neq \emptyset$. Indeed, in that case incidence occurs at least for one pair of the elements $\bar{a} \in \bar{A}, \bar{c} \in \bar{C}$ and we have $\mathbf{s}\bar{A} \cap \mathbf{s}\bar{C} \supset \bar{a} \cap \bar{c} \neq \emptyset$. To simplify the notation, we put $\mathbf{s}\bar{A} = A$, $\mathbf{s}\bar{C} = C$, so that $A \cap C \neq \emptyset$.

The decompositions $\bar{A}$, $\bar{C}$ are semi-coupled if and only if the intersections $\bar{A} \sqcap C$, $\bar{C} \sqcap A$ are equal: $\bar{A} \sqcap C = \bar{C} \sqcap A$.

Proof. a) Suppose $\bar{A}$ and $\bar{C}$ are semi-coupled. Then, with regard to $A \cap C \neq \emptyset$, we have: $\bar{A} \sqcap C \neq \emptyset \neq \bar{C} \sqcap A$. Let $\bar{a}' \in \bar{A} \sqcap C$ be an arbitrary element; evidently $\bar{a}' = \bar{a} \cap C$, $\bar{a}$ standing for a convenient element of $\bar{A}$. Since $\bar{a}' \subset C$, $\bar{a}$ is incident with at least one and therefore, by the above assumption, exactly one element $\bar{c} \in \bar{C}$; $\bar{a}$ is obviously the only element of $\bar{A}$ which is incident with $\bar{c}$. We see that: $\bar{a}' = \bar{a} \cap \bar{c} = \bar{c} \cap A \in \bar{C} \sqcap A$. Thus we have $\bar{A} \sqcap C \subset \bar{C} \sqcap A$. Naturally, there simultaneously holds the relation $\supset$ and, consequently, the equality of both decompositions.

b) Suppose $\bar{A} \sqcap C = \bar{C} \sqcap A$. Let $\bar{a} \in \bar{A}$ be an arbitrary element. The element $\bar{a}$ is either not incident with any element of $\bar{C}$ or is incident with at least one. If it is incident with the elements $\bar{c}_1, \bar{c}_2 \in \bar{C}$, then we have: $\bar{a} \cap (\bar{c}_1 \cup \bar{c}_2) \subset \bar{a} \cap C \in \bar{A} \sqcap C = \bar{C} \sqcap A$ and, consequently, there exists an element $\bar{c} \in \bar{C}$ for which there holds $\bar{a} \cap (\bar{c}_1 \cup \bar{c}_2) = A \cap \bar{c}$. Since any two different elements of a decomposition are disjoint, there follows $\bar{c}_1 = \bar{c}_2 = \bar{c}$. We see that every element of $\bar{A}$ is incident with at most one element of $\bar{C}$ and, obviously, there also holds that every element of $\bar{C}$ is incident with at most one element of $\bar{A}$. From $A \cap C \neq \emptyset$ it is clear that at least for one pair of the elements $\bar{a} \in \bar{A}$, $\bar{c} \in \bar{C}$ the incidence really occurs and the proof is accomplished.

The decompositions $\bar{A}, \bar{C}$ *are semi-coupled if and only if the closures* $(\mathrm{H}\bar{A} =)\, \bar{C} \sqsubset \bar{A}$, $(\mathrm{H}\bar{C} =)\, \bar{A} \sqsubset \bar{C}$ *are coupled.*

Proof. a) Suppose $\bar{A}, \bar{C}$ are semi-coupled. Then, on taking account of $A \cap C \neq \emptyset$, we first have: $\mathrm{H}\bar{A} \neq \emptyset \neq \mathrm{H}\bar{C}$. Let us now consider an element $\bar{a} \in \mathrm{H}\bar{A}$. It is incident with at least one and, by the above assumption, exactly one element $\bar{c} \in \bar{C}$. The element $\bar{c}$ evidently belongs to the closure $\mathrm{H}\bar{C}$, hence $\bar{c} \in \mathrm{H}\bar{C}$, and is the only element of $\mathrm{H}\bar{C}$ which is incident with $\bar{a}$. It follows that every element of $\mathrm{H}\bar{A}$ is incident with exactly one element of $\mathrm{H}\bar{C}$. Since, analogously, every element of $\mathrm{H}\bar{C}$ is incident with exactly one element of $\mathrm{H}\bar{A}$, the closures $\mathrm{H}\bar{A}$ and $\mathrm{H}\bar{C}$ are coupled.

b) Suppose the closures $\mathrm{H}\bar{A}$, $\mathrm{H}\bar{C}$ are coupled. Then an arbitrary element $\bar{a} \in \bar{A}$ is either not incident with any element of $\bar{C}$ or is incident with at least one element of $\bar{C}$. In the latter case, $\bar{a}$ belongs to the closure $\mathrm{H}\bar{A}$ and, by the above assumption, it is incident with exactly one element $\bar{c} \in \mathrm{H}\bar{C}$. Except the elements of $\mathrm{H}\bar{C}$, no element of $\bar{C}$ is incident with $\bar{a}$. Consequently, every element of $\bar{A}$ is incident with at most one element of $\bar{C}$. For similar reasons, every element of $\bar{C}$ is incident with at most one element of $\bar{A}$. Therefore the decompositions $\bar{A}, \bar{C}$ are semi-coupled and the proof is complete.

The decompositions $\bar{A}, \bar{C}$ *are coupled if and only if there simultaneously holds*

$$\bar{A} \sqcap C = \bar{C} \sqcap A, \tag{1}$$

$$A = \mathbf{s}(C \sqcap \bar{A}), \quad C = \mathbf{s}(A \sqcap \bar{C}). \tag{2}$$

Proof. a) Suppose $\bar{A}, \bar{C}$ are coupled. Then every element of $\bar{A}$ and $\bar{C}$ is incident with at most and, at the same time, at least one element of $\bar{C}$ and $\bar{A}$, respectively. Consequently, on taking account of the above result, there holds (1) and, simultaneously, by 2.6.6, the first (second) equality (2).

b) Suppose the equalities (1), (2) are true. By means of the same theorems as in a), we can verify that $\bar{A}, \bar{C}$ are coupled.

If $\bar{A}, \bar{C}$ are coupled, then every element of $\bar{A}$ or $\bar{C}$ is incident with at least and, simultaneously, at most one element of $\bar{C}$ or $\bar{A}$, respectively. Consequently, there holds: $\bar{A} = \bar{C} \sqsubset \bar{A}$, $\bar{C} = \bar{A} \sqsubset \bar{C}$ (2.6.6).

Let us now assume that $\bar{A} = \bar{C} \sqcap \bar{A}$, $\bar{C} = \bar{A} \sqsubset \bar{C}$. Then, of course, our assumption: $A \cap C \neq \emptyset$ is satisfield as well.

Suppose $\bar{B}$ is an arbitrary common covering of the decompositions $\bar{A} \sqcap C, \bar{C} \sqcap A$ of the set $A \cap C$. By means of $\bar{B}$ we define, first, the decomposition $\bar{\bar{A}}$ $(\bar{\bar{C}})$ on $\bar{A}$ $(\bar{C})$ as follows: Each element of $\bar{\bar{A}}$ $(\bar{\bar{C}})$ consists of all the elements $\bar{a} \in \bar{A}$ $(\bar{c} \in \bar{C})$ that are incident with the same element of $\bar{B}$. Furthermore, by means of $\bar{\bar{A}}$ $(\bar{\bar{C}})$ we define the decomposition $\mathring{A}$ $(\mathring{C})$ in G: $\mathring{A}$ $(\mathring{C})$ is the covering of $\bar{A}$ $(\bar{C})$ enforced by $\bar{\bar{A}}$ $(\bar{\bar{C}})$.

Accordingly, there holds $\bigcup \bar{a} \in \mathring{A}$ $(\bigcup \bar{c} \in \mathring{C})$ if and only if $\bigcup (\bar{a} \cap C) \in \bar{B}$ $\big(\bigcup (\bar{c} \cap A) \in \bar{B}\big)$.

Thus we have, by means of the decomposition $\bar{B}$, constructed certain coverings $\mathring{A}$ and $\mathring{C}$ of $\bar{A}$ and $\bar{C}$, respectively. The coverings $\mathring{A}$, $\mathring{C}$ are said to be *enforced* by the common covering $\bar{B}$ of the decompositions $\bar{A} \sqcap C$, $\bar{C} \sqcap A$. Note that the construction is based upon the relations $\bar{A} = \bar{C} \sqsubset \bar{A}$, $\bar{C} = \bar{A} \sqsubset \bar{C}$.

Obviously: $\mathbf{s}\mathring{A} = \mathbf{s}\bar{A}$ $(= A)$, $\mathbf{s}\mathring{C} = \mathbf{s}\bar{C}$ $(= C)$.

Now we shall prove that *the decompositions $\mathring{A}$, $\mathring{C}$ are coupled and intersect each other in the decomposition $\bar{B}$* so that $\mathring{A} \sqcap \mathring{C} = \bar{B}$.

Proof. The equality $\bar{A} = \bar{C} \sqsubset \bar{A}$ yields $\mathring{A} = \mathring{C} \sqsubset \mathring{A}$ and, similarly, $\mathring{C} = \mathring{A} \sqsubset \mathring{C}$. To prove the theorem, it is sufficient to verify that

$$\mathring{A} \sqcap C = \mathring{C} \sqcap A = \bar{B}.$$

Indeed, if these equalities are satisfied, then, by the above result, the decompositions $\mathring{A}$, $\mathring{C}$ are coupled and, on taking account of 2.3, we have: $\mathring{A} \sqcap \mathring{C} = (\mathring{A} \sqcap C) \sqcap (\mathring{C} \sqcap A) = \bar{B} \sqcap \bar{B} = \bar{B}$.

To every element $\mathring{a}' \in \mathring{A} \sqcap C$ there exist elements $\mathring{a} = \bigcup \bar{a}$, $\mathring{a} \in \mathring{A}$, $\bar{a} \in \bar{A}$ such that $\mathring{a}' = \mathring{a} \cap C = (\bigcup \bar{a}) \cap C = \bigcup (\bar{a} \cap C) \in \bar{B}$, whence $\mathring{A} \sqcap C \subset \bar{B}$. Conversely, every element $\bar{b} \in \bar{B}$ has the form: $\bar{b} = \bigcup (\bar{a} \cap C)$ where $\bar{a} \in \bar{A}$, $\mathring{a} = \bigcup \bar{a} \in \mathring{A}$ and there holds $\bar{b} = \bigcup (\bar{a} \cap C) = (\bigcup \bar{a}) \cap C = \mathring{a} \cap C \in \mathring{A} \sqcap C$, whence $\bar{B} \subset \mathring{A} \sqcap C$. So we have $\mathring{A} \sqcap C = \bar{B}$ and, for analogous reasons, even $\mathring{C} \sqcap A = \bar{B}$.

4.2. Adjoint decompositions

Suppose $\bar{A}$, $\bar{C}$ are decompositions and B, D subsets of G. Let $B \in \bar{A}$, $D \in \bar{C}$ and $B \cap D \neq \emptyset$. We shall again make use of the notation: $A = \mathbf{s}\bar{A}$, $C = \mathbf{s}\bar{C}$.

By the above assumptions there holds $B \in D \sqsubset \bar{A}$, $D \in B \sqsubset \bar{C}$ and, on taking account of $B \subset A$, $D \subset C$, we have

$$\emptyset \neq B \cap D \subset (B \cap C),\ (D \cap A).$$

Consequently (2.6.5),

$$D \sqsubset \bar{A} \sqcap C, \quad B \sqsubset \bar{C} \sqcap A$$

are decompositions in G.

If there holds:

$$\mathbf{s}(D \sqsubset \bar{A} \sqcap C) = \mathbf{s}(B \sqsubset \bar{C} \sqcap A),$$

then the decompositions $\bar{A}$, $\bar{C}$ are said to be *adjoint with regard to the sets* B, D; we also say that $\bar{A}$ $(\bar{C})$ *is adjoint to* $\bar{C}$ $(\bar{A})$ *with regard to* B, D.

On taking account of the equalities

$$D \sqsubset \bar{A} \sqcap C = (D \cap A) \sqsubset (\bar{A} \sqcap C),$$
$$B \sqsubset \bar{C} \sqcap A = (B \cap C) \sqsubset (\bar{C} \sqcap A),$$

the formula (1) may be replaced by:

$$\mathbf{s}\big((D \cap A) \sqsubset (\bar{A} \sqcap C)\big) = \mathbf{s}\big((B \cap C) \sqsubset (\bar{C} \sqcap A)\big). \tag{1'}$$

For example, the decompositions $\bar{A}, \bar{C}$ are adjoint with regard to B, D if $\bar{A}$ is the greatest or the least decomposition of A.

Let us now assume that $\bar{A}, \bar{C}$ are adjoint with regard to B, D. Then:

$$\bar{A}_1 = C \sqsubset \bar{A}, \quad \bar{A}_2 = D \sqsubset \bar{A},$$
$$\bar{C}_1 = A \sqsubset \bar{C}, \quad \bar{C}_2 = B \sqsubset \bar{C}$$

are decompositions in G. Denote: $A_1 = \mathbf{s}\bar{A}_1$, $A_2 = \mathbf{s}\bar{A}_2$, $C_1 = \mathbf{s}\bar{C}_1$, $C_2 = \mathbf{s}\bar{C}_2$. Then we have:

$$\bar{A} \supset \bar{A}_1 \supset \bar{A}_2 \supset \{B\}, \quad A \supset A_1 \supset A_2 \supset B,$$
$$\bar{C} \supset \bar{C}_1 \supset \bar{C}_2 \supset \{D\}, \quad C \supset C_1 \supset C_2 \supset D.$$

We shall show that *there exist coupled coverings $\mathring{A}$, $\mathring{C}$ of the decompositions $\bar{A}_1, \bar{C}_1$ such that $A_2 \in \mathring{A}$, $C_2 \in \mathring{C}$*. These coverings are determined by the construction described in part a) of the following proof. *The sets A_2, C_2 are incident.*

Proof. a) Every element of $\bar{A}_1$ (C_1) lies in $\bar{A}$ ($\bar{C}$) and is incident with C (A) and, therefore, with some element of $\bar{C}$ ($\bar{A}$) incident with A (C); this element of $\bar{C}$ ($\bar{A}$) is, of course, contained in $\bar{C}_1$ ($\bar{A}_1$). Hence:

$$\bar{A}_1 = \bar{C}_1 \sqsubset \bar{A}_1, \quad \bar{C}_1 = \bar{A}_1 \sqsubset \bar{C}_1.$$

It is also easy to realize that $A_1 \cap C_1 = A \cap C$. We observe that $A_1 \sqcap C_1$ $\bar{C}_1, \sqcap \bar{A}_1$ are decompositions on $A \cap C$. Let $\bar{U}$ be their least common covering so that $\bar{U} = [A_1 \sqcap \bar{C}_1, C_1 \sqcap \bar{A}_1]$. Now the decompositions $\mathring{A}$, $\mathring{C}$ are defined as the coverings of $\bar{A}_1, \bar{C}_1$, enforced by $\bar{U}$. So we have $\mathring{A} \sqcap \mathring{C} = \bar{U}$ and every element of $\mathring{A}$ ($\mathring{C}$) is the sum of all the elements of $\bar{A}_1$ ($\bar{C}_1$) which are incident with one element of $\bar{U}$.

b) There holds $A_2 \in \mathring{A}$ and $C_2 \in \mathring{C}$. In fact, as $B \in \bar{A}$, $D \in \bar{C}$, we have

$$C \cap B \in C \sqcap \bar{A}, \quad A \cap D \in A \sqcap \bar{C}$$

and, since $\bar{A}, \bar{C}$ are adjoint with regard to B, D, there holds (1′). So we have, by 3.7.1, $\bar{u} \in \bar{U}$ where $\bar{u}$ is the set (1′). The individual elements of $\mathring{A}$ and $\mathring{C}$, respectively, are the sums of all the elements of $\bar{A}_1$ and $\bar{C}_1$, incident with one element of $\bar{U}$. To prove the relations $A_2 \in \mathring{A}$ and $C_2 \in \mathring{C}$, we only need to show that A_2 and C_2 are the sums of all the elements of $\bar{A}_1$ and $\bar{C}_1$, respectively, incident with $\bar{u}$.

We see, first, that there holds:

$$\bar{u} = \mathbf{s}(D \sqsubset \bar{A} \sqcap C) = \mathbf{s}(\bar{A}_2 \sqcap C) = A_2 \cap C.$$

An arbitrary element of $\bar{A}_1$ lies in $\bar{A}$ and is incident with the set C; it simultaneously lies in $\bar{A}_2$ if and only if it is incident with the set D and, consequently, with the set $A_2 \cap C = \bar{u}$. Hence, it is exactly the elements of $\bar{A}_1$ which lie in $\bar{A}_2$ that are incident with $\bar{u}$; their sum is, as we see, A_2. Similarly, from

$$\bar{\bar{u}} = \mathbf{s}(B \sqsubset \bar{C} \sqcap A) = \mathbf{s}(\bar{C}_2 \sqcap A) = C_2 \cap A$$

there follows that the sum of the elements of $\bar{C}_1$ which are incident with $\bar{\bar{u}}$ is C_2.

c) From $\emptyset \neq B \cap D \subset A_2 \cap C_2$ we have $A_2 \cap C_2 \neq \emptyset$.

The notion of adjoint decompositions may be extended to adjoint chains of decompositions.

Suppose $(\emptyset \neq)\, B \subset A \subset G$, $(\emptyset \neq)\, D \subset C \subset G$ and let

$$([\bar{K}] =)\; \bar{K}_1 \to \cdots \to \bar{K}_\alpha,$$
$$([\bar{L}] =)\;\; \bar{L}_1 \to \cdots \to \bar{L}_\beta$$

be chains of decompositions in G from A to B and from C to D.

The chains $[\bar{K}]$, $[\bar{L}]$ are called *adjoint* if: 1) their ends coincide, i.e., $A = C$, $B = D$; 2) every two members $\bar{K}_\gamma$, $\bar{L}_\delta$ are adjoint with regard to the sets $\mathbf{s}\bar{K}_{\gamma+1}$, $\mathbf{s}\bar{L}_{\delta+1}$; γ and δ run over $1, \ldots, \alpha$ and $1, \ldots, \beta$, respectively, and $\mathbf{s}\bar{K}_{\alpha+1} = B$, $\mathbf{s}\bar{L}_{\beta+1} = D$.

4.3. Modular decompositions

In this chapter we shall deal with special decompositions lying *on* G.

Suppose $\bar{X}$, $\bar{A}$, $\bar{B}$ are decompositions on G and let $\bar{X} \geqq \bar{A}$.

The reader has certainly noticed (see 3.7.2, 3) that the decomposition $(\bar{X}, [\bar{A}, \bar{B}])$ is a covering of the decomposition $[\bar{A}, (\bar{X}, \bar{B})]$ but that these two decompositions need not be equal.

If they are equal, i.e., if there holds

$$[\bar{A}, (\bar{X}, \bar{B})] = (\bar{X}, [\bar{A}, \bar{B}]),$$

then the decomposition $\bar{B}$ is called *modular with regard to* $\bar{X}$, $\bar{A}$ (in this order). If, e.g., $\bar{X} = \bar{A}$ or $\bar{X} = \bar{G}_{\max}$, then $\bar{B}$ is modular with regard to $\bar{X}$, $\bar{A}$.

Let now $\bar{X}, \bar{Y}$ and $\bar{A}, \bar{B}$ stand for decompositions on G such that $\bar{X} \geqq \bar{A}, \bar{Y} \geqq \bar{B}$ and suppose $\bar{B}$ and $\bar{A}$ are modular with regard to $\bar{X}$, $\bar{A}$ and $\bar{Y}$, $\bar{B}$, respectively.

Then there holds:

$$(\mathring{A} =)\, [\bar{A}, (\bar{X}, \bar{B})] = (\bar{X}, [\bar{A}, \bar{B}]),$$
$$(\mathring{B} =)\, [\bar{B}, (\bar{Y}, \bar{A})] = (\bar{Y}, [\bar{B}, \bar{A}]),$$

where the decompositions on either side of the first as well as the second formula are denoted $\mathring{A}$ and $\mathring{B}$, respectively.

We see, first, that *there holds*

$$\overline{X} \geqq \mathring{A} \geqq \overline{A}, \quad \overline{Y} \geqq \mathring{B} \geqq \overline{B},$$

so that the decompositions $\mathring{A}$, $\mathring{B}$ interpolate the decompositions $\overline{X}$, $\overline{A}$ or $\overline{Y}$, $\overline{B}$, respectively, in the sense of the above formulae.

Next, *there holds*:

$$[\mathring{A}, \mathring{B}] = [\overline{A}, \overline{B}], \ [\overline{X}, \mathring{B}] = [\overline{X}, \overline{B}], \ [\overline{Y}, \mathring{A}] = [\overline{Y}, \overline{A}], \tag{1}$$

$$(\mathring{A}, \mathring{B}) = (\overline{X}, \mathring{B}) = (\overline{Y}, \mathring{A}) = \big((\overline{X}, \overline{Y}), [\overline{A}, \overline{B}]\big). \tag{2}$$

These relations can easily be deduced from the properties of the least common covering and the greatest common refinement of two decompositions. For example, the first equality (1) by means of $(\overline{X}, \overline{B}) \leqq \overline{B} \leqq [\overline{B}, (\overline{Y}, \overline{A})]$, $(\overline{Y}, \overline{A}) \leqq \overline{A} \leqq [\overline{A}, \overline{B}]$ as follows:

$$\begin{aligned} [\mathring{A}, \mathring{B}] &= \big[[\overline{A}, (\overline{X}, \overline{B})], \ [\overline{B}, (\overline{Y}, \overline{A})]\big] \\ &= \big[\overline{A}, [(\overline{X}, \overline{B}), [\overline{B}, (\overline{Y}, \overline{A})]]\big] = \big[\overline{A}, [\overline{B}, (\overline{Y}, \overline{A})]\big] \\ &= \big[[\overline{A}, \overline{B}], (\overline{Y}, \overline{A})\big] = [\overline{A}, \overline{B}]. \end{aligned}$$

The other equalities may be deduced analogously.

The mentioned properties of modular decompositions can be specified as "global", since they concern decompositions as a whole without regard to the individual elements of which they consist. Besides these "global" properties, the modular decompositions also have the following "local" property, important to our purposes:

For any two incident elements $\bar{x} \in \overline{X}$, $\bar{y} \in \overline{Y}$ *the closures* $(\bar{x} \cap \bar{y}) \sqsubset \mathring{A}$, $(\bar{x} \cap \bar{y}) \sqsubset \mathring{B}$ *are coupled.*

Proof. Suppose $\bar{x} \in \overline{X}$, $\bar{y} \in \overline{Y}$ are arbitrary incident elements. Consider an element $\mathring{a} \in (\bar{x} \cap \bar{y}) \sqsubset \mathring{A}$ and show that it is incident with exactly one element $\mathring{b} \in (\bar{x} \cap \bar{y}) \sqsubset \mathring{B}$. In fact, since the element $\mathring{a} \in \mathring{A}$ is incident with the set $\bar{x} \cap \bar{y}$ and, according to the assumption, there holds $\overline{X} \geqq \mathring{A}$, we have: $\mathring{a} \subset \bar{x}$, $\bar{y} \cap \mathring{a} \neq \emptyset$. Hence, in particular, $\bar{y} \cap \mathring{a}$ is an element of the decomposition $(\overline{Y}, \mathring{A})$. As $(\overline{X}, \mathring{B}) = (\overline{Y}, \mathring{A})$, there exists an element $\mathring{b} \in \mathring{B}$ such that $\bar{x} \cap \mathring{b} = \bar{y} \cap \mathring{a}$. We see that $\mathring{b}$ is incident with $\mathring{a}$ so that $\mathring{b} \cap \mathring{a} \neq \emptyset$. As $\mathring{b}$ is also incident with $\bar{x} \cap \bar{y}$, we have $\mathring{b} \in (\bar{x} \cap \bar{y}) \sqsubset \mathring{B}$. Consequently, the element $\mathring{a}$ is incident at least with the element $\mathring{b}$ of the closure $(\bar{x} \cap \bar{y}) \sqsubset \mathring{B}$. But in the latter there are no further elements incident with $\mathring{a}$ because every element incident with $\mathring{a}$ forms a part of $\bar{y}$, cuts the set $\bar{y} \cap \mathring{a} = \bar{x} \cap \mathring{b}$ and therefore coincides with $\mathring{b}$. For analogous reasons, every element of $(\bar{x} \cap \bar{y}) \sqsubset \mathring{B}$ is incident with exactly one element of $(\bar{x} \cap \bar{y}) \sqsubset \mathring{A}$ and the proof is accomplished.

4.4. Exercises

1. Two finite coupled decompositions have the same number of elements.
2. On taking account of the last theorem of 4.3, show that there holds:

$$((\bar{x} \cap \bar{y}) \sqsubset \mathring{A}) \sqcap \mathbf{s}((\bar{x} \cap \bar{y}) \sqsubset \mathring{B}) = ((\bar{x} \cap \bar{y}) \sqsubset \mathring{B}) \sqcap \mathbf{s}((\bar{x} \cap \bar{y}) \sqsubset \mathring{A})$$
$$= (\bar{x} \cap \bar{y}) \sqcap [\bar{A}, \bar{B}].$$

5. Complementary (commuting) decompositions

Further particular situations generated by decompositions on the set G arise from the so-called complementary or commuting decompositions. As the latter play an important part in the following deliberations, we shall discuss them in a special chapter.

5.1. The notion of complementary (commuting) decompositions

Let $\bar{A}$, $\bar{B}$, $\bar{C}$ stand for arbitrary decompositions on G.

By the definition of the least common covering $[\bar{A}, \bar{B}]$, every element $\bar{u} \in [\bar{A}, \bar{B}]$ is the sum of certain elements $\bar{a} \in \bar{A}$ and, at the same time, the sum of certain elements $\bar{b} \in \bar{B}$. The *decomposition* $\bar{A}$ *is* called *complementary to* or *commuting with the decomposition* $\bar{B}$ if every element $\bar{a} \in \bar{A}$ is incident with each element $\bar{b} \in \bar{B}$ that lies in the same element $\bar{u} \in [\bar{A}, \bar{B}]$ as $\bar{a}$.

If. for example, $\bar{A}$ is a covering of $\bar{B}$, then $\bar{A}$ is complementary to $\bar{B}$. The new notion generalizes the concept of a covering.

There holds:

a) $\bar{A}$ *is complementary to* $\bar{A}$.
b) *If* $\bar{A}$ *is complementary to* $\bar{B}$, *then* $\bar{B}$ *is complementary to* $\bar{A}$.

Indeed, a) is obviously true. To prove b), let us accept the assumption but reject the assertion. Then there exists an element $\bar{b} \in \bar{B}$, lying in a certain element $\bar{u} \in [\bar{B}, \bar{A}]$, which is not incident with every element of $\bar{A}$ that lies in $\bar{u}$. Consequently, $\bar{b}$ is not incident with an element $\bar{a} \in \bar{A}$ lying in $\bar{u}$. Hence, $\bar{a}$ is not incident with all the elements of $\bar{B}$ lying in $\bar{u}$, which contradicts our assumption that $\bar{A}$ is complementary to $\bar{B}$ and the proof is accomplished.

With regard to b), we generally speak about complementary (commuting) decompositions without stressing which is complementary to (commuting with) which.

The following example proves the fact that *if* $\bar{A}$, $\bar{B}$ *and, at the same time,* $\bar{B}$, $\bar{C}$ *are complementary, then* $\bar{A}$, $\bar{C}$ *need not be complementary.*

Suppose $G = \{a_1, a_2, a_3, a_4, a_5, a_6\}$ is a set consisting of six elements. Denote, furthermore,

$$\bar{a}_1 = \{a_1, a_2\}, \quad \bar{a}_2 = \{a_3, a_4\}, \quad \bar{a}_3 = \{a_5, a_6\};$$

$$\bar{b}_1 = \{a_1, a_3, a_5\}, \quad \bar{b}_2 = \{a_2, a_4, a_6\};$$

$$\bar{c}_1 = \{a_1, a_2, a_3\}, \quad \bar{c}_2 = \{a_4, a_5, a_6\},$$

so that we have following decompositions on G:

$$\bar{A} = \{\bar{a}_1, \bar{a}_2, \bar{a}_3\}, \quad \bar{B} = \{\bar{b}_1, \bar{b}_2\}, \quad \bar{C} = \{\bar{c}_1, \bar{c}_2\}.$$

Every element $\bar{a}_\alpha$ is incident with every element $\bar{b}_\beta$ and every element $\bar{b}_\beta$ is incident with every element $\bar{c}_\gamma$ ($\alpha = 1, 2, 3$; $\beta, \gamma = 1, 2$). So we have $[\bar{A}, \bar{B}] = \bar{G}_{\max}$, $[\bar{B}, \bar{C}] = \bar{G}_{\max}$ and it is clear that $\bar{A}$, $\bar{B}$ and, at the same time, $\bar{B}$, $\bar{C}$ are complementary. Moreover, both elements $\bar{c}_1$, $\bar{c}_2$ are incident with $\bar{a}_2$ so that $[\bar{A}, \bar{C}] = \bar{G}_{\max}$ but the elements $\bar{a}_1$, $\bar{c}_2$, for example, are not incident. Hence $\bar{A}$, $\bar{C}$ are not complementary.

.2. Characteristic properties

Suppose, again, that $\bar{A}$, $\bar{B}$, $\bar{C}$ are decompositions on G.

If every two elements $\bar{a} \in \bar{A}$, $\bar{b} \in \bar{B}$ *lying in the same element of a common covering* $\bar{C}$ *of the decompositions* $\bar{A}$, $\bar{B}$ *are incident, then* $\bar{C} = [\bar{A}, \bar{B}]$ *and therefore the decompositions* $\bar{A}$, $\bar{B}$ *are complementary.*

Indeed, let $\bar{C}$ stand for a common covering of $\bar{A}$, $\bar{B}$ and let $\bar{c} \in \bar{C}$. Then $\bar{c}$ is the sum of certain elements of the decomposition $[\bar{A}, \bar{B}]$. Let $\bar{u}$, $\bar{v}$ be elements of $[\bar{A}, \bar{B}]$, lying in $\bar{c}$. Every element $\bar{a}_1 \in \bar{A}$ lying in $\bar{u}$ is incident with some element $\bar{b} \in \bar{B}$ which must, therefore, lie in $\bar{u}$ and, consequently, in $\bar{c}$. If $\bar{A}$, $\bar{B}$ have the above property, then $\bar{b}$ is incident with every element $\bar{a}_2 \in \bar{A}$ lying in $\bar{v}$ so that the two-membered sequence $\bar{a}_1$, $\bar{a}_2$ forms a binding $\{\bar{A}, \bar{B}\}$ from $\bar{a}_1$ to $\bar{a}_2$. Hence $\bar{v} = \bar{u}$ as well as $\bar{c} = \bar{u}$ and, furthermore, $\bar{C} \subset [\bar{A}, \bar{B}]$. Since every element of $[\bar{A}, \bar{B}]$ lies in an element of $\bar{C}$, there also holds the relation $\supset$, hence even the equality and the proof is complete.

The decompositions $\bar{A}$, $\bar{B}$ *are complementary if and only if for every two elements* $\bar{a}_1, \bar{a}_2 \in \bar{A}$ *lying in the same element* $\bar{u} \in [\bar{A}, \bar{B}]$ *there holds* $\bar{a}_1 \sqsubset \bar{B} = \bar{a}_2 \sqsubset \bar{B}$.

Proof. a) Suppose the decompositions $\bar{A}, \bar{B}$ are complementary. If an element $\bar{b} \in \bar{B}$ is incident with $\bar{a}_1$, then it lies in $\bar{u}$ and is, therefore, incident with $\bar{a}_2$. Hence $\bar{a}_1 \sqsubset \bar{B} \subset \bar{a}_2 \sqsubset \bar{B}$ and, analogously, $\bar{a}_2 \sqsubset \bar{B} \subset \bar{a}_1 \sqsubset \bar{B}$.

b) Suppose $\bar{a}_1 \sqsubset \bar{B} = \bar{a}_2 \sqsubset \bar{B}$. Let the elements $\bar{a} \in \bar{A}$, $\bar{b} \in \bar{B}$ lie in the same element $\bar{u} \in [\bar{A}, \bar{B}]$. The element $\bar{b}$ is incident with an element $\bar{x} \in \bar{A}$ and the latter lies in $\bar{u}$. So we have $\bar{b} \in \bar{x} \sqsubset \bar{B} = \bar{a} \sqsubset \bar{B}$ and, consequently, $\bar{a}$ and $\bar{b}$ are incident.

5.3. Further properties

Suppose $\bar{A}$, $\bar{B}$ are complementary decompositions on G.

For every two elements $\bar{a} \in \bar{A}$, $\bar{u} \in [\bar{A}, \bar{B}]$ *where* $\bar{a} \subset \bar{u}$ *there holds* $\bar{u} = \mathbf{s}(\bar{a} \sqsubset \bar{B})$.

In fact, let $\bar{a} \in \bar{A}$, $\bar{u} \in [\bar{A}, \bar{B}]$ be arbitrary elements such that $\bar{a} \subset \bar{u}$. Every point $u \in \bar{u}$ lies in a certain element $\bar{b} \in \bar{B}$ which is, of course, a part of $\bar{u}$. Since the decompositions $\bar{A}, \bar{B}$ are complementary, the elements $\bar{a}$, $\bar{b}$ are incident and, therefore, $\bar{b}$ is an element of the closure $\bar{a} \sqsubset \bar{B}$, namely $\bar{b} \in \bar{a} \sqsubset \bar{B}$. There follows $u \in \bar{b} \subset \mathbf{s}(\bar{a} \sqsubset \bar{B})$ and $\bar{u} \subset \mathbf{s}(\bar{a} \sqsubset \bar{B})$. Furthermore, every point $a \in \mathbf{s}(\bar{a} \sqsubset \bar{B})$ lies in a certain element $\bar{b} \in \bar{B}$ incident with $\bar{a}$ and $\bar{b}$ is a part of $\bar{u}$. Consequently, $a \in \bar{u}$ as well as $\mathbf{s}(\bar{a} \sqsubset \bar{B}) \subset \bar{u}$ and the above statement is correct.

Every decomposition $\bar{C}$ *on* G *that satisfies* $[\bar{A}, \bar{B}] \geqq \bar{C} \geqq \bar{A}$ *is complementary to* $\bar{B}$.

In fact, suppose $\bar{C}$ is a decomposition on G, satisfying the above relations. Then (3.7.2a; 3.4): $[\bar{A}, \bar{B}] \geqq [\bar{C}, \bar{B}] \geqq [\bar{A}, \bar{B}]$, so that (3.2): $[\bar{C}, \bar{B}] = [\bar{A}, \bar{B}]$. Consider arbitrary elements $\bar{c} \in \bar{C}$, $\bar{b} \in \bar{B}$ lying in the same element $\bar{u} \in [\bar{C}, \bar{B}]$. Since $[\bar{C}, \bar{B}] = [\bar{A}, \bar{B}]$, the elements $\bar{c}$, $\bar{b}$ are subsets of the same element $\bar{u} \in [\bar{A}, \bar{B}]$. From $\bar{C} \geqq \bar{A}$ there follows that $\bar{c}$ is the sum of some elements $\bar{a} \in \bar{A}$. As $\bar{A}, \bar{B}$ are complementary, $\bar{c}$, $\bar{b}$ are incident. Therefore $\bar{C}$, $\bar{B}$ are complementary.

Furthermore, there holds:

If the decomposition $\bar{X}$ *on* G *is a covering of* $\bar{A}$, *i.e.*, $\bar{X} \geqq \bar{A}$, *then* $\bar{A}$ *is complementary to* $(\bar{X}, \bar{B})$.

If the decomposition $\bar{Z}$ *on* G *is a refinement of* $\bar{A}$, *i.e.*, $\bar{Z} \leqq \bar{A}$, *then* $\bar{A}$ *is complementary to* $[\bar{Z}, \bar{B}]$.

Proof. Suppose $\bar{X} \geqq \bar{A}$. Consider an element $\bar{u} \in [\bar{A}, (\bar{X}, \bar{B})]$. We are to show that every two elements $\bar{a} \in \bar{A}$, $\bar{b}' \in (\bar{X}, \bar{B})$ contained in $\bar{u}$ are incident, so that $\bar{a} \cap \bar{b}' \neq \emptyset$. Indeed, by 3.7.2a and for convenient elements $\bar{x} \in \bar{X}$, $\bar{w} \in [\bar{A}, \bar{B}]$ we have $\bar{u} \subset \bar{x} \cap \bar{w}$; moreover, with regard to $\bar{b}'$ and for convenient $\bar{x}' \in \bar{X}$, $\bar{b} \in \bar{B}$, there holds $\bar{b}' = \bar{x}' \cap \bar{b}$. From $\bar{x}' \cap \bar{b} \subset \bar{u} \subset \bar{x} \cap \bar{w}$ there follows $\bar{x}' = \bar{x}$ and $\bar{b} \subset \bar{w}$.

Furthermore: $\bar{a} \subset \bar{u} \subset \bar{x} \cap \bar{w}$. Since $\bar{a}, \bar{b} \subset \bar{w}$ and the decompositions $\bar{A}, \bar{B}$ are complementary, there holds $\bar{a} \cap \bar{b} \neq \emptyset$ and since $\bar{a} \subset \bar{x}$, we have:

$$\bar{a} \cap \bar{b} = (\bar{a} \cap \bar{x}) \cap \bar{b} = \bar{a} \cap (\bar{x} \cap \bar{b}) = \bar{a} \cap \bar{b}'.$$

Consequently: $\bar{a} \cap \bar{b}' \neq \emptyset$.

b) Suppose $\bar{Z} \leqq \bar{A}$. Then we have (3.7.2a): $[\bar{B}, \bar{A}] \geqq [\bar{Z}, \bar{B}] \geqq \bar{B}$ and, by the above (second) statement, the assertion is correct.

5.4. Modularity

Let again $\bar{A}, \bar{B}$ stand for complementary decompositions on G.

If $\bar{X} \geqq \bar{A}$, then $\bar{B}$ is modular with respect to $\bar{X}, \bar{A}$.

Proof. Suppose $\bar{X}$ is a covering of $\bar{A}$, i.e., $\bar{X} \geqq \bar{A}$. Taking account of 3.7.2, our object is to show that $(\bar{X}, [\bar{A}, \bar{B}]) \leqq [\bar{A}, (\bar{X}, \bar{B})]$. Consider an element $\bar{u}' \in \in (\bar{X}, [\bar{A}, \bar{B}])$ so that $\bar{u}' = \bar{x} \cap \bar{u}$ for convenient elements $\bar{x} \in \bar{X}$, $\bar{u} \in [\bar{A}, \bar{B}]$. The element $\bar{u}$ is the sum of certain elements of $\bar{A}$ some of which, let us denote them $\bar{a}$, are incident with $\bar{x}$ whereas others, if there are any, are disjoint with $\bar{x}$. Since $\bar{X} \geqq \bar{A}$, there applies to every $\bar{a}$ the relation $\bar{x} \supset \bar{a}$. Hence $\bar{u}'$ is the sum of all the elements $\bar{a}$ and we have $\bar{u}' = \bigcup \bar{a}$. It remains to be shown that any two elements $\bar{a}$ may be connected in $(\bar{X}, \bar{B})$. Let, therefore, $\bar{a}_1$, $\bar{a}_2$ be such elements, so that $\bar{a}_1, \bar{a}_2 \subset \bar{x} \cap \bar{u}$. Since $\bar{A}, \bar{B}$ are complementary and $\bar{a}_1$, $\bar{a}_2$ lie in $\bar{u}$, there exists an element $\bar{b} \in \bar{B}$ which lies in $\bar{u}$ and is incident with $\bar{a}_1, \bar{a}_2$: $\bar{a}_1 \cap \bar{b} \neq \emptyset, \bar{a}_2 \cap \bar{b} \neq \emptyset$; as, moreover, $\bar{a}_1, \bar{a}_2$ lie in $\bar{x}$, we have $\bar{a}_1 \cap \bar{b} = \bar{a}_1 \cap (\bar{x} \cap \bar{b})$, $\bar{a}_2 \cap \bar{b} = \bar{a}_2 \cap (\bar{x} \cap \bar{b})$. It is easy to see that the elements $\bar{a}_1$, $\bar{a}_2$ are incident with $\bar{x} \cap \bar{b} \in (\bar{X}, \bar{B})$ so that the two-membered sequence $\bar{a}_1$, $\bar{a}_2$ is a binding $\{\bar{A}, (\bar{X}, \bar{B})\}$ from $\bar{a}_1$ to $\bar{a}_2$ and the proof is accomplished.

The above theorem cannot be converted. In fact, let us show that for two decompositions $\bar{A}_0$, $\bar{B}_0$ on the set G the following statement is correct: *if $\bar{B}_0$ is modular with regard to any covering of $\bar{A}_0$ and to $\bar{A}_0$ itself, then $\bar{A}_0$, $\bar{B}_0$ need not be complementary.*

Assuming the set G to consist of four elements: a_1, a_2, a_3, a_4, i.e., $G = \{a_1, a_2, a_3, a_4\}$, let $\bar{A}_0$, $\bar{B}_0$ be decompositions on G consisting of the elements:

$$\bar{a}_1 = \{a_1, a_2\}, \quad \bar{a}_2 = \{a_3, a_4\};$$

$$\bar{b}_1 = \{a_1\}, \quad \bar{b}_2 = \{a_2, a_3\}, \quad \bar{b}_3 = \{a_4\},$$

hence

$$\bar{A}_0 = \{\bar{a}_1, \bar{a}_2\}, \quad \bar{B}_0 = \{\bar{b}_1, \bar{b}_2, \bar{b}_3\}.$$

Then there holds $[\bar{A}_0, \bar{B}_0] = \{G\}$ and we see that, e.g., the elements $\bar{a}_1$ and $\bar{b}_3$ have no points in common; consequently, $\bar{A}_0$, $\bar{B}_0$ are not complementary. On the whole, there exist two coverings of $\bar{A}_0$, namely: $\bar{X}_1 = \bar{A}_0$, $\bar{X}_2 = \bar{G}_{\max}$ and $\bar{B}_0$ is modular with regard to both $\bar{X}_1, \bar{A}_0$ and $\bar{X}_2, \bar{A}_0$ (4.3).

From the above theorem we realize that the figures generated by the decompositions $\bar{X} \geqq \bar{A}, \bar{Y} \geqq \bar{B}$ have all the properties of modular decompositions described in (4.3). In particular, for

$$\mathring{A} = (\bar{X}, [\bar{A}, \bar{B}]) = [\bar{A}, (\bar{X}, \bar{B})],$$
$$\mathring{B} = (\bar{Y}, [\bar{B}, \bar{A}]) = [\bar{B}, (\bar{Y}, \bar{A})]$$

there hold the formulae (1), (2) given in 4.3. $\mathring{A}, \mathring{B}$ have even further properties based on the fact that $\bar{A}, \bar{B}$ are complementary. Let us just remark that *$\mathring{A}, \mathring{B}$ are complementary*, as the reader might verify by means of the formula $[\mathring{A}, \mathring{B}] = [\bar{A}, \bar{B}]$.

5.5. Local properties

Let again $\bar{A}, \bar{B}$ stand for complementary decompositions on G and $\bar{X}, \bar{Y}$ for coverings of $\bar{A}, \bar{B}$ so that $\bar{X} \geqq \bar{A}, \bar{Y} \geqq \bar{B}$. Let $\mathring{A}, \mathring{B}$ have the same meaning as in 5.4.

Let, moreover, $a \in G$ be an arbitrary point and $\bar{x} \in \bar{X}$, $\bar{a} \in \bar{A}$, $\bar{y} \in \bar{Y}$, $\bar{b} \in \bar{B}$ the elements of $\bar{X}, \bar{A}, \bar{Y}, \bar{B}$ containing a.

First, owing to the modularity of $\bar{A}, \bar{B}$, *the closures* $(\bar{x} \cap \bar{y}) \sqsubset \mathring{A}$, $(\bar{x} \cap \bar{y}) \sqsubset \mathring{B}$ *are coupled.*

Next, consider the following decompositions in G:

$$\bar{X}^a = \bar{x} \sqsubset \bar{A} \; (= \bar{A} \sqcap \bar{x}), \quad \bar{Y}^a = \bar{y} \sqsubset \bar{B} \; (= \bar{B} \sqcap \bar{y}).$$

We observe that the decomposition $\bar{X}^a$ lies on $\bar{x}$ and $\bar{a} \in \bar{X}^a$; analogously, the decomposition $\bar{Y}^a$ lies on $\bar{y}$ and $\bar{b} \in \bar{Y}^a$.

We shall prove that $\bar{X}^a$ *and* $\bar{Y}^a$ *are adjoint with regard to* $\bar{a}, \bar{b}$. To that purpose we must show that

$$\mathbf{s}(\bar{b} \sqsubset \bar{X}^a \sqcap \bar{y}) = \mathbf{s}(\bar{a} \sqsubset \bar{Y}^a \sqcap \bar{x}).$$

Indeed, let $\mathring{a} \in \mathring{A}$, $\mathring{b} \in \mathring{B}$ denote elements containing the point a. Since, by 5.3, the decompositions $\bar{A}$, $(X, \bar{B})$ are complementary and $\mathring{A}$ is their least common covering, we have (by 5.3)

$$\mathring{a} = \mathbf{s}\big((\bar{b} \cap \bar{x}) \sqsubset \bar{A}\big).$$

On taking account of $\bar{X} \geqq \bar{A}$, we see that the closure $(\bar{b} \cap \bar{x}) \sqsubset \bar{A}$ consists of exactly those elements of $\bar{A}$ that lie in $\bar{x}$ and are incident with $\bar{b}$ so that $(\bar{b} \cap \bar{x}) \sqsubset \bar{A} = \bar{b} \sqsubset \bar{X}^a$. Hence $\mathbf{s}(\bar{b} \sqsubset \bar{X}^a) = \mathring{a}$ and, moreover,

$$\mathbf{s}(\bar{b} \sqsubset \bar{X}^a \sqcap \bar{y}) = \mathbf{s}(\bar{b} \sqsubset \bar{X}^a) \cap \bar{y} = \mathring{a} \cap \bar{y} \in (\bar{Y}, \mathring{A}),$$

the last relation following from $\mathring{a} \cap \bar{y} \supset \{a\} \neq \emptyset$. Thus the set $\mathbf{s}(\bar{b} \sqsubset \bar{X}^a \sqcap \bar{y})$ is an element of $(\overline{Y}, \mathring{A})$ and, in fact, the element containing a. In a similar way we can verify that the set $\mathbf{s}(\bar{a} \sqsubset \overline{Y} \sqcap \bar{x})$ is the element of $(\overline{X}, \mathring{B})$, containing a. From this and from $(\overline{X}, \mathring{B}) = (\overline{Y}, \mathring{A})$ there follows the equality we were to prove.

5.6. Exercises

1. If the decompositions $\bar{A}, \bar{B}$ are complementary, then the formulae $(\mathring{A}, \mathring{B}) = (\overline{X}, \mathring{B}) = (\overline{Y}, \mathring{A}) = ((\overline{X}, \overline{Y}), [\bar{A}, \bar{B}])$, valid for modular decompositions $\overline{X} \geqq \bar{A}$, $\overline{Y} \geqq \bar{B}$ (see 4.3. (2)), may be completed by $(\mathring{A}, \mathring{B}) = [(\overline{X}, \bar{B}), (\overline{Y}, \bar{A})]$. In that case the decompositions $(\overline{X}, \bar{B})$, $(\overline{Y}, \bar{A})$ are complementary as well.

2. Show that in a set of four elements there exist, beside the pairs consisting of a covering and a refinement, only the following pairs of complementary decompositions: a) pairs of decompositions consisting of two elements each of which comprises only two points of the set; b) pairs of disjoint decompositions each of which contains three elements.

6. Mappings of sets

The theory of decompositions in sets considered in the previous chapters is the set-basis of the theory of groupoids and groups we intend to develop. But the results we have hitherto arrived at are only one part of the means necessary to attain our object. The other part consists of the theory of the mappings of sets, dealt with in the following chapters. The reader will certainly welcome the fact that the preceding, at times rather complicated, deliberations will now again be replaced by simpler ones.

6.1. Mappings into a set

In everyday life we often come across phenomena connected with the mathematical concept of mapping. Such phenomena are, in the simplest case, of the following kind: We have two nonempty sets G, G^* and between their elements a certain relation by which there corresponds, to each element of G, exactly one element of G^*. For example:

[1] Between the spectators at a certain performance and the tickets issued for the latter there exists the relation that each of the spectators is present on the ground of exactly one ticket.

[2] Between the pupils of a certain school and its classes there is the relation that each of the pupils belongs to exactly one class.

[3] The number n of certain objects is determined by way of associating each object with exactly one integer $1, 2, \ldots, n$; this is generally done by taking each of the objects, one by one, in hand and marking it, actually or only in mind, with one of the integers $1, 2, \ldots, n$.

Let G, G^* stand for nonempty sets. By a *mapping of the set G into G^** we understand a correspondence between the elements of both sets such that to each element of G there corresponds exactly one element of G^*; in other words, a relation by which each element of G is mapped exactly into one element of G^*.

A mapping of the set G into G^* is also called a *function* the domain of which is the set G and the range a part of G^*.

Consider an arbitrary mapping $\boldsymbol{g}$ of the set G into G^*. The mapping $\boldsymbol{g}$ associates, with each element $a \in G$, a certain element $a^* \in G^*$. The element a is called an *inverse image of* a^* and the element a^* the *image of* a under the mapping $\boldsymbol{g}$; we write $a^* = \boldsymbol{g}(a)$ or only $a^* = \boldsymbol{g}a$. Sometimes we also say that a^* is the *value of the function* $\boldsymbol{g}$ *in* a. Another way of notation is $\begin{pmatrix} a \\ a^* \end{pmatrix}$; the symbol $\begin{pmatrix} a & b & \ldots \\ a^* & b^* & \ldots \end{pmatrix}$ expresses $a^* = \boldsymbol{g}a$, $b^* = \boldsymbol{g}b, \ldots$

If A is a subset of G and A^* the subset of G^* consisting of the images of the individual elements of A, we write $A^* = \boldsymbol{g}(A)$ or only $A^* = \boldsymbol{g}A$. If $A \neq \emptyset$, then we can associate, with every element $a \in A$, the element $\boldsymbol{g}a \in G^*$ and thus obtain a mapping of the set A into G^*. It is called the *partial mapping* (*function*) *determined by* $\boldsymbol{g}$ and denoted $\boldsymbol{g}_A$.

By the definition of a mapping of G into G^* there corresponds, to an arbitrary element $a \in G$, exactly one image $a^* \in G^*$. Accordingly, such mappings are called *single-valued.*

In our study we shall sometimes meet with several mappings $\boldsymbol{g}, \boldsymbol{h}, \ldots$ simultaneously. In such cases we mark the concepts connected with the single mappings by a prefix, for example: $\boldsymbol{g}$-, $\boldsymbol{h}$-, ... and speak about $\boldsymbol{g}$-images, $\boldsymbol{h}$-inverse images, etc.

If two mappings $\boldsymbol{g}, \boldsymbol{h}$ of the set G into G^* are such that $\boldsymbol{g}a = \boldsymbol{h}a$ for each element $a \in G$, we call them *equal* and write $\boldsymbol{g} = \boldsymbol{h}$. In the opposite case we call them *different* and write $\boldsymbol{g} \neq \boldsymbol{h}$.

6.2. Mappings onto a set

By the definition of a mapping $\boldsymbol{g}$ of the set G into G^*, each element of G has, under the mapping $\boldsymbol{g}$, an image but, conversely, each element of G^* need not have an inverse image. If each element of G^* has an inverse image, then $\boldsymbol{g}$ is said to be a mapping of G onto G^*; we also say that the function $\boldsymbol{g}$ maps the set G onto G^*. If $\emptyset \neq A \subset G$, then $\boldsymbol{g}_A$ is evidently a mapping of the set A onto the set $\boldsymbol{g}A$.

From the above examples, the second [2] as well as the third [3] is a mapping onto a set: to each class there belongs at least one pupil associated with it under the mentioned mapping; if we have n objects and are to determine their number, then each object is marked by one of the numbers $1, 2, \ldots, n$. Example [1], on the other hand, is a mapping onto a set only if we assume that the house is quite full. In the opposite case, there have still remained some tickets for which there are no spectators.

6.3. Simple (one-to-one) mappings

In the notion of a mapping of the set G into G^* there is a further asymmetry with regard to both sets: Under the mapping $\boldsymbol{g}$ each element of G has exactly one image in G^* whereas, conversely, the same element on G^* may have several, even an infinite number of, inverse images in G.

If each element of G^* has, under $\boldsymbol{g}$, at most one inverse image, then $\boldsymbol{g}$ is called a *simple* or *one-to-one mapping of the set G into G^*.*

It is clear that $\boldsymbol{g}$ is a simple mapping of the set G onto the set G^* if and only if each element of G^* has exactly one inverse image.

The above example [3] is a simple mapping onto a set; [2] is an example of a simple mapping onto a set only if (in theory) each class has only one pupil; [1] is an example of a simple mapping onto a set only if the house is full and no tickets have remained.

6.4. Inverse mappings. Equivalent sets. Ordered finite sets

The concept of a simple mapping of a set onto a set is connected with two important notions: the notion of the inverse mapping and the notion of equivalent sets.

1. *Inverse mapping.* Suppose $\boldsymbol{g}$ is a simple mapping of the set G onto G^*. Then we can define a mapping of G^* onto G, denoted by $\boldsymbol{g}^{-1}$ and called the *inverse mapping* with regard to $\boldsymbol{g}$, in the following way: Each element $a^* \in G^*$ is, under $\boldsymbol{g}^{-1}$, associated with its $\boldsymbol{g}$-inverse image $a \in G$.

In example [1], provided the house is full and no tickets have remained, there corresponds, under $\boldsymbol{g}^{-1}$, to each ticket the spectator who owns it.

Obviously, the inverse mapping is simple and its inverse, $(\boldsymbol{g}^{-1})^{-1}$, is again the mapping $\boldsymbol{g}$, hence $(\boldsymbol{g}^{-1})^{-1} = \boldsymbol{g}$.

2. *Equivalent sets.* Given two nonempty sets G, G^*, there need not exist any mapping of G *onto* G^* as we see, for example, in the case when G consists of one element and G^* of two elements; therefore even a simple mapping of a set onto another does not necessarily exist.

Note that if there exists a simple mapping $\boldsymbol{g}$ of G onto G^*, then there also exists a simple mapping, $\boldsymbol{g}^{-1}$, in the opposite direction, i.e., a mapping of G^* onto G.

If there exists a simple mapping $\boldsymbol{g}$ of G onto G^*, then the set G *is said to be equivalent to* G^*. Then, of course, the set G^* is also equivalent to G. With regard to this symmetry, we speak about equivalent sets G, G^* without differentiating which is equivalent to which. The equivalence of sets is expressed by the formulae: $G^* \simeq G$ or $G \simeq G^*$.

For example, every set A consisting of n (> 0) elements and the set $1, 2, \ldots, n$ are equivalent because, if the elements of A are denoted, let us say, $a_1, a_2, \ldots, a_n$ (it makes no difference for which element each symbol stands), then we have a simple mapping of A onto the set $(1, 2, \ldots, n)$, namely:

$$\begin{pmatrix} a_1 & a_2 & \ldots & a_n \\ 1 & 2 & \ldots & n \end{pmatrix}.$$

3. *Ordered finite sets.* If the set A consists of n (> 0) elements and a simple mapping of A onto the set $\{1, 2, \ldots, n\}$ is given, then A is said to be an *ordered set* and the mapping is called an *ordering of A*. An ordering of A is obtained, for example, by way of ranging its elements in a certain order, i.e., a certain element $a_1 \in A$ is marked as first, the next one as second, etc. and the last: $a_n \in A$ as the nth. Then A is said to be the ordered set of elements $a_1, a_2, \ldots, a_n$. This notion therefore depends on the order in which the names of the individual elements are quoted or written.

By the *inversely ordered set* we mean the ordered set $\{a_1', \ldots, a'_{n-1}, a_n'\}$ where $a_1' = a_n, \ldots, a_n' = a_1$.

6.5. The decompositions of sets, corresponding to mappings

Let $\boldsymbol{g}$ stand for a mapping of the set G onto G^*. We have already noticed that an element $a^* \in G^*$ may have, under $\boldsymbol{g}$, several inverse images.

Consider the system $\bar{G}$ of all subsets $\bar{a}$ of G each of which is formed by all the inverse images under $\boldsymbol{g}$ of an element $a^* \in G^*$. Each element of $\bar{G}$ is therefore a subset of G, consisting of all the points mapped, under $\boldsymbol{g}$, onto the same point of G^*. Since G^* contains at least one element a^*, the system $\bar{G}$ is not empty because it contains the set $\bar{a}$ consisting of the inverse images of a^*. As $\boldsymbol{g}$ is a mapping of G onto G^*, each element of G^* has at least one inverse image, hence the set $\bar{a}$ of the inverse of each element $a^* \in G^*$ is not empty. $\bar{G}$ is therefore a nonempty system of nonempty subsets of G.

Moreover, it is easy to see that the system $\bar{G}$ is disjoint, i.e., every two of its elements are disjoint, and that it covers G (each element $a \in G$ has exactly one image $a^* \in G^*$ and therefore lies in exactly one element $\bar{a} \in \bar{G}$, namely in the set of the inverse images of a^*). Consequently, *the system $\bar{G}$ of all subsets of G, each of* *which is formed by all the inverse images under the mapping* $\boldsymbol{g}$ *of some element of* G^*, *is a decomposition of the set* G. We say that this decomposition *corresponds* or *belongs to the mapping* $\boldsymbol{g}$.

In the above example [2], the corresponding decomposition consists of single sets of pupils belonging to the same class.

Note, in particular, the following extreme cases: If the set G^* consists of one element only, then the corresponding decomposition $\overline{G}$ is $\overline{G}_{\max}$. If $\boldsymbol{g}$ is a simple mapping, then the corresponding decomposition is $\overline{G}_{\min}$.

6.6. Mappings of sets into and onto themselves

The above deliberations do not exclude that G^* may be identical with G. If $G^* = G$, then we speak about a mapping of the set G *into* or *onto itself*.

Associating, for example, with every natural number n the number $n + 1$, we obtain a mapping of the set of all natural numbers into itself.

The simplest mapping of the set G onto itself is obtained by associating, with every element $a \in G$, again the element a; it is the so-called *identical mapping of* G, denoted $\boldsymbol{e}$.

A simple mapping of the set G onto itself is called a *permutation of* G. Permutations of finite sets are the object of a more detailed study in Chapter 8.

6.7. Composition of mappings

The concept of a composite mapping. Let G, H, K stand for arbitrary nonempty sets, $\boldsymbol{g}$ denote a mapping of the set G into H and $\boldsymbol{h}$ a mapping of the set H into K. Then there corresponds, under the mapping $\boldsymbol{g}$, to every element $a \in G$ a certain element $\boldsymbol{g}a \in H$ and to $\boldsymbol{g}a$ there corresponds, under the mapping $\boldsymbol{h}$, an element $\boldsymbol{h}(\boldsymbol{g}a) \in K$. Associating with every element $a \in G$ the element $\boldsymbol{h}(\boldsymbol{g}a) \in K$, we have a mapping of the set G into K. It is called the *composite mapping of* $\boldsymbol{g}$ *and* $\boldsymbol{h}$ (in this order) and is denoted by $\boldsymbol{hg}$. As a mapping of the set G into K, $\boldsymbol{hg}$ has the property that, for $a \in G$, there holds $(\boldsymbol{hg})a = \boldsymbol{h}(\boldsymbol{g}a)$.

Let us note some particular cases. If $\boldsymbol{g}$ maps the set G onto H and $\boldsymbol{h}$ maps the set H onto K, then $\boldsymbol{hg}$ is obviously a mapping of the set G onto K.

If both $\boldsymbol{g}$ and $\boldsymbol{h}$ are simple mappings, then $\boldsymbol{hg}$ is simple as well because, in that case, any two different elements $a, b \in G$ have two different $\boldsymbol{g}$-images: $\boldsymbol{g}a, \boldsymbol{g}b \in H$ and the latter have two different $\boldsymbol{h}$-images: $\boldsymbol{hg}a, \boldsymbol{hg}b \in K$.

Furthermore, it is clear that if the set K is identical with G so that $\boldsymbol{h}$ is a mapping of the set H into G, then $\boldsymbol{hg}$ is a mapping of the set G into itself; if $\boldsymbol{g}$ maps the set G onto H and $\boldsymbol{h}$ the set H onto G, then $\boldsymbol{hg}$ is a mapping of the set G onto itself; in particular, if the mapping $\boldsymbol{g}$ is simple and $\boldsymbol{h} = \boldsymbol{g}^{-1}$, then $\boldsymbol{hg}$ is the identical mapping of the set G.

Note, moreover, that if the sets H and K are both identical with G so that both $\boldsymbol{g}$ and $\boldsymbol{h}$ are mappings of G into itself, then even $\boldsymbol{hg}$ is a mapping of G into itself; if both $\boldsymbol{g}$ and $\boldsymbol{h}$ map G onto itself, then even $\boldsymbol{hg}$ maps G onto itself.

A simple mapping $\boldsymbol{g}$ of the set G onto itself is called *involutory* if the composite mapping $\boldsymbol{gg}$ is the identical mapping of G: $\boldsymbol{gg} = \boldsymbol{e}$. The inverse mapping $\boldsymbol{g}^{-1}$ of any involutory mapping $\boldsymbol{g}$ obviously equals $\boldsymbol{g}$, i.e., $\boldsymbol{g}^{-1} = \boldsymbol{g}$.

Finally, let us note that for the identical mapping $\boldsymbol{e}$ of the set G and for an arbitrary mapping $\boldsymbol{g}$ of G into itself there holds: $\boldsymbol{eg} = \boldsymbol{ge} = \boldsymbol{g}$.

Example of a composite mapping: If $\boldsymbol{g}$ denotes the mapping considered in the above example [1] and $\boldsymbol{h}$ stands for the mapping of the set of tickets into the set of colours associated with the tickets, then the composite mapping $\boldsymbol{hg}$ associates, with every spectator, a certain colour, namely the colour of his ticket.

The associative law for the composition of mappings. Let us now consider three mappings $\boldsymbol{g}, \boldsymbol{h}, \boldsymbol{k}$, where $\boldsymbol{k}$ stands for a mapping of the set K into some set L (without excluding the case that L is identical with one of the sets G, H, K). An important property of the composition of mappings consists in that there holds:

$$\boldsymbol{k}(\boldsymbol{hg}) = (\boldsymbol{kh})\boldsymbol{g},$$

called the *associative law for the composition of mappings.*

The above equality expresses that every element of G has, under both the mappings $\boldsymbol{k}(\boldsymbol{hg})$ and $(\boldsymbol{kh})\boldsymbol{g}$, the same image lying, of course, in the set L.

To prove this, let us consider the image of an element $a \in G$ under the mapping $\boldsymbol{k}(\boldsymbol{hg})$. The $\boldsymbol{k}(\boldsymbol{hg})$-image of a is the image of the element $(\boldsymbol{hg})a$ under the mapping $\boldsymbol{k}$ and is therefore obtained by associating, with the element $\boldsymbol{g}a \in H$, its $\boldsymbol{h}$-image $\boldsymbol{h}(\boldsymbol{g}a) \in K$ and then, with the latter, its $\boldsymbol{k}$-image $\boldsymbol{k}(\boldsymbol{hg})a \in L$. But the $\boldsymbol{k}$-image of the element $\boldsymbol{h}(\boldsymbol{g}a)$ is, by the definition of the mapping $\boldsymbol{kh}$, the same as the $(\boldsymbol{kh})\boldsymbol{g}$-image of the element a. Consequently, the above equality is true.

Instead of $\boldsymbol{k}(\boldsymbol{hg})$ or $(\boldsymbol{kh})\boldsymbol{g}$ we simply write $\boldsymbol{khg}$.

6.8. The equivalence theorems

Let us now introduce three theorems called equivalence theorems. They can, owing to their simplicity, be justly regarded as describing the properties of certain equivalent sets. Their value is due to the fact that they express the set-structure of important situations connected with the so-called theorems of isomorphism we shall deal with in the theory of groupoids and groups.

1. The first equivalence theorem. *If there exists a mapping of the set G onto the set G^*, then G^* is equivalent to a certain decomposition lying on G and vice versa. The mapping of the decomposition $\bar{G}$ belonging to a mapping $\boldsymbol{g}$ of the set G onto G^* under which there corresponds, to every element $\bar{a} \in \bar{G}$, the $\boldsymbol{g}$-image of the points lying in $\bar{a}$, is simple.*

Indeed, if there exists a mapping $\boldsymbol{g}$ of the set G onto G^*, then the set G^* is equivalent to the decomposition $\bar{G}$ belonging to $\boldsymbol{g}$. A simple mapping of $\bar{G}$ onto G^* is obtained by associating, with every element $\bar{a} \in \bar{G}$, the $\boldsymbol{g}$-image of the points $a \in G$ lying in $\bar{a}$. If, conversely, there exists a simple mapping $\boldsymbol{i}$ of a decomposition $\bar{G}$ of the set G onto the set G^*, then the composite mapping $\boldsymbol{ij}$ maps the set G onto G^*; $\boldsymbol{j}$ denotes the mapping of the set G onto the decomposition $\bar{G}$, associating with each point $a \in G$ that element $\bar{a} \in \bar{G}$ which contains a: $a \in \bar{a} = \boldsymbol{j}a \in \bar{G}$. The decomposition of G belonging to the mapping $\boldsymbol{ij}$ is $\bar{G}$.

2. Second theorem. *Every two coupled decompositions* $\bar{A}$, $\bar{B}$ *in* G *are equivalent*, i.e., $\bar{A} \simeq \bar{B}$. *The mapping of the decomposition* $\bar{A}$ *onto* $\bar{B}$ *under which there corresponds, to every element* $\bar{a} \in \bar{A}$, *the element* $\bar{b} \in \bar{B}$ *incident with* $\bar{a}$, *is simple.*

An important case of this theorem (see 4.1) concerns the equivalence of the closure and the intersection of a subset $X \subset G$ and a decomposition $\bar{Y}$ in G: *If* $X \cap \mathbf{s}\bar{Y} \neq \emptyset$, *then there holds* $X \sqsubset \bar{Y} \simeq \bar{Y} \sqcap X$. *The mapping given by the incidence of the elements is simple.*

3. Third theorem. *A decomposition* $\bar{\bar{B}}$ *of some decomposition* $\bar{B}$ *of the set G and the covering* $\bar{A}$ *of* $\bar{B}$, *enforced by* $\bar{\bar{B}}$, *are equivalent sets*, i.e., $\bar{\bar{B}} \simeq \bar{A}$. *The mapping of the decomposition* $\bar{\bar{B}}$ *onto* $\bar{A}$ *under which there corresponds, to every element* $\bar{\bar{b}} \in \bar{\bar{B}}$, *the sum* $\bar{a} \in \bar{A}$ *of the elements of* $\bar{B}$ *lying in* $\bar{\bar{b}}$, *is simple.*

6.9. Mappings of sequences and α-grade structures

In this chapter we shall deal with some more complicated notions based on the concept of the equivalence of sets.

1. *Mappings of sequences.* Let α ($\geqq 1$) be a positive integer. Consider wo arbitrary α-membered sequences:

$$(a) = (a_1, \ldots, a_\alpha), \quad (b) = (b_1, \ldots, b_\alpha).$$

a) By a *mapping* $\boldsymbol{a}$ *of the sequence* (a) *onto the sequence* (b) we naturally understand a simple mapping (6.10.2) of the set formed by the members of (a) onto the set of the members of (b). Under any mapping $\boldsymbol{a}$ of the sequence (a) onto the sequence (b) there corresponds, therefore, to each member a_γ of (a), exactly one member $b_\delta = \boldsymbol{a}a_\gamma$ of (b) and, simultaneously, to members of different indices there correspond, in (b), members with different indices as well. Every mapping $\boldsymbol{a}$ of the sequence (a) onto (b) is uniquely determined by a certain permutation $\boldsymbol{p}$ of the set $\{1, \ldots, \alpha\}$ in the sense of the formula: $\boldsymbol{a}a_\gamma = b_{\boldsymbol{p}\gamma}$ ($\gamma = 1, \ldots, \alpha$). The function inverse of a mapping of the sequence (a) onto (b) is, of course, a mapping of the sequence (b) onto (a).

It is clear that there exist (even in the number $\alpha!$) mappings of the sequence (a) onto (b) as well as mappings in the opposite direction. We see that the sequences (a) and (b) are equivalent. Every two finite sequences of the same length are equivalent.

b) Suppose the members $a_1, \ldots, a_\alpha$ of the sequence (a) as well as the members $b_1, \ldots, b_\alpha$ of the sequence (b) are nonempty sets.

The *sequence* (b) *is* said to be *strongly equivalent to the sequence* (a) if the following situation occurs: There exists a mapping $\boldsymbol{a}$ of the sequence (a) onto (b) such that, to each member a_γ of (a), there corresponds a simple mapping $\boldsymbol{a}_\gamma$ of a_γ onto the member $b_\delta = \boldsymbol{a} a_\gamma$ of (b).

If (b) is strongly equivalent to (a), then (a) has, obviously, the same property with regard to (b). On taking account of this symmetry, we call the sequences (a) and (b) strongly equivalent.

c) Assume the members $a_1, \ldots, a_\alpha$ of the sequence (a) as well as the members $b_1, \ldots, b_\alpha$ of (b) to be decompositions in the set G.

The *sequence* (b) *is* said to be *semi-coupled* or *loosely coupled* (*coupled*) *with* (a) in the following situation: There exists a mapping $\boldsymbol{a}$ of the sequence (a) onto (b) such that every member a_γ of (a) is semi-coupled (coupled) with its $\boldsymbol{a}$-image $b_\delta = \boldsymbol{a} a_\gamma$ in (b).

If (b) is semi-coupled (coupled) with (a), then the sequence (a) has evidently the same property with regard to (b). In that case, on taking account of this symmetry, we call the sequences (a) and (b) semi-coupled (coupled).

Suppose (b) is semi-coupled with (a) and let $\boldsymbol{a}$ stand for a mapping of the sequence (a) onto (b), determining the pairs of semi-coupled (coupled) members. Consider a member a_γ of (a) and its $\boldsymbol{a}$-image $b_\delta = \boldsymbol{a} a_\gamma$ in (b). Then the closures $\mathrm{H}a_\gamma = b_\delta \sqsubset a_\gamma$, $\mathrm{H}b_\delta = a_\gamma \sqsubset b_\delta$ are nonempty and coupled (4.1). According to the second equivalence theorem (6.8), the mapping $\boldsymbol{a}_\gamma$ of the closure $\mathrm{H}a_\gamma$ onto the closure $\mathrm{H}b_\delta$, determined by the incidence of the elements, is simple. In particular, if the sequence (b) is coupled with (a), we have $\mathrm{H}a_\gamma = a_\gamma$, $\mathrm{H}b_\delta = b_\delta$. We see that *two coupled sequences are always strongly equivalent.*

2. *Mappings of α-grade structures.* Let α $(\geqq 1)$ be a positive integer and $\big((A) =\big)$ $(A_1, \ldots, A_\alpha)$, $\big((B) =\big)$ $(B_1, \ldots, B_\alpha)$ stand for arbitrary sequences of nonempty sets. Let, moreover, $\tilde{A}$ be an α-grade structure with regard to the sequence (A) and $\tilde{B}$ a structure of the same kind with regard to (B) (1.9).

Note that any element $\bar{\bar{a}} \in \tilde{A}$ $(\bar{\bar{b}} \in \tilde{B})$ is an α-membered sequence $\bar{\bar{a}} = (\bar{a}_1, \ldots, \bar{a}_\alpha)$ $\big(\bar{\bar{b}} = (\bar{b}_1, \ldots, \bar{b}_\alpha)\big)$ every member $\bar{a}_\gamma$ $(\bar{b}_\gamma)$ of which is a nonempty part of the set A_γ (B_γ); $(\gamma = 1, \ldots, \alpha)$.

Suppose there is a simple mapping $\boldsymbol{f}$ of the structure $\tilde{A}$ onto $\tilde{B}$.

a) The *mapping* $\boldsymbol{f}$ *is* called *a strong equivalence-mapping of the structure* $\tilde{A}$ *onto* $\tilde{B}$, briefly, a *strong equivalence of* $\tilde{A}$ *onto* $\tilde{B}$ in the following situation:

There exists a permutation $\boldsymbol{p}$ of the set $\{1, \ldots, \alpha\}$ with the following effect: To every member $\bar{a}_\gamma$ of any element $\bar{\bar{a}} = (\bar{a}_1, \ldots, \bar{a}_\alpha) \in \tilde{A}$ there exists a simple mapping $\boldsymbol{a}_\gamma$ of $\bar{a}_\gamma$ onto the member $\bar{b}_{\boldsymbol{p}\gamma}$ of the sequence $\boldsymbol{f}\bar{\bar{a}} = \bar{\bar{b}} = (\bar{b}_1, \ldots, \bar{b}_\alpha) \in \tilde{B}$; $(\gamma = 1, \ldots, \alpha)$.

We observe that the inverse function $\boldsymbol{f}^{-1}$ of an arbitrary strong equivalence $\boldsymbol{f}$ of the structure $\tilde{A}$ onto $\tilde{B}$ is a strong equivalence in the opposite direction, i.e., a strong equivalence of $\tilde{B}$ onto $\tilde{A}$.

If there exists a strong equivalence of $\tilde{A}$ onto $\tilde{B}$, then we say that the *structure $\tilde{B}$ is strongly equivalent to $\tilde{A}$*. The notion of strong equivalence is, of course, symmetric with regard to both structures; therefore we also speak about strongly equivalent structures $\tilde{A}$, $\tilde{B}$.

b) Let us now assume that the sequences (A), (B) consist of the decompositions $\bar{A}_1, \ldots, \bar{A}_\alpha$ and $\bar{B}_1, \ldots, \bar{B}_\alpha$ in the set G. Then every element $\bar{\bar{a}} = (\bar{a}_1, \ldots, \bar{a}_\alpha) \in \tilde{A}$ $(\bar{\bar{b}} = (\bar{b}_1, \ldots, \bar{b}_\alpha) \in \tilde{B})$ is an α-membered sequence every member of which, $\bar{a}_\gamma$ $(\bar{b}_\gamma)$, is a decomposition in G, namely, a part of the decomposition $\bar{A}_\gamma$ $(\bar{B}_\gamma)$; $(\gamma = 1, \ldots, \alpha)$.

The mapping $\boldsymbol{f}$ of the structure $\tilde{A}$ onto $\tilde{B}$ is called *equivalence-mapping connected with semi-coupling* or *equivalence-mapping connected with loose coupling* (*equivalence-mapping connected with coupling*), briefly *equivalence connected with semi-coupling* or *with loose coupling* (*with coupling*) if the following situation occurs:

There exists a permutation $\boldsymbol{p}$ of the set $\{1, \ldots, \alpha\}$ with the following effect: Every member $\bar{a}_\gamma$ of any element $\bar{\bar{a}} = (\bar{a}_1, \ldots, \bar{a}_\alpha) \in \tilde{A}$ is semi-coupled (coupled) with the member $\bar{b}_{\boldsymbol{p}\gamma}$ of the element $\boldsymbol{f}\bar{\bar{a}} = \bar{\bar{b}} = (\bar{b}_1, \ldots, \bar{b}_\alpha) \in \tilde{B}$ $(\gamma = 1, \ldots, \alpha)$.

It is easy to see that the inverse function $\boldsymbol{f}^{-1}$ of an arbitrary equivalence connected with semi-coupling (coupling) of the structure $\tilde{A}$ onto the structure $\tilde{B}$ is an equivalence-mapping of $\tilde{B}$ onto $\tilde{A}$ which is of the same type.

Let $\boldsymbol{f}$ denote an equivalence connected with a loose coupling (coupling) of the structure $\tilde{A}$ onto $\tilde{B}$. Consider arbitrary members $\bar{a}_\gamma$, $\bar{b}_\delta$ $(\delta = \boldsymbol{p}\gamma)$ which are in the above relation so that $\bar{a}_\gamma$ is in $\bar{\bar{a}}$, $\bar{b}_\delta$ is in $\boldsymbol{f}\bar{\bar{a}} = \bar{\bar{b}}$ and the decompositions $\bar{a}_\gamma$, $\bar{b}_\delta$ are semi-coupled (coupled). Then the closures $\mathrm{H}\bar{a}_\gamma = \bar{b}_\delta \sqsubset \bar{a}_\gamma$, $\mathrm{H}\bar{b}_\delta = \bar{a}_\gamma \sqsubset \bar{b}_\delta$ are non-empty and coupled (4.1). By the second equivalence theorem (6.8), the mapping $\boldsymbol{a}_\gamma$ of the closure $\mathrm{H}\bar{a}_\gamma$ onto $\mathrm{H}\bar{b}_\delta$, given by the incidence of the elements, is simple. In particular, if $\boldsymbol{f}$ is an equivalence connected with coupling, we have $\mathrm{H}\bar{a}_\gamma = \bar{a}_\gamma$, $\mathrm{H}\bar{b}_\delta = \bar{b}_\delta$. We observe that *every equivalence of the structure $\tilde{A}$ onto $\tilde{B}$, connected with coupling, is a strong equivalence.*

If there exists an equivalence connected with semi-coupling (coupling) of the structure $\tilde{A}$ onto $\tilde{B}$, we say that $\tilde{B}$ *is equivalent to and semi-coupled* or *loosely coupled* (*coupled*) *with* $\tilde{A}$. These notions are obviously symmetric with regard to both structures; for that reason we speak about equivalent and semi-coupled or equivalent and loosly coupled (equivalent and coupled) structures $\tilde{A}$, $\tilde{B}$. Especially, *every two equivalent and coupled structures $\tilde{A}$, $\tilde{B}$ are strongly equivalent.*

6.10. Exercises

1. Consider some simple real functions (for example $y = ax + b$ or $y = x^2$ and similar) as particular cases of the above concept of a function.

2. If the sets G and G^* are finite and of the same order, then: a) every mapping of G onto G^* is simple; b) every simple mapping of G into G^* is a mapping *onto* G^*.

3. Assume $A \subset G$ and let $\boldsymbol{g}[A]$ denote the mapping of G into the set $\{0, 1\}$, defined als follows: For $a \in G$ there is $\boldsymbol{g}[A]a = 1$ or 0 according as a lies or does not lie in A. Prove that the following relations are true:
   a) $\boldsymbol{g}[A \cap B]a = (\boldsymbol{g}[A]a) \cdot (\boldsymbol{g}[B]a) =$ the least of the numbers $\boldsymbol{g}[A]a, \boldsymbol{g}[B]a$;
   b) $\boldsymbol{g}[A \cup B]a =$ the greatest of the numbers $\boldsymbol{g}[A]a, \boldsymbol{g}[B]a$;
   c) if $A \cap B = \emptyset$, then $\boldsymbol{g}[A \cup B]a = \boldsymbol{g}[A]a + \boldsymbol{g}[B]a$.

4. Let $\boldsymbol{f}[a]$ denote the mapping of a straight line onto itself, defined as follows: to every point of the straight line with the coordinate x there corresponds the point with the coordinate $x' = x + a$, a standing for a real number. Similarly, let $\boldsymbol{g}[a]$ be the mapping of the straight line onto itself, given by the formula $x' = -x + a$. The distance between two arbitrary points x_1, x_2 of the straight line, i.e., the number[1] $|x_1 - x_2|$ and the distance between their images under both mappings $\boldsymbol{f}[a]$ and $\boldsymbol{g}[a]$ are equal. Under the mapping $\boldsymbol{f}[a]$, no point of the straight line is mapped onto itself unless $a = 0$ and then we have the identical mapping of the straight line onto itself; under the mapping $\boldsymbol{g}[a]$ exactly one point is mapped onto itself. For the composition of the mappings $\boldsymbol{f}[a]$ and $\boldsymbol{g}[a]$ there hold the following formulae:

$$\boldsymbol{f}[b]\boldsymbol{f}[a] = \boldsymbol{f}[a + b], \quad \boldsymbol{g}[b]\boldsymbol{f}[a] = \boldsymbol{g}[-a + b],$$
$$\boldsymbol{f}[b]\boldsymbol{g}[a] = \boldsymbol{g}[a + b], \quad \boldsymbol{g}[b]\boldsymbol{g}[a] = \boldsymbol{f}[-a + b].$$

Remark. The mappings $\boldsymbol{f}[a]$ and $\boldsymbol{g}[a]$ are called *Euclidean motions on a straight line.*

5. Let $\boldsymbol{f}[\alpha; a, b]$ denote the mapping of a plane onto itself, defined in the following way: to every point in the plane, with the coordinates x, y, there corresponds the point with the coordinates x', y', where

$$x' = x \,.\, \cos\alpha + y \,.\, \sin\alpha + a,$$
$$y' = -x \,.\, \sin\alpha + y \,.\, \cos\alpha + b,$$

α, a, b denoting real numbers. Similarly, let $\boldsymbol{g}[\alpha; a, b]$ be the mapping of the plane onto itself, given by the formulae

$$x' = x \cdot \cos\alpha + y \cdot \sin\alpha + a,$$
$$y' = x \cdot \sin\alpha - y \cdot \cos\alpha + b.$$

The distance between two arbitrary points x_1, y_1 and x_2, y_2 in the plane, namely, the number $\sqrt{[(x_1 - x_2)^2 + (y_1 - y_2)^2]}$ and the distance between their images in both mappings $\boldsymbol{f}[\alpha; a, b]$ and $\boldsymbol{g}[\alpha; a, b]$ are equal. Under the mapping $\boldsymbol{f}[\alpha; a, b]$, where α is a multiple of 2π, no point in the plane is mapped onto itself except in the case: $a = b = 0$ and then we have the identical mapping of the plane onto itself. If α is not a multiple of 2π, then

---

[1]) If x is an arbitrary number, then $|x|$ denotes the absolute value of x, namely, the non-negative number of both x and $-x$.

exactly one point in the plane is mapped onto itself. Under the mapping $\boldsymbol{g}[\alpha; a, b]$ no point in the plane is mapped onto itself unless the numbers α, a, b are connected by the relation:

$$a \,.\, \cos \frac{1}{2} \alpha + b \,.\, \sin \frac{1}{2} \alpha = 0;$$

in that case all the points in the plane that are mapped onto themselves form a straight line. For the composition of the mappings $\boldsymbol{f}[\alpha; a, b]$, $\boldsymbol{g}[\alpha; a, b]$ there hold the following formulae:

$$\begin{aligned}
\boldsymbol{f}[\beta; c, d]\ \boldsymbol{f}[\alpha; a, b] &= \boldsymbol{f}[\alpha + \beta;\ a \,.\, \cos \beta + b \,.\, \sin \beta + c, \\
&\qquad -a \,.\, \sin \beta + b \,.\, \cos \beta + d], \\
\boldsymbol{g}[\beta; c, d]\ \boldsymbol{f}[\alpha; a, b] &= \boldsymbol{g}[\alpha + \beta;\ a \,.\, \cos \beta + b \,.\, \sin \beta + c, \\
&\qquad a \,.\, \sin \beta - b \,.\, \cos \beta + d], \\
\boldsymbol{f}[\beta; c, d]\ \boldsymbol{g}[\alpha; a, b] &= \boldsymbol{g}[\alpha - \beta;\ a \,.\, \cos \beta + b \,.\, \sin \beta + c, \\
&\qquad -a \,.\, \sin \beta + b \,.\, \cos \beta + d], \\
\boldsymbol{g}[\beta; c, d]\ \boldsymbol{g}[\alpha; a, b] &= \boldsymbol{f}[\alpha - \beta;\ a \,.\, \cos \beta + b \,.\, \sin \beta + c, \\
&\qquad a \,.\, \sin \beta - b \,.\, \cos \beta + d].
\end{aligned}$$

Remark. The mappings $\boldsymbol{f}[\alpha; a, b]$ and $\boldsymbol{g}[\alpha; a, b]$ are called *Euclidean motions in a plane*.

6. Every α-membered (infinite) sequence on a set A is the set formed from the images of the elements of the set $\{1, \ldots, \alpha\}$ ($\{1, 2, \ldots\}$) onto A under a convenient mapping of the latter onto the set A (1.7).

7. For the equivalence of nonempty sets A, B, C the following statements are correct: a) $A \simeq A$ (reflexivity); b) from $A \simeq B$ there follows $B \simeq A$ (symmetry); c) from $A \simeq B$, $B \simeq C$ there follows $A \simeq C$ (transitivity) (6.4).

8. Let $\boldsymbol{g}$, $\boldsymbol{h}$ denote mappings of the set G into itself and $\bar{G}_{\boldsymbol{g}}$, $\bar{G}_{\boldsymbol{h}}$, $\bar{G}_{\boldsymbol{hg}}$ be decompositions on G, corresponding to the mappings $\boldsymbol{g}$, $\boldsymbol{h}$, $\boldsymbol{hg}$. Show that the following relations apply:
   a) $\boldsymbol{hg}G \subset \boldsymbol{h}G$, $\bar{G}_{\boldsymbol{hg}} \geqq \bar{G}_{\boldsymbol{g}}$,
   b) the equality $\boldsymbol{hg}G = \boldsymbol{h}G$ yields $\boldsymbol{g}G \sqsubset \bar{G}_{\boldsymbol{h}} = \bar{G}_{\boldsymbol{h}}$ and vice versa,
   c) the equality $\bar{G}_{\boldsymbol{hg}} = \bar{G}_{\boldsymbol{g}}$ yields $\boldsymbol{g}G \sqcap \bar{G}_{\boldsymbol{h}} = (\overline{\boldsymbol{g}G})_{\min}$ and vice versa. ($(\overline{\boldsymbol{g}G})_{\min}$ is the least decomposition of the set $\boldsymbol{g}G$.)

9. Any two adjoint chains of decompositions in G have a coupled refinement. (Prove it by means of the construction described in 4.2.)

7. Mappings of decompositions

Let $\boldsymbol{g}$ denote a mapping of the set G onto a set G^*. Thus every element $a \in G$ is, under $\boldsymbol{g}$, mapped onto a certain element $a^* \in G^*$; a^* is the image of the element a under the mapping $\boldsymbol{g}$. To the mapping $\boldsymbol{g}$ there corresponds a certain decomposition $\bar{G}$ on G; each element of $\bar{G}$ consists of all $\boldsymbol{g}$-inverse images of the same point in G^*. The decomposition $\bar{G}$ is equivalent to the set G^*.

7.1. Extended mappings

The mapping $\boldsymbol{g}$ determines a mapping $\bar{\boldsymbol{g}}$ of the system of all subsets of G into the system of all subsets of G^*, the so-called *extended mapping*. $\bar{\boldsymbol{g}}$ is defined in the way that, for $\emptyset \neq A \subset G$, $\bar{\boldsymbol{g}}A \subset G^*$ is the set of the $\boldsymbol{g}$-images of all the points lying in A; moreover, we put $\bar{\boldsymbol{g}}\emptyset = \emptyset$. In particular, for $\bar{a} \in \bar{G}$, the set $\bar{\boldsymbol{g}}\bar{a}$ consists of a single point of G^*, namely, of the $\boldsymbol{g}$-image of the points of G lying in $\bar{a}$.

To simplify the notation, we generally write $\boldsymbol{g}$ instead of $\bar{\boldsymbol{g}}$. The symbol $\boldsymbol{g}$ is thus applied to the points of G, e.g. $a \in G$, and then the result $\boldsymbol{g}a$ denotes the image of the point a under the original mapping $\boldsymbol{g}$. The symbol $\boldsymbol{g}$ is also applied to subsets of G, e.g. $A \subset G$, in which case the result $\boldsymbol{g}A$ denotes the image of the subset A under the extended mapping $\bar{\boldsymbol{g}}$.

This rule is observed even for systems of subsets of G: If $\tilde{A}$ is a nonempty system of subsets of G, then we generally denote the system of the $\bar{\boldsymbol{g}}$-images of the indivdual elements of $\tilde{A}$ by the symbol $\boldsymbol{g}\tilde{A}$.

Fori example, if $\bar{A}$ is a decomposition of G, then $\boldsymbol{g}\bar{A}$ denotes the system of the $\bar{\boldsymbol{g}}$-images of the elements of $\bar{A}$. If, in particular, $\boldsymbol{g}\bar{A}$ is a decomposition on G^*, then the extended mapping $\bar{\boldsymbol{g}}$ defines the partial mapping $\boldsymbol{g}_{\bar{A}}$ of the decomposition $\bar{A}$ onto the decomposition $\boldsymbol{g}\bar{A}$ under which there corresponds, to every element $\bar{a} \in \bar{A}$, its image $\boldsymbol{g}\bar{a} \in \boldsymbol{g}\bar{A}$.

Let A and B stand for arbitrary subsets of G.

It is obvious that $A \subset B$ yields $\boldsymbol{g}A \subset \boldsymbol{g}B$.

Let us prove the following theorem:

The equality $\boldsymbol{g}A = \boldsymbol{g}B$ is true if and only if every element of $\bar{G}$, incident with one of the subsets A, B, is also incident with the other.

Proof. a) Suppose $\boldsymbol{g}A = \boldsymbol{g}B$. If an element $\bar{g} \in \bar{G}$ is incident with, for example, A, then there exists an element $a \in A$ such that $\bar{g}$ is the set of all the $\boldsymbol{g}$-inverse images of $\boldsymbol{g}a$. Since $\boldsymbol{g}a \in \boldsymbol{g}A = \boldsymbol{g}B$, there exists an element $b \in B$ such that $\boldsymbol{g}b = \boldsymbol{g}a$, so that $b \in \bar{g}$ and, consequently, $\bar{g}$ is incident with B.

b) Let every element of $\bar{G}$, incident with one of the sets A, B, be also incident with the other. Then, e.g., for $a^* \in \boldsymbol{g}A$, the element $\bar{g} \in \bar{G}$ which consists of all the $\boldsymbol{g}$-inverse images of a^* is incident with A and therefore, by the assumption, even with B. Hence there exists an element $b \in B$ such that $a^* = \boldsymbol{g}b \in \boldsymbol{g}B$ and we have $\boldsymbol{g}A \subset \boldsymbol{g}B$. At the same time there holds, of course, the relation $\boldsymbol{g}B \subset \boldsymbol{g}A$ and we have $\boldsymbol{g}A = \boldsymbol{g}B$.

The above theorem can, naturally, also be expressed by saying that *the equality $\boldsymbol{g}A = \boldsymbol{g}B$ applies if and only if $A \sqsubset \bar{G} = B \sqsubset \bar{G}$.*

Let $\tilde{A}$ stand for a system of subsets of G.

If all the elements of $\tilde{A}$ have, under the extended mapping $\boldsymbol{g}$, the same image $A^* \subset G^*$ so that, for $A \in \tilde{A}$, there holds $\boldsymbol{g}A \subset A^*$, then even the set $\mathbf{s}\tilde{A}$ is mapped onto A^*, i.e., $\boldsymbol{g}(\mathbf{s}\tilde{A}) = A^*$.

Indeed, first of all, for every element $A \in \tilde{A}$ there holds $A \subset \mathbf{s}\tilde{A}$ whence $A^* = \mathbf{g}A \subset \mathbf{g}(\mathbf{s}\tilde{A})$. Moreover, every element $\mathbf{a} \in \mathbf{s}\tilde{A}$ lies in a certain subset $A \in \tilde{A}$ and we have: $\mathbf{g}a \in \mathbf{g}A = A^*$ which yields $\mathbf{g}(\mathbf{s}\tilde{A}) \subset A^*$ and the proof is accomplished.

7.2. Theorems on mappings of decompositions

Let $\bar{A}$ denote a decomposition on G.

The system $\mathbf{g}\bar{A}$ of the subsets of G^* evidently covers the set G^*. But this system is not necessarily a decomposition of the set G^* because the $\mathbf{g}$-images of two different elements of $\bar{A}$ may be incident without coinciding.

The following theorem states a necessary and sufficient condition under which the decomposition $\bar{A}$ is mapped, under $\mathbf{g}$, onto a decomposition of G^*.

$\mathbf{g}\bar{A}$ is a decomposition of the set G^* if and only if the decompositions $\bar{A}$, $\bar{G}$ are complementary.

Proof. a) Suppose $\mathbf{g}\bar{A}$ is a decomposition on G^*. Let the elements $\bar{a} \in \bar{A}$, $\bar{g} \in \bar{G}$ lie in the same element $\bar{u} \in [\bar{A}, \bar{G}]$. We are to show that $\bar{a} \cap \bar{g} \neq \emptyset$. Let $\bar{b} \in \bar{A}$ stand for an arbitrary element incident with $\bar{g}$. Then $\bar{b} \subset \bar{u}$, hence there exists a binding $\{\bar{A}, \bar{B}\}$ from $\bar{a}$ to $\bar{b}$:

$$(\bar{a} =)\quad \bar{a}_1, \ldots, \bar{a}_\alpha \quad (= \bar{b}).$$

By the definition of a binding, every two of its neighbouring elements $\bar{a}_\beta$, $\bar{a}_{\beta+1}$ $(\beta = 1, \ldots, \alpha - 1)$ are incident with an element of the decomposition $\bar{G}$ and thus both images $\mathbf{g}\bar{a}_\beta$, $\mathbf{g}\bar{a}_{\beta+1}$ are incident. Since $\mathbf{g}\bar{A}$ is a decomposition on G^*, we have $\mathbf{g}\bar{a}_\beta = \mathbf{g}\bar{a}_{\beta+1}$ and thus even $\mathbf{g}\bar{a} = \mathbf{g}\bar{b}$. Consequently, $\bar{a} \sqsubset \bar{G} = \bar{b} \sqsubset \bar{G}$. As $\bar{g} \in \bar{b} \sqsubset \bar{G}$, we have $\bar{g} \in \bar{a} \sqsubset \bar{G}$ so that $\bar{a} \cap \bar{g} \neq \emptyset$.

b) Let the decompositions $\bar{A}$, $\bar{G}$ be complementary. Our object now is to show that, for $\bar{a}, \bar{b} \in \bar{A}$, the sets $\mathbf{g}\bar{a}$, $\mathbf{g}\bar{b}$ either are disjoint or coincide. If the sets $\mathbf{g}\bar{a}$, $\mathbf{g}\bar{b}$ are not disjoint, then there exist points $a \in \bar{a}$, $b \in \bar{b}$ such that $\mathbf{g}a = \mathbf{g}b \in \mathbf{g}\bar{a} \cap \mathbf{g}\bar{b}$. Then the element $\bar{g} \in \bar{G}$, consisting of all the $\mathbf{g}$-inverse images of the element $\mathbf{g}a$, is incident with both the elements $\bar{a}$, $\bar{b}$ and the latter therefore lie in the same element of the decomposition $[\bar{A}, \bar{G}]$. Since the decompositions $\bar{A}$, $\bar{G}$ are complementary, there holds $\bar{a} \sqsubset \bar{G} = \bar{b} \sqsubset \bar{G}$ which yields $\mathbf{g}\bar{a} = \mathbf{g}\bar{b}$.

Let again $\bar{A}$, $\bar{G}$ be complementary.

By the above theorem, $\mathbf{g}\bar{A}$ is a decomposition on G. The extended mapping $\mathbf{g}$ determines the partial mapping of the decomposition $\bar{A}$ onto $\mathbf{g}\bar{A}$ under which there corresponds, of course, to every element $\bar{a} \in \bar{A}$, its image $\mathbf{g}\bar{a} \in \mathbf{g}\bar{A}$. By the mapping $\mathbf{g}$ of the decomposition $\bar{A}$ onto $\mathbf{g}\bar{A}$ we shall, in what follows, understand this partial mapping.

To the mapping $\boldsymbol{g}$ of $\bar{A}$ onto $\boldsymbol{g}\bar{A}$ there naturally corresponds a certain decomposition $\bar{\bar{A}}$ of $\bar{A}$. Its elements consist of all the elements of $\bar{A}$ that have, under the extended mapping $\boldsymbol{g}$, the same image.

We shall show that *the covering of the decomposition $\bar{A}$ enforced by $\bar{\bar{A}}$ is the least common covering $[\bar{A}, \bar{G}]$ of the decompositions $\bar{A}, \bar{G}$.*

Indeed, consider an arbitrary element $\bar{\bar{a}} \in \bar{\bar{A}}$. We are to show that the set $\boldsymbol{s}\bar{\bar{a}}$ is an element of the decomposition $[\bar{A}, \bar{G}]$. Let $\bar{a} \in \bar{\bar{a}}$ be an arbitrary element and $\bar{u} \in [\bar{A}, \bar{G}]$ the element of $[\bar{A}, \bar{G}]$, containing $\bar{a}$; consequently, we have $\bar{a} \subset \boldsymbol{s}\bar{\bar{a}} \cap \bar{u}$. Every element $\bar{x} \in \bar{\bar{a}}$ has, under the extended mapping $\boldsymbol{g}$, the same image as $\bar{a}$, hence $\bar{a} \sqsubset \bar{G} = \bar{x} \sqsubset \bar{G}$; it follows that the element $\bar{x}$ may be connected with the element $\bar{a}$ in the decomposition $\bar{G}$ and therefore lies in the element $\bar{u}$. Thus we have verified that $\boldsymbol{s}\bar{\bar{a}} \subset \bar{u}$. Conversely, for any element $\bar{x} \in \bar{A}$ lying in $\bar{u}$ there holds $\bar{a} \sqsubset \bar{G} = \bar{x} \sqsubset \bar{G}$; consequently, the element $\bar{x}$ has, under the extended mapping $\boldsymbol{g}$, the same image as $\bar{a}$, thus $\bar{x} \subset \bar{u}$ and we have $\bar{x} \subset \boldsymbol{s}\bar{\bar{a}}$. Hence $\bar{u} \subset \boldsymbol{s}\bar{\bar{a}}$ and the proof is accomplished.

Associating, with every element $\bar{u} \in [\bar{A}, \bar{G}]$, the element $\bar{\bar{a}} \in \bar{\bar{A}}$ which contains all the elements of $\bar{A}$ lying in $\bar{u}$, we obtain a simple mapping of the decomposition $[\bar{A}, \bar{G}]$ onto $\bar{\bar{A}}$ (6.8); associating, with every element $\bar{\bar{a}} \in \bar{\bar{A}}$, the element $\bar{a}^* \in \boldsymbol{g}\bar{A}$ which is the image of every element $\bar{a} \in \bar{A}$ lying in $\bar{\bar{a}}$, we obtain a simple mapping of the decomposition $\bar{\bar{A}}$ onto $\boldsymbol{g}\bar{A}$ (6.8). Composing these simple mappings, we get a simple mapping of the decomposition $[\bar{A}, \bar{G}]$ onto $\boldsymbol{g}\bar{A}$ (6.7). Under this mapping there corresponds, to every element $\bar{u} \in [\bar{A}, \bar{G}]$, a certain element $\bar{a}^* \in \boldsymbol{g}\bar{A}$; the element $\bar{a}^*$ is the image, under the extended mapping $\boldsymbol{g}$, of every element of $\bar{A}$ lying in the element $\bar{\bar{a}} \in \bar{\bar{A}}$ which contains all the elements of $\bar{A}$ lying in $\bar{u}$. Since $\bar{u} = \boldsymbol{s}\bar{\bar{a}}$ and for $\bar{a} \in \bar{\bar{a}}$ we have $\boldsymbol{g}\bar{a} = \bar{a}^*$, we conclude, with respect to the last theorem in 7.1, that the element $\bar{u}$ has, under the extended mapping $\boldsymbol{g}$, the image $\bar{a}^*$, i.e., $\boldsymbol{g}\bar{u} = \bar{a}^*$.

Thus we have the following result:

If a decomposition $\bar{A}$ on G is mapped, under $\boldsymbol{g}$, onto some decomposition $\bar{A}^*$ on G, then the decompositions $[\bar{A}, \bar{G}]$ and $\bar{A}^*$ are equivalent, i.e., $[\bar{A}, \bar{G}] \simeq \bar{A}^*$; *a simple mapping of the decomposition $[\bar{A}, \bar{G}]$ onto $\bar{A}^*$ is obtained by associating, with each element of $[\bar{A}, \bar{G}]$, its image under the extended mapping $\boldsymbol{g}$.*

Consequently, *every covering of the decomposition $\bar{G}$ is equivalent to its image under $\boldsymbol{g}$; the mapping under which every element of the covering is associated with its own image is simple.*

7.3. Exercises

1. Let $\boldsymbol{g}$ be a mapping of the set G onto G^* and A, B stand for arbitrary subsets of G. Show that the following relations are true: $\boldsymbol{g}(A \cup B) = \boldsymbol{g}A \cup \boldsymbol{g}B$; $\boldsymbol{g}(A \cap B) \subset \boldsymbol{g}A \cap \boldsymbol{g}B$.

2. Assuming the situation described in exercise 1., let $\overline{G}$ be the decomposition on G corresponding to the mapping $\boldsymbol{g}$. Show that the equality $\boldsymbol{g}(A \cap B) = \boldsymbol{g}A \cap \boldsymbol{g}B$ applies if and only if there holds $(A \cap B) \sqsubset \overline{G} = (A \sqsubset \overline{G}) \cap (B \sqsubset \overline{G})$.

3. Let $\boldsymbol{g}$ be a mapping of the set G onto G^* and $\{\bar{a}, \bar{b}, \ldots\}$ stand for a decomposition on G. Then $\{\boldsymbol{g}\bar{a}, \boldsymbol{g}\bar{b}, \ldots\}$ is a decomposition on G^* if and only if $\{\bar{a}, b, \ldots\}$ is a covering of the decomposition corresponding to $\boldsymbol{g}$.

4. Suppose $\boldsymbol{g}$ is a simple mapping of the set G onto G^*. Let, moreover, $A \subset G$ be a nonempty subset and $\bar{A}, \bar{B}$ stand for decompositions in (on) G. In this situation there holds:
   a) the extended mapping $\bar{\boldsymbol{g}}$ of the system of all the nonempty parts of G onto the system of all the nonempty parts of G^* is simple;
   b) the sets $A, \boldsymbol{g}A$ are equivalent, i.e., $A \simeq \boldsymbol{g}A$;
   c) $\boldsymbol{g}\bar{A}$ is a decomposition in (on) the set G^*;
   d) the decompositions $\bar{A}, \boldsymbol{g}\bar{A}$ are equivalent, i.e., $\bar{A} \simeq \boldsymbol{g}\bar{A}$;
   e) if the decompositions $\bar{A}, \bar{B}$ are equivalent or loosely coupled or coupled, then the decompositions $\boldsymbol{g}\bar{A}, \boldsymbol{g}\bar{B}$ have, in each case, the same property.

8. Permutations

In this chapter we shall deal with simple mappings of finite sets onto themselves; they play an important role in algebra, particularly, in the theory of groups.

8.1. Definition

By a *permutation of the set* G we mean a simple mapping of the set G onto itself (6.6).

In this section we shall restrict our considerations to permutations of *finite* sets.

Let G denote an arbitrary set consisting of a finite number $n(\geqq 1)$ of elements. From the assumption that G is finite it follows that every simple mapping $\boldsymbol{p}$ of the set G into itself is a permutation of G (6.10.2).

Let the elements of G be denoted by the letters $a, b, \ldots, m$. Then we can uniquely associate, with every permutation $\boldsymbol{p}$ of the set G, a symbol of the form:

$$\begin{pmatrix} a & b & \ldots & m \\ a^* & b^* & \ldots & m^* \end{pmatrix},$$

where $a^*, b^*, \ldots, m^*$ are the letters denoting the elements $\boldsymbol{p}a, \boldsymbol{p}b, \ldots, \boldsymbol{p}m$. Since $\boldsymbol{p}G = G$, the letters $a^*, b^*, \ldots, m^*$ are again $a, b, \ldots, m$ written in a certain order.

Conversely, every symbol of the above form, where the letters $a^*, b^*, \ldots, m^*$ are again the letters $a, b, \ldots, m$ written in a certain order, determines a certain permutation of the set G, namely, the permutation under which every element denoted by a letter x in the first row is mapped onto the element denoted by the letter lying under x in the second row. Note that the same permutation $\boldsymbol{p}$ may similarly be expressed by other symbols if the letters $a, b, \ldots, m$ in the first row are written in a different order but under each of them there remains the same letter as before. The identical mapping of G is, naturally, a particular permutation of G, the so-called *identical permutation*; it is denoted by the symbol $\begin{pmatrix} a\,b \ldots m \\ a\,b \ldots m \end{pmatrix}$ or any of the other symbols, e.g., $\begin{pmatrix} b\,a \ldots m \\ b\,a \ldots m \end{pmatrix}$.

8.2. Examples of permutations

Let us first introduce a few simple examples of permutations of sets containing $n = 1, 2, 3, 4$ elements.

1. $n = 1$. Let G be a set consisting of a single point a in a plane. In that case there exists, of course, exactly one permutation of G, namely, the identical permutation $\begin{pmatrix} a \\ a \end{pmatrix}$.

2. $n = 2$. Let G be a set consisting of two arbitrary points a, b in a plane. If a, b are rotated, in the plane, in one or the other direction, about the center of the line segment with the end-points a, b through an angle α, then the point a shifts to a certain point a' and the point b to b' and we have a simple mapping of the set G onto the set $\{a', b'\}$. If α equals 0°, 180°, then the set $\{a', b'\}$ is identical with G and we have the following permutations of the set G: $\begin{pmatrix} a\,b \\ a\,b \end{pmatrix}$, $\begin{pmatrix} a\,b \\ b\,a \end{pmatrix}$, respectively.

3. $n = 3$. Suppose G is a set of three points on a plane: a, b, c, forming the vertices of an equilateral triangle. If the points a, b, c are rotated, in the plane, in one or the other direction, about the center of the triangle through an angle α, then the point a shifts to a certain point a', the point b to b', the point c to c' and we have a simple mapping of the set G onto the set $\{a', b', c'\}$. If α equals 0°, 120°, 240°, then the set $\{a', b', c'\}$ is identical with G and we have the following permutations of G:

$$\begin{pmatrix} a & b & c \\ a & b & c \end{pmatrix}, \quad \begin{pmatrix} a & b & c \\ b & c & a \end{pmatrix}, \quad \begin{pmatrix} a & b & c \\ c & a & b \end{pmatrix},$$

respectively. Further permutations of G are obtained by associating, with the points a, b, c, the points symmetric with regard to some axis of symmetry of the triangle in question. The latter has altogether three axes of symmetry; each of them passes through one vertex and bisects the opposite side. Associating, with each of the points a, b, c, the point symmetric with regard to the axis of symmetry passing through a, we obtain the permutation

$$\begin{pmatrix} a & b & c \\ a & c & b \end{pmatrix};$$

in a similar way we obtain further permutations:

$$\begin{pmatrix} a & b & c \\ c & b & a \end{pmatrix}, \quad \begin{pmatrix} a & b & c \\ b & a & c \end{pmatrix}.$$

So we have found, in this case, altogether 6 permutations, namely:

$$\begin{pmatrix} a & b & c \\ a & b & c \end{pmatrix}, \quad \begin{pmatrix} a & b & c \\ b & c & a \end{pmatrix}, \quad \begin{pmatrix} a & b & c \\ c & a & b \end{pmatrix}, \quad \begin{pmatrix} a & b & c \\ a & c & b \end{pmatrix}, \quad \begin{pmatrix} a & b & c \\ c & b & a \end{pmatrix}, \quad \begin{pmatrix} a & b & c \\ b & a & c \end{pmatrix}.$$

4. $n = 4$. Now let G be a set of four points in a plane: a, b, c, d, forming the vertices of a square. Rotating the points a, b, c, d, in the plane, in one or the other direction about the center of the square through an angle α, we again obtain a simple mapping of the set G onto the set of certain points a', b', c', d' in the plane; if $\alpha = 0°$, $90°$ $180°$, $270°$, then we get the following permutations of the set G, respectively:

$$\begin{pmatrix} a & b & c & d \\ a & b & c & d \end{pmatrix}, \quad \begin{pmatrix} a & b & c & d \\ b & c & d & a \end{pmatrix}, \quad \begin{pmatrix} a & b & c & d \\ c & d & a & b \end{pmatrix}, \quad \begin{pmatrix} a & b & c & d \\ d & a & b & c \end{pmatrix}.$$

Further permutations of the set G are found, again, by associating, with the points a, b, c, d, the points symmetric with regard to some axis of symmetry of the mentioned square. The latter has althogether four axes of symmetry; two of them pass through two diagonal vertices and the other two bisect the two opposite sides. Associating, with each of the points a, b, c, d, the point symmetric with regard to the axis of symmetry passing through the vertices a, c, we obtain the permutation

$$\begin{pmatrix} a & b & c & d \\ a & d & c & b \end{pmatrix};$$

in a similar way we obtain further permutations:

$$\begin{pmatrix} a & b & c & d \\ c & b & a & d \end{pmatrix}, \quad \begin{pmatrix} a & b & c & d \\ b & a & d & c \end{pmatrix}, \quad \begin{pmatrix} a & b & c & d \\ d & c & b & a \end{pmatrix}.$$

Thus we have found, in this case, altogether 8 permutations, namely:

$$\begin{pmatrix} a & b & c & d \\ a & b & c & d \end{pmatrix}, \quad \begin{pmatrix} a & b & c & d \\ b & c & d & a \end{pmatrix}, \quad \begin{pmatrix} a & b & c & d \\ c & d & a & b \end{pmatrix}, \quad \begin{pmatrix} a & b & c & d \\ d & a & b & c \end{pmatrix},$$

$$\begin{pmatrix} a & b & c & d \\ a & d & c & b \end{pmatrix}, \quad \begin{pmatrix} a & b & c & d \\ c & b & a & d \end{pmatrix}, \quad \begin{pmatrix} a & b & c & d \\ b & a & d & c \end{pmatrix}, \quad \begin{pmatrix} a & b & c & d \\ d & c & b & a \end{pmatrix}.$$

8.3. The number of permutations

Let us now resume our study of the permutations on a set G of $n(\geqq 1)$ elements $a, b, \ldots, m$.

How many permutations of G are there altogether? To answer this question, let us first note the fact that, under an arbitrary permutation $\boldsymbol{p}$ of G, the element a is mapped onto a certain element $\boldsymbol{p}a$ of G; if $n > 1$ then, moreover, the element b is mapped onto an element $\boldsymbol{p}b$ different from $\boldsymbol{p}a$, the element c onto an element $\boldsymbol{p}c$ different from $\boldsymbol{p}a, \boldsymbol{p}b$, etc., up to the element m mapped onto an element $\boldsymbol{p}m$ different from the elements $\boldsymbol{p}a, \boldsymbol{p}b, \boldsymbol{p}c, \ldots$. Conversely, associating with the element a any element $a^* \in G$ and, if $n > 1$, with the element b any element $b^* \in G$ different from a^* and with the element c any element $c^* \in G$ different from a^*, b^* and so on up to the element $m^* \in G$ different from the elements $a^*, b^*, c^*, \ldots$, we obtain a certain permutation

$$\begin{pmatrix} a & b & c & \ldots & m \\ a^* & b^* & c^* & \ldots & m^* \end{pmatrix}$$

of the set G. The number of the permutations is exactly the same as the number of the possibilities of the above associations. But with the element a we may associate an element $a^* \in G$ in n ways: first, the element a itself, then the element b and so on, until, the nth time, the element m; if $n > 1$ we may, moreover, associate with the element b an element $b^* \in G$ different from a^* in altogether $n - 1$ ways and, similarly, with the element c some element $c^* \in G$ different from a^*, b^* in altogether $n - 2$ ways, and so on up to the element m with which we may associate some element $m^* \in G$ different from $a^*, b^*, c^*, \ldots$, exactly in one way. So we have altogether $n(n - 1)(n - 2) \ldots 1$ possibilities and the answer to the above question is that there *exist exactly* $1.2.3 \ldots n$ *permutations of the set* G. This number is generally denoted by the symbol $n!$. For example, for every set consisting of $n = 1, 2, 3, 4, 5, 6, 7, 8, 9, 10$ elements there exist exactly $n! = 1, 2, 6, 24, 120, 720, 5040, 40320, 362880, 3628800$ permutations. The permutations we have found in the above examples of 1, 2, 3 points in a plane are evidently all that there exist but, in case of 4 points in a plane, there exist, beside the 8 permutations we have found, $2.8 = 16$ further permutations.

8.4. Properties of permutations

1. *Inverse permutations.* Let us now proceed to a more detailed study of the properties of permutations. Suppose $\boldsymbol{p}$ is a permutation of the set G. Since $\boldsymbol{p}$ is a simple mapping, there exists an inverse permutation $\boldsymbol{p}^{-1}$ of $\boldsymbol{p}$ of G. It is easy to see that the symbol of $\boldsymbol{p}^{-1}$ is obtained by interchanging the two rows in the symbol $\boldsymbol{p}$. For instance, the permutations inverse of the above 8 permutations of four points in

a plane are:

$$\begin{pmatrix} a & b & c & d \\ a & b & c & d \end{pmatrix}, \quad \begin{pmatrix} a & b & c & d \\ d & a & b & c \end{pmatrix}, \quad \begin{pmatrix} a & b & c & d \\ c & d & a & b \end{pmatrix}, \quad \begin{pmatrix} a & b & c & d \\ b & c & d & a \end{pmatrix},$$

$$\begin{pmatrix} a & b & c & d \\ a & d & c & b \end{pmatrix}, \quad \begin{pmatrix} a & b & c & d \\ c & b & a & d \end{pmatrix}, \quad \begin{pmatrix} a & b & c & d \\ b & a & d & c \end{pmatrix}, \quad \begin{pmatrix} a & b & c & d \\ d & c & b & a \end{pmatrix}.$$

2. *Invariant elements.* An arbitrary point $x \in G$ is mapped, under the permutation $\boldsymbol{p}$, onto an element $\boldsymbol{p}x$ which is or is not identical with x. In the first case, $\boldsymbol{p}x = x$, we say that *the permutation* $\boldsymbol{p}$ *leaves the element* x *invariant* (*unchanged*) or that the element x is invariant under the permutation $\boldsymbol{p}$. It is obvious that, under the permutation $\boldsymbol{p}$ and the inverse permutation $\boldsymbol{p}^{-1}$, the same elements of the set G are invariant. For instance, the above permutations of four points in a plane leave the following elements invariant: a, b, c, d; none; none; none; a, c; b, d; none; none.

3. *Cyclic* (or *circular*) *permutations.* An element x and the permutation $\boldsymbol{p}$ uniquely determine the sequence of elements of G: $x, \boldsymbol{p}x, \boldsymbol{p}(\boldsymbol{p}x), \boldsymbol{p}(\boldsymbol{p}(\boldsymbol{p}x)), \ldots$, in which every element except the first is the $\boldsymbol{p}$-image of the preceding one. Instead of $x, \boldsymbol{p}x$ we sometimes write $\boldsymbol{p}^0x, \boldsymbol{p}^1x$ and, for brevity, instead of $(\boldsymbol{p}\boldsymbol{p}x), \boldsymbol{p}(\boldsymbol{p}(\boldsymbol{p}x))$ we generally put $\boldsymbol{p}^2x, \boldsymbol{p}^3x, \ldots$.

The permutation $\boldsymbol{p}$ is called *cyclic* or *circular* if there exists an element $x \in G$ and a positive integer k such that, in the sequence $x, \boldsymbol{p}x, \boldsymbol{p}^2x, \boldsymbol{p}^3x, \ldots, \boldsymbol{p}^{k-1}x$, no two elements are identical but the image $\boldsymbol{p}^kx$ of $\boldsymbol{p}^{k-1}x$ is again the element x and if, moreover, all the other elements of G—if there are any—remain invariant under $\boldsymbol{p}$. The permutation $\boldsymbol{p}$ can be more precisely described as a cyclic (circular) permutation with regard to the elements $x, \boldsymbol{p}x, \boldsymbol{p}^2x, \ldots, \boldsymbol{p}^{k-1}x$.

The ordered set of elements $x, \boldsymbol{p}x, \boldsymbol{p}^2x, \ldots, \boldsymbol{p}^{k-1}x$ is called a *cycle of the permutation* $\boldsymbol{p}$ or, more precisely, a *k-membered cycle* or a *k-cycle of* $\boldsymbol{p}$. If, in particular, $k = n$, i.e., if every element of G lies in this cycle we say that $\boldsymbol{p}$ is a *pure cyclic* (*circular*) *permutation.*

Let the permutation $\boldsymbol{p}$ be cyclic with regard to the elements $x, \boldsymbol{p}x, \boldsymbol{p}^2x, \ldots, \boldsymbol{p}^{k-1}x$. Then the permutation $\boldsymbol{p}$ is usually expressed by a simple symbol: the elements $x, \boldsymbol{p}x, \boldsymbol{p}^2x, \cdots, \boldsymbol{p}^{k-1}x$ are written in this order, next to each other, in parentheses. The inverse of the permutation $\boldsymbol{p}$, i.e. $\boldsymbol{p}^{-1}$, maps every element of the sequence $x, \boldsymbol{p}x, \boldsymbol{p}^2x, \ldots, \boldsymbol{p}^{k-1}x$ except the first onto the preceding one, the element x onto $\boldsymbol{p}^{k-1}x$ and the other elements of G—if there are any—remain invariant; consequently, $\boldsymbol{p}^{-1}$ is cyclic with regard to $\boldsymbol{p}^{k-1}x, \ldots, \boldsymbol{p}^2x, \boldsymbol{p}x, x$. If we change the symbols of the elements of G by denoting the elements $x, \boldsymbol{p}x, \boldsymbol{p}^2x, \ldots, \boldsymbol{p}^{k-1}x$ by the letters $a, b, c, \ldots, j$, respectively, and the other elements of G—if there are any—by other arbitrarily chosen letters, then the simplified symbol of $\boldsymbol{p}$ is: $(a, b, c, \ldots, j)$. The permutation may, of course, be expressed by any other symbol: $(b, c, \ldots, j, a)$, $(c, \ldots, j, a, b)$, etc., altogether in k ways. Then the symbol of the inverse permutation is, for example, $(j, \ldots, c, b, a)$.

The simplest cyclic permutations are those with regard to one single element; by the above definition, every permutation of this kind is the identical permutation of G and, consequently, may be expressed by any symbol (a), (b), ..., (m).

Every cyclic permutation of G with regard to two elements is called a *transposition.*

For instance, in the above permutations of the set of $n = 1, 2, 3, 4$ points in a plane we have the following cyclic permutations:

for $n = 1$: (a);

for $n = 2$: (a), (a, b);

for $n = 3$: (a), (a, b), (a, c), (b, c), (a, b, c), (a, c, b);

for $n = 4$: (a), (a, c), (b, d), (a, b, c, d), (a, d, c, b).

4. *Invariant subsets and decompositions.* Now, let again $\boldsymbol{p}$ stand for an arbitrary permutation of the set G. Any nonempty subset $A \subset G$ is mapped, under the extended mapping $\boldsymbol{p}$, onto a subset $\boldsymbol{p}A \subset G$ which is or is not a part of A. In the first case, if $\boldsymbol{p}A \subset A$, then $\boldsymbol{p}A = A$. In fact, by the definition of the partial mapping $\boldsymbol{p}_A$, there holds $\boldsymbol{p}A = \boldsymbol{p}_A A$; moreover, as $\boldsymbol{p}$ is a simple mapping of the finite set A into itself, it is a permutation of the set A; so we have $\boldsymbol{p}_A A = A$.

If $\boldsymbol{p}A = A$, we say that *the permutation* $\boldsymbol{p}$ *leaves the subset* A *invariant* or that *the subset* A *is invariant under the permutation* $\boldsymbol{p}$.

The subset A is invariant under the permutation $\boldsymbol{p}$ if each of its elements is invariant under $\boldsymbol{p}$. If $\boldsymbol{p}$ leaves the subset A invariant, then the same evidently holds for the inverse permutation $\boldsymbol{p}^{-1}$. For example, the above permutations of four points in the plane leave the following proper subsets of the set $\{a, b, c, d\}$ invariant: all; none; $\{a, c\}$, $\{b, d\}$; none; $\{a\}$, $\{c\}$, $\{b, d\}$; $\{b\}$, $\{d\}$, $\{a, c\}$; $\{a, b\}$, $\{c, d\}$; $\{a, d\}$, $\{b, c\}$. Note that, if $\boldsymbol{p}$ is the cyclic permutation $(a, b, c, \ldots, j)$, then every subset $A \subset G$ containing the elements $a, b, c, \ldots, j$ is invariant under $\boldsymbol{p}$, the partial permutation $\boldsymbol{p}_A$ is cyclic as well and is expressed by the same symbol $(a, b, c, \ldots, j)$.

Suppose $\bar{G} = \{\bar{a}, \bar{b}, \ldots, \bar{m}\}$ is a decomposition of the set G. If $\bar{G}$ has the property that, under the extended mapping $\boldsymbol{p}$, the image of every element of $\bar{G}$ is again an element of $\bar{G}$, we say that *the permutation* $\boldsymbol{p}$ *leaves the decomposition* $\bar{G}$ *invariant* or that *the permutation* $\bar{G}$ *is invariant under the permutation* $\boldsymbol{p}$. It is clear that, if the permutation $\boldsymbol{p}$ leaves the decomposition $\bar{G}$ invariant, the same holds for the inverse permutation $\boldsymbol{p}^{-1}$.

Let us, in particular, consider the case when every element of $\bar{G}$ is invariant under $\boldsymbol{p}$ so that $\boldsymbol{p}\bar{a} = \bar{a}$, $\boldsymbol{p}\bar{b} = \bar{b}$, ..., $\boldsymbol{p}\bar{m} = \bar{m}$. Then, $\bar{x}$ being an element of $\bar{G}$, the partial mapping $\boldsymbol{p}_{\bar{x}}$ is a permutation of $\bar{x}$. The partial permutations $\boldsymbol{p}_{\bar{a}}, \boldsymbol{p}_{\bar{b}}, \ldots, \boldsymbol{p}_{\bar{m}}$ uniquely determine the permutation $\boldsymbol{p}$ in the sense that the $\boldsymbol{p}$-image of every element $x \in G$ is the same under the partial permutation $\boldsymbol{p}_{\bar{x}}$ of the element $\bar{x} \in \bar{G}$ containing x. Under the inverse permutation $\boldsymbol{p}^{-1}$ every element of $\bar{G}$ is invariant as well and $\boldsymbol{p}^{-1}$ is determined by the inverse permutations $\boldsymbol{p}_{\bar{a}}^{-1}$; $\boldsymbol{p}_{\bar{b}}^{-1}, \ldots, \boldsymbol{p}_{\bar{m}}^{-1}$. Conversely, let $\bar{G} = \{\bar{a}, \bar{b}, \ldots, \bar{m}\}$ be an arbitrary decomposition on G and choose, on each of its elements $\bar{x}$, an arbitrary permutation $\boldsymbol{p}_{\bar{x}}$; define, on G, the permutation $\boldsymbol{p}$ by associating with every element $x \in G$ its $\boldsymbol{p}_{\bar{x}}$-image where $x \in \bar{x}$; then every element of $\bar{G}$ is invariant under $\boldsymbol{p}$ and $\boldsymbol{p}_{\bar{a}}, \boldsymbol{p}_{\bar{b}}, \ldots, \boldsymbol{p}_{\bar{m}}$ are the determining partial permutations of $\boldsymbol{p}$.

8.5. The determination of permutations by pure cyclic permutations

Now we shall show that *an arbitrary permutation* $\boldsymbol{p}$ *of any set* G *consisting of* $n(\geq 1)$ *elements is determined by a finite number of pure cyclic permutations*, in other words, that there exists a decomposition $\bar{G} = \{\bar{a}, \bar{b}, \ldots, \bar{m}\}$ of G such that each element of G is invariant under $\boldsymbol{p}$ and the partial permutations are pure cyclic permutations of the elements $\bar{a}, \bar{b}, \ldots, \bar{m}$.

The proof will be based on the method of complete induction.[1]) For $n = 1$ our statement is correct because, in that case, $\boldsymbol{p}$ is the identical permutation of G and the greatest decomposition of G has the above property. It remains to be shown that, if our statement holds for every set consisting of at most $n - 1$ elements, n standing for an integer > 1, then it also holds for any set consisting of n elements. Let G stand for a set of n elements and $\boldsymbol{p}$ for a permutation of G. Let, moreover, a denote an element of G. Consider the sequence of the elements $a, \boldsymbol{p}a, \boldsymbol{p}^2a, \ldots, \boldsymbol{p}^na$ of G, each of which is the $\boldsymbol{p}$-image of the preceding element. The number of these elements is $n + 1$ so that at least one element occurs in G at least twice. Proceeding, in the mentioned sequence, from the first element a successively to the subsequent elements, we arrive, *for the first time*, at:

a) a certain element $\boldsymbol{p}^ja$, j denoting a number $0, \ldots, n - 1$ that occurs among the elements $\boldsymbol{p}^{j+1}a, \ldots, \boldsymbol{p}^na$ at least once more;

b) the element $\boldsymbol{p}^{j+k}a$, k being a number $1, \ldots, n - j$ which is identical with the element $\boldsymbol{p}^ja$, so that $\boldsymbol{p}^ja = \boldsymbol{p}^{j+k}a$.

If $\boldsymbol{p}^ja$ is not the first element a, i.e., if $j > 0$, then both elements $\boldsymbol{p}^{j-1}a$ and $\boldsymbol{p}^{j+k-1}a$ are mapped, under $\boldsymbol{p}$, onto the same element $\boldsymbol{p}^ja$ and, since $\boldsymbol{p}$ is a simple mapping, there holds $\boldsymbol{p}^{j-1}a = \boldsymbol{p}^{j+k-1}a$; but that is not possible because, in the sequence $a, \boldsymbol{p}a$,

---

1) The method of complete induction is based on the following theorem: *If one associates, with every positive integer* n, *a certain statement* $\boldsymbol{g}n$ *such that:* (1) *the statement* $\boldsymbol{g}1$ *is correct,* (2) *for every* $n > 1$ *for which the statements* $\boldsymbol{g}1, \ldots, \boldsymbol{g}(n-1)$ *are correct, even* $\boldsymbol{g}n$ *is correct, then all the statements are correct.* In fact, in the opposite case the incorrect statements are associated with certain positive integers one of which, let us denote it by m, is the least. By the assumption (1), there holds $m > 1$; by the definition of m, the statements $\boldsymbol{g}1, \ldots, \boldsymbol{g}(m-1)$ are correct, whereas the statement $\boldsymbol{g}m$ is incorrect, but that contradicts the assumption (2).

An analogous theorem applies to the statements associated with integers greater than or equal to an integer k.

$\boldsymbol{p}^2a, \ldots, \boldsymbol{p}^n a$, the element $\boldsymbol{p}^j a$ is not preceded by any element occurring once more whereas, according to the above equality, $\boldsymbol{p}^{j-1}a$ is such an element. Thus we have ascertained that $j = 0$. By the definition of the number k, we have $\boldsymbol{p}^k a = a$ but none of the elements $\boldsymbol{p}a, \ldots, \boldsymbol{p}^{k-1}a$ is a. If any two of the elements $a, \boldsymbol{p}a, \ldots, \boldsymbol{p}^{k-1}a$ are equal, i.e., if for some integers r, s satisfying the inequalities $0 \leqq r < s \leqq k-1$, there holds $\boldsymbol{p}^r a = \boldsymbol{p}^s a$, then we have $\boldsymbol{p}^{k-s}(\boldsymbol{p}^r a) = \boldsymbol{p}^{k-s}(\boldsymbol{p}^s a)$, i.e., $\boldsymbol{p}^{k-s+r}a = \boldsymbol{p}^k a = a$; but this contradicts the fact that none of the elements $\boldsymbol{p}a, \ldots, \boldsymbol{p}^{k-1}a$ is a because $1 \leqq k - s + r \leqq k - 1$ and therefore $\boldsymbol{p}^{k-s+r}a$ is one of these elements. Thus we have verified that no two elements $a, \boldsymbol{p}a, \ldots, \boldsymbol{p}^{k-1}a$ are equal.

Let $\bar{a}$ stand for the set of the elements $a, \boldsymbol{p}a, \ldots, \boldsymbol{p}^{k-1}a$. We observe that the subset $\bar{a} \subset G$ is invariant under the permutation $\boldsymbol{p}$ and that the partial permutation $\boldsymbol{p}_{\bar{a}}$ is a pure cyclic permutation of $\bar{a}$. If $k = n$, i.e., if $\bar{a} = G$, then $\boldsymbol{p}_{\bar{a}} = \boldsymbol{p}$ and the greatest decomposition of G has the above property. Let us now consider the case $k < n$. In that case the set G contains, besides $a, \boldsymbol{p}a, \ldots, \boldsymbol{p}^{k-1}a$, further elements the number of which is, at most, $n - 1$; the set of these elements will be denoted by H. Under the partial mapping $\boldsymbol{p}_H$, the image of every element $x \in H$ is again an element of H because, in the opposite case, there holds $\boldsymbol{p}x = \boldsymbol{p}^l a$, l standing for one of the numbers $0, \ldots, k-1$ and, consequently, $x = \boldsymbol{p}^{l-1}a$ and $x = \boldsymbol{p}^{k-1}a$ if $l > 0$ and $l = 0$, respectively; but in both cases this contradicts the assumption $x \in H$. The permutation $\boldsymbol{p}_H$ is therefore a mapping of the set H into itself and, since it is simple and H has only a finite number of elements, $\boldsymbol{p}_H$ is a permutation of H. If our statement holds for every set of, at most, $n - 1$ elements, then there exists a decomposition $\overline{H} = \{\bar{b}, \ldots, \overline{m}\}$ of the set H such that every element of H is invariant under the permutation $\boldsymbol{p}_H$ and the partial permutations of the elements $\bar{b}, \ldots, \overline{m}$, determined by $\boldsymbol{p}_H$ are pure cyclic permutations. Since $\boldsymbol{p}_H$ maps every element of H onto the same element as $\boldsymbol{p}$, the partial mappings $\boldsymbol{p}_{\bar{b}}, \ldots, \boldsymbol{p}_{\overline{m}}$ of $\bar{b}, \ldots, \overline{m}$, determined by $\boldsymbol{p}$, are exactly these pure cyclic permutations. The system of the sets $\overline{G} = \{\bar{a}, \bar{b}, \ldots, \overline{m}\}$ is obviously a decomposition of G and we see that each element $\bar{a}, \bar{b}, \ldots, \overline{m}$ is invariant under $\boldsymbol{p}$ and that $\boldsymbol{p}_{\bar{a}}, \boldsymbol{p}_{\bar{b}}, \ldots, \boldsymbol{p}_{\overline{m}}$ are pure cyclic permutations of $\bar{a}, \bar{b}, \ldots, \overline{m}$, which completes the proof.

8.6. The method of determining the pure cyclic permutations forming a given permutation

Given a permutation $\boldsymbol{p}$ of the set G consisting of $n \geqq 1$ elements, the pure cyclic permutations by which it is determined are obtained as follows: Starting from an arbitrary element $a \in G$ we first determine the cycle $a, \boldsymbol{p}a, \ldots, \boldsymbol{p}^{k-1}a$; then, if $k < n$, we choose an element $b \in G$ which is not in this cycle and determine the next cycle $b, \boldsymbol{p}b, \ldots, \boldsymbol{p}^{l-1}b$; furthermore, if $k + l < n$, we choose an element $c \in G$ which is not in any of the preceding cycles and determine the cycle beginning with the element c; in this way we proceed. To express the permutation $\boldsymbol{p}$ we then write, in a certain order, side by side, the symbols of the individual pure cyclic permutations. From this we obtain the symbol of the inverse permutation

p^{-1} by way of reversing, in each cycle, the order of the letters. For example, the above permutations of the set of $n = 1, 2, 3, 4$ points in a plane is determined by pure cyclic permutations as follows:

if $n = 1$: (a);

if $n = 2$: $(a)(b)$, (a, b);

if $n = 3$: $(a)(b)(c)$, (a, b, c), (a, c, b), $(a)\ (b, c)$, $(a, c)(b)$, $(a, b)(c)$;

if $n = 4$: $(a)(b)(c)(d)$, (a, b, c, d), $(a, c)(b, d)$, (a, d, c, b), $(a)(c)(b, d)$, $(a, c)(b)(d)$, $(a, b)(c, d)$, $(a, d)(b, c)$.

The inverse permutations of the latter are:

if $n = 1$: (a);

if $n = 2$: $(a)(b)$, (a, b);

if $n = 3$: $(a)(b)(c)$, (c, b, a), (b, c, a), $(a)(b, c)$, $(a, c)(b)$, $(a, b)(c)$;

if $n = 4$: $(a)(b)(c)(d)$, (d, c, b, a), $(a, c)(b, d)$, (b, c, d, a), $(a)(c)(b, d)$, $(a, c)(b)(d)$, $(a, b)(c, d)$, $(a, d)(b, c)$.

8.7. Composition of permutations

1. *The concept of the composition of permutations.* The permutations of the set G may, of course, be composed according to the rule of composing mappings. Let $\boldsymbol{p}$, $\boldsymbol{q}$ denote arbitrary permutations of G. The mapping $\boldsymbol{qp}$ composed of the permutations $\boldsymbol{p}$, $\boldsymbol{q}$ is again a permutation of G. The symbol of the latter is obtained by writing, under each letter x denoting some element of G, the letter of the element $\boldsymbol{q}(\boldsymbol{p}x)$. If the permutations $\boldsymbol{p}$, $\boldsymbol{q}$ are expressed in usual two-lined symbols, then the letter denoting the element $\boldsymbol{q}(\boldsymbol{p}x)$ is found as follows: First we find the letter denoting the element $\boldsymbol{p}x$ which lies, in the symbol of $\boldsymbol{p}$, under x, and then the letter denoting the element $\boldsymbol{q}(\boldsymbol{p}x)$ which lies, in the symbol of $\boldsymbol{q}$, under the letter denoting $\boldsymbol{p}x$. If, for instance, $n = 3$ and $\boldsymbol{p}$, $\boldsymbol{q}$ are given by the symbols

$$\begin{pmatrix} a & b & c \\ b & c & a \end{pmatrix}, \quad \begin{pmatrix} a & b & c \\ a & c & b \end{pmatrix},$$

then the symbol of $\boldsymbol{qp}$ is

$$\begin{pmatrix} a & b & c \\ c & b & a \end{pmatrix}.$$

Analogously we proceed if $\boldsymbol{p}$, $\boldsymbol{q}$ are expressed by the pure cyclic permutations by which they are determined. For example, if $n = 3$ and $\boldsymbol{p}$, $\boldsymbol{q}$ are given by the symbols (a, b, c), $(a)(b, c)$, then $\boldsymbol{qp}$ is expressed by $(a, c)(b)$.

2. *Interchangeable permutations.* Note that the result of composing two permutations of G may depend on the order in which they are composed, i.e., the permutation $\boldsymbol{qp}$ composed of $\boldsymbol{p}, \boldsymbol{q}$ may be different from the permutation $\boldsymbol{pq}$ composed of $\boldsymbol{q}, \boldsymbol{p}$. In the above example there holds $\boldsymbol{qp} \neq \boldsymbol{pq}$, for $\boldsymbol{qp}$ is the permutation (a, c), whereas $\boldsymbol{pq}$ is (a, b). If the permutations $\boldsymbol{p}, \boldsymbol{q}$ are such that the results of their composition does not depend on their order, i.e., if $\boldsymbol{qp} = \boldsymbol{pq}$, then they are called *interchangeable.* E.g., the identical permutation of the set G and any other permutation of G are interchangeable.

3. *Associative law for the composition of permutations.* To any permutations $\boldsymbol{p}, \boldsymbol{q}, \boldsymbol{r}$ of the set G there, of course, applies the associative law

$$\boldsymbol{r}(\boldsymbol{qp}) = (\boldsymbol{rq})\boldsymbol{p};$$

the permutation of G lying on either side of this equality is briefly denoted by $\boldsymbol{rqp}$.

4. *The inverse of a composed permutation.* By means of the associative law we can easily show that *the inverse of the composed permutation* $\boldsymbol{qp}$ *is* $\boldsymbol{p}^{-1}\boldsymbol{q}^{-1}$, i.e., that there holds

$$(\boldsymbol{qp})^{-1} = \boldsymbol{p}^{-1}\,\boldsymbol{q}^{-1}.$$

In fact, let x denote an arbitrary element of G. Taking account of the definition of $\boldsymbol{p}^{-1}\boldsymbol{q}^{-1}$ and the associative law, we have $(\boldsymbol{p}^{-1}\boldsymbol{q}^{-1})(\boldsymbol{qp}x) = \boldsymbol{p}^{-1}(\boldsymbol{q}^{-1}(\boldsymbol{qp}x)) = \boldsymbol{p}^{-1}((\boldsymbol{q}^{-1}\boldsymbol{q})\boldsymbol{p}x)$ and, furthermore, $\boldsymbol{p}^{-1}((\boldsymbol{q}^{-1}\boldsymbol{q})\boldsymbol{p}x) = \boldsymbol{p}^{-1}(\boldsymbol{e}(\boldsymbol{p}x)) = \boldsymbol{p}^{-1}((\boldsymbol{ep})x) = \boldsymbol{p}^{-1}(\boldsymbol{p}x) = (\boldsymbol{p}^{-1}\boldsymbol{p})x = \boldsymbol{e}x = x$, where $\boldsymbol{e}$ denotes the identical permutation of G. Consequently. the permutation $\boldsymbol{p}^{-1}\boldsymbol{q}^{-1}$ maps the element $\boldsymbol{qp}x$ onto x and our statement is correct.

8.8. Exercises

1. Give an example of a simple mapping of an infinite set (let us say, the set of all natural numbers) into itself which is not a permutation.
2. Write down the symbols of all the permutations of a set consisting of four elements and express the single permutations by means of pure cyclic permutations.
3. Say by which rule you would proceed if you were to write down the symbols of all the permutations of a set consisting of $n(\geqq 1)$ elements so as not to forget any of them.
4. A regular n-gon ($n \geqq 3$) in a plane has altogether n axes of symmetry. Rotating the vertices about the center of the n-gon through the angles of $0°$, $\left(\frac{360}{n}\right)^\circ$, $\left(2.\frac{360}{n}\right)^\circ, \ldots, \left((n-1).\frac{360}{n}\right)^\circ$ and, furthermore, associating with them the vertices symmetric with regard to the single axes of symmetry we obtain, altogether, $2n$ permutations of the set of vertices; let us denote the set of these permutations M_n. Prove that M_n has the following properties: 1. If $\boldsymbol{p} \in M_n$, $\boldsymbol{q} \in M_n$, then even $\boldsymbol{qp} \in M_n$; 2. $\boldsymbol{e} \in M_n$; 3. if $\boldsymbol{p} \in M_n$, then $\boldsymbol{p}^{-1} \in M_n$.
5. Any two cyclic permutations of a set of $n(\geqq 1)$ elements whose cycles have no common elements are interchangeable.

9. General (multiple-valued) mappings

9.1. Basic notions and properties

The notion of a mapping of the set G into the set G^* may be generalized by the following definition:

A *general (multiple-valued) mapping of G into G^** is a relation between the elements of both sets by which there corresponds, to every element of G, at least one element of G^*.

Let $\boldsymbol{g}$ be a general mapping of G into G^*. Then each element $a \in G$ has at least one, generally more, maybe even an infinite number of images in G^*; the set of these images is denoted by $\boldsymbol{g}a$.

If every element $a^* \in G^*$ is contained in the set of the images of some element $a \in G$, then $\boldsymbol{g}$ is said to be a general mapping of G *onto* G^*.

In that case $\boldsymbol{g}$ determines a certain general mapping of G^* onto G, called the *inverse of* $\boldsymbol{g}$ and denoted by $\boldsymbol{g}^{-1}$. It is defined in the way that to any element $a^* \in G^*$ there corresponds every element $a \in G$ whose set of images under $\boldsymbol{g}$ contains a^*. By this definition both $a^* \in \boldsymbol{g}a$ and $a \in \boldsymbol{g}^{-1}a^*$ are simultaneously valid, that is to say, if $a^* \in \boldsymbol{g}a$ then $a \in \boldsymbol{g}^{-1}a^*$, and conversely.

It is easy to show that *the mapping inverse of $\boldsymbol{g}^{-1}$ is the original mapping $\boldsymbol{g}$*, i.e., $(\boldsymbol{g}^{-1})^{-1} = \boldsymbol{g}$. Indeed, the relation $a^* \in \boldsymbol{g}a$ yields $a \in \boldsymbol{g}^{-1}a^*$ whence $a^* \in (\boldsymbol{g}^{-1})^{-1}a$, so that $\boldsymbol{g}a \subset (\boldsymbol{g}^{-1})^{-1}a$; vice versa, from $a^* \in (\boldsymbol{g}^{-1})^{-1}a$ follows $a \in \boldsymbol{g}^{-1}a^*$ whence $a^* \in \boldsymbol{g}a$ and so $(\boldsymbol{g}^{-1})^{-1}a \subset \boldsymbol{g}a$. So we have $(\boldsymbol{g}^{-1})^{-1}a = \boldsymbol{g}a$ and the proof is complete.

On taking account of the above property of the inverse mapping, we call both $\boldsymbol{g}$ and $\boldsymbol{g}^{-1}$ inverse without discerning which is inverse of which.

If $\boldsymbol{g}$ is single-valued, then $\boldsymbol{g}^{-1}$ is, as a rule, general. In that case the set of the images $\boldsymbol{g}^{-1}a^* \subset G$ of an element $a^* \in \boldsymbol{g}G$ consists of all the points of G that are, under the function $\boldsymbol{g}$, mapped onto a^*; $\boldsymbol{g}^{-1}a^*$ is therefore an element of the decomposition in G corresponding to $\boldsymbol{g}$.

The concept of a general mapping may serve as the basis of an extensive theory which, naturally, also comprises results concerning the single-valued mappings considered above. From this theory we shall now introduce a few details about the composition of general mappings.

The notion of a composite mapping which we have, in 6.7, introduced for single-valued mappings may be directly extended to general mappings. Let G, H, K denote nonempty sets, $\boldsymbol{g}$ a general mapping of G into H and $\boldsymbol{h}$ a general mapping of H into K. Then the mapping $\boldsymbol{hg}$, composed of the functions $\boldsymbol{g}$ and $\boldsymbol{h}$ (in this order), is defined by associating, with every element $a \in G$, all the $\boldsymbol{h}$-images of the individual elements lying in $\boldsymbol{g}a$. Obviously, there holds $\boldsymbol{hg}a \subset K$. In particular, the set of the images $\boldsymbol{g}^{-1}\boldsymbol{g}a$ of an element $a \in G$ consists of elements $x \in G$ such that $\boldsymbol{g}x \cap \boldsymbol{g}a \neq 0$.

Finally, let us remark that the notion of a general mapping of G into G^* may be extended by assuming that there exist elements of G with which no element of G^* is associated. Such a general mapping is called a *relation of the set G into the set G^**.

In the following study of general mappings we shall restrict our attention to the case when G and G^* coincide and we have to deal with general mappings of G *onto* itself.

9.2. Congruences

A general mapping $\boldsymbol{g}$ of the set G onto itself is called a *congruence on G* if it has the following properties:

a) For $a \in G$ there holds $a \in \boldsymbol{g}a$;

b) if for $a, b, c \in G$ there holds $b \in \boldsymbol{g}a$, $c \in \boldsymbol{g}b$, then $c \in \boldsymbol{g}a$.

These properties are expressed by saying that $\boldsymbol{g}$ is *reflexive* or *transitive*, respectively.

A congruence on a set is therefore a general mapping of a set onto itself which is both reflexive and transitive.

Suppose $\boldsymbol{g}$ is a congruence.

The relation $b \in \boldsymbol{g}a$ is expressed by saying that b is congruent to a under $\boldsymbol{g}$.

It is easy to realize that *the inverse general mapping $\boldsymbol{g}^{-1}$ is also a congruence*. In fact, the mapping $\boldsymbol{g}^{-1}$ is obviously reflexive. Moreover, from $b \in \boldsymbol{g}^{-1}a$, $c \in \boldsymbol{g}^{-1}b$ there follows $a \in \boldsymbol{g}b$, $b \in \boldsymbol{g}c$, hence $a \in \boldsymbol{g}c$ and we have $c \in \boldsymbol{g}^{-1}a$, so that $\boldsymbol{g}^{-1}$ is even transitive.

The congruence $\boldsymbol{g}^{-1}$ is, of course, called *the inverse of $\boldsymbol{g}$*. The congruence inverse of $\boldsymbol{g}^{-1}$ is $\boldsymbol{g}$.

If we have, for example, arbitrary decompositions $\bar{A}, \bar{B}$ on G and associate, with every element $\bar{a} \in \bar{A}$, all the elements of $\bar{A}$ that may be connected with $\bar{a}$ in $\bar{B}$, then we have a congruence $\boldsymbol{g}$ on $\bar{A}$ (see 3.1 a, b). In this case, every element $\bar{b} \in \bar{A}$ that may be connected with $\bar{a}$ in $\bar{B}$ is congruent to $\bar{a}$. In the inverse congruence $\boldsymbol{g}^{-1}$ there correspond, to every element $\bar{a} \in \bar{A}$, all the elements of $\bar{A}$ with which $\bar{a}$ may be connected in $\bar{B}$; the latter are, according to 3.1 c, precisely those elements that can be connected with $\bar{a}$ in $\bar{B}$. Consequently, in this particular case, the inverse congruence $\boldsymbol{g}^{-1}$ is identical with $\boldsymbol{g}$, i.e., $\boldsymbol{g}^{-1} = \boldsymbol{g}$.

Some other examples of congruences: Let us associate, with every decomposition $\bar{A}$ of the set G, all its coverings (refinements). In both cases we obtain a congruence on the set of all decompositions of G (see 2.4 a, b). Every decomposition of G which is a covering (refinement) of $\bar{A}$ is congruent to $\bar{A}$. Either of the congruences is the inverse of the other.

Of particular importance are the symmetric and antisymmetric congruences.

9.3. Symmetric congruences

A congruence $\boldsymbol{g}$ on the set G is called *symmetric* if:

From $b \in \boldsymbol{g}a$ there follows $a \in \boldsymbol{g}b$. (S)

This property expresses the symmetry of the congruence $\boldsymbol{g}$: Of every two elements in G either not one is or both are contained in the set of images of the other. If $b \in \boldsymbol{g}a$, then we write $b \equiv a$ $(\boldsymbol{g})$, briefly $b \equiv a$. Then, of course, we also have $a \equiv b$ and say that *the elements a, b are congruent.*

For instance, the congruence on the decomposition $\bar{A}$, given in 9.2, first example, is symmetric according to 3.1c.

Let $\boldsymbol{g}$ be an arbitrary symmetric congruence on G.

The congruence $\boldsymbol{g}$ has the remarkable property that *the system of all subsets of G, each of which consists of all the elements congruent to some element of G, is a decomposition of G.* The latter is said to *belong* or to *correspond* to $\boldsymbol{g}$; its elements are called *classes of the congruence $\boldsymbol{g}$.*

The proof of this statement is an easy generalization of the proof in 3.4 where we have shown that the system $\bar{\bar{A}}$ of subsets of $\bar{A}$ is a decomposition on $\bar{A}$; we leave it to the reader to carry it out himself.

It is also easy to see that *any two points of G lying in the same class of $\boldsymbol{g}$ are congruent, whereas any two that do not lie in the same class are not.* A subset of G which has exactly one point of G in common with each element of the decomposition corresponding to $\boldsymbol{g}$ forms a *system of representatives of the congruence $\boldsymbol{g}$* in the sense that every element of G is congruent to exactly one of the representatives.

Vice versa, *if we have, on the set G, an arbitrary decomposition $\bar{A}$, then there exists a congruence on G such that the corresponding decomposition is $\bar{A}$.* This congruence is defined in the way that each point $a \in G$ is congruent to any point lying in the same element of $\bar{A}$ as a, whereas the other points of G are not congruent to a.

Between the study of symmetric congruences and the study of decompositions of sets there is no essential difference.

Finally, we shall show that *the inverse congruence $\boldsymbol{g}^{-1}$ is equal to $\boldsymbol{g}$*, i.e., $\boldsymbol{g}^{-1} = \boldsymbol{g}$. Indeed, $b \in \boldsymbol{g}^{-1}a$ yields $a \in \boldsymbol{g}b$ and hence, on taking account of (S), there holds $b \in \boldsymbol{g}a$ and we have $\boldsymbol{g}^{-1}a \subset \boldsymbol{g}a$; conversely: by (S), $b \in \boldsymbol{g}a$ yields $a \in \boldsymbol{g}b$ and therefore also $b \in \boldsymbol{g}^{-1}a$ so that $\boldsymbol{g}a \subset \boldsymbol{g}^{-1}a$. Hence $\boldsymbol{g}^{-1}a = \boldsymbol{g}a$, which was to be proved.

Note that every congruence on G that coincides with its own inverse is symmetric.

Symmetric congruences are also called *equivalences.*

9.4. Antisymmetric congruences

1. *Basic concepts and properties.* A congruence $\boldsymbol{g}$ on the set G is called *antisymmetric* if:

From $b \in \boldsymbol{g}a$, $a \in \boldsymbol{g}b$ there follows $a = b$. (AS)

This property expresses the antisymmetry of $\boldsymbol{g}$: Of any two different elements of G either not one or exactly one is congruent to the other. If b is congruent to a, i.e., if $b \in \boldsymbol{g}a$, we write $a \leqq b$ $(\boldsymbol{g})$ or $b \geqq a$ $(\boldsymbol{g})$, briefly $a \leqq b$ or $b \geqq a$.

If the congruence $\boldsymbol{g}$ is antisymmetric, then its inverse, $\boldsymbol{g}^{-1}$ is antisymmetric as well, for $b \in \boldsymbol{g}^{-1}a$, $a \in \boldsymbol{g}^{-1}b$ yield $a \in \boldsymbol{g}b$, $b \in \boldsymbol{g}a$ and, consequently, by (AS), there holds $a = b$.

For example, by 2.4c, both congruences on the system of all decompositions on G considered in 9.2 are antisymmetric; as we have already said, either of them is the inverse of the other.

Antisymmetric congruences are also called *partial orderings*; partial orderings which are inverse of each other are called *dual*.

2. *The least upper bound (join) and the greatest lower bound (meet) of two elements.* Remarkable notions based on the concept of antisymmetric congruence are those of the least upper bound and the greatest lower bound of two elements.

Let there be given, on G, an antisymmetric congruence $\boldsymbol{g}$.

The *least upper bound* or the *join of a two-membered sequence of elements $a, b \in G$ with regard to $\boldsymbol{g}$*, briefly, the *least upper bound* or the *join of a, b* is the element $c \in G$ such that $a \leqq c$, $b \leqq c$ and, furthermore, $c \leqq x$ for every $x \in G$ satisfying $a \leqq x$, $b \leqq x$. Any two-membered sequence of elements may have at most one join because, if c, c' are joins, we have $c \leqq c'$ and, simultaneously, $c' \leqq c$ so that, with respect to (AS), there holds $c = c'$. The join of the elements a, b need not exist at all; if it does, it is denoted by $a \cup b$.

Analogously we define the *greatest lower bound* or the *meet of a two-membered sequence of elements $a, b \in G$ with regard to $\boldsymbol{g}$*, briefly, the *greatest lower bound* or the *meet of $a, b \in G$*; it is the element $c \in G$ such that $c \leqq a$, $c \leqq b$ and, furthermore $x \leqq c$ for every $x \in G$ satisfying $x \leqq a$, $x \leqq b$. There may exist at most one meet b; if it does, it is denoted by $a \cap b$.

Comparing the definitions of the join and the meet we observe that *the join (meet) of a, b with regard to $\boldsymbol{g}$, if it exists, is the meet (join) of a, b with regard to $\boldsymbol{g}^{-1}$.*

We leave it to the reader to verify that, *for every three elements $a, b, c \in G$, the following formulae are true whenever the included joins and meets exist*:

a) $a \cup b = b \cup a$, a′) $a \cap b = b \cap a$,

b) $a \cup a = a$, b′) $a \cap a = a$,

c) $a \cup (b \cup c) = (a \cup b) \cup c$, c′) $a \cap (b \cap c) = (a \cap b) \cap c$,

d) $a \cup (a \cap b) = a$, d′) $a \cap (a \cup b) = a$.

Since there hold a) and a′) we generally speak about the join and the meet of two elements without drawing any distinction as to their arrangement.

To give an example of a join and a meet, let us note the antisymmetric congruence on the system of all decompositions of G under which there correspond, to each decomposition of G, all its coverings or refinements. Every two decompositions $\bar{A}$, $\bar{B}$ of G have the join $[\bar{A}, \bar{B}]$ or $(\bar{A}, \bar{B})$ and the meet $(\bar{A}, \bar{B})$ or $[\bar{A}, \bar{B}]$.

9.5. Exercises

1. Let the set G be mapped, under the single-valued functions $\boldsymbol{a}, \boldsymbol{b}$, onto the set A or B, respectively, and let its decompositions corresponding to these mappings be equal. Show that, in that case, $\boldsymbol{f} = \boldsymbol{b}\boldsymbol{a}^{-1}$ is a single-valued and simple mapping of A onto B and $\boldsymbol{f}^{-1} = \boldsymbol{a}\boldsymbol{b}^{-1}$ the inverse mapping of B onto A. Hence, in this case the sets A, B are equivalent.
2. Let n denote an arbitrary positive integer. Associating, with every integer a, each number $a + \nu n$ where $\nu = \ldots, -2, -1, 0, 1, 2, \ldots$, we obtain a symmetric congruence on the set of all integers. The corresponding decomposition consists of n classes; the numbers $0, 1, \ldots, n-1$ form a system of representatives of the congruence.
3. Associating, with every positive integer, each of its positive multiples (each of its positive divisors), we obtain an antisymmetric congruence on the set of all positive integers. Every two positive integers have, with regard to this congruence, a join formed by their least common multiple (greatest common divisor) and a meet formed by their greatest common divisor (least common multiple). Either of the congruences is the inverse of the other.
4. Associating, with every part of G, each of its supersets (subsets), we obtain an antisymmetric congruence on the set of all parts of G. Every two parts of G have, with regard to this congruence, a join formed by their sum (intersection) and a meet formed by their intersection (sum). Either of the congruences is the inverse of the other.
5. If $\boldsymbol{g}$ is an antisymmetric congruence on G and some elements $a, b \in G$ have the join $a \cup b$, then:

   a) $\boldsymbol{g}(a \cup b) = \boldsymbol{g}a \cap \boldsymbol{g}b$ (the right-hand side denotes, of course, the intersection of $\boldsymbol{g}a$, $\boldsymbol{g}b$),
   b) $\boldsymbol{g}^{-1}(a \cup b) \supset \boldsymbol{g}^{-1}a \cup \boldsymbol{g}^{-1}b$.

10. Series of decompositions of sets

In this chapter we shall develop a theory of the so-called series of decompositions of sets. We shall make use of many results arrived at in the previous considerations and concerning decompositions and mappings of sets. The mentioned theory describes the set-structure of the appropriate sections of the theory of groupoids and groups and admits of a better understanding of the results of the theory of groups arrived at by classical methods. A study of the series of decompositions of sets has, moreover, proved most useful in connection with mappings onto sets of sequences and the domain of scientific classifications.

10.1. Basic concepts

Let $\bar{A} \geqq \bar{B}$ stand for arbitrary decompositions of the set G.

A *series of decompositions of the set G from $\bar{A}$ to $\bar{B}$* (briefly, a *series of decompositions from $\bar{A}$ to $\bar{B}$*) is a finite sequence of the decompositions $\bar{A}_1, \ldots, \bar{A}_\alpha$ on G, of length $\alpha (\geqq 1)$, with the following properties: 1. The first member of the sequence is the decomposition $\bar{A}$, the last member is $\bar{B}$, hence $\bar{A}_1 = \bar{A}$, $\bar{A}_\alpha = \bar{B}$. 2. Every decomposition is a refinement of the one directly preceding it, so that

$$(\bar{A} =)\ \bar{A}_1 \geqq \cdots \geqq \bar{A}_\alpha\ (= \bar{B}).$$

Such a series is briefly denoted $(\bar{A})$. The decompositions $\bar{A}_1, \ldots, \bar{A}_\alpha$ are called *members of* $(\bar{A})$; $\bar{A}_1$ is the *initial* and $\bar{A}_\alpha$ the *final* member of $(\bar{A})$. By the *length of* $(\bar{A})$ we mean the number α of the members of $(\bar{A})$.

For example, a decomposition $\bar{A}$ on G forms a series of length 1; its initial as well as final member coincides with $\bar{A}$.

Suppose $\big((\bar{A}) = \big)\ \bar{A}_1 \geqq \cdots \geqq \bar{A}_\alpha$ is a series of decompositions from $\bar{A}$ to $\bar{B}$.

A member of $(\bar{A})$ is called *essential* if it is either the initial member $\bar{A}_1$ or a proper refinement of the member directly preceding it. In the opposite case it is *inessential*. If $(\bar{A})$ contains at least one inessential member $\bar{A}_{\gamma+1}$, then it is called (because $\bar{A}_{\gamma+1} = \bar{A}_\gamma$) a *series with iteration*. If all the members of $(\bar{A})$ are essential, then $\bar{A}$ is said to be *without iteration*. The number α' of essential members of $(\bar{A})$ is the *reduced length of* $(\bar{A})$. There evidently holds $1 \leqq \alpha' \leqq \alpha$ and the equality $\alpha' = \alpha$ is characteristic of series without iteration. If any iterations in $(\bar{A})$ occur, then $(\bar{A})$ may be reduced by omitting all the inessential members, that is to say, shortened to a series $(\bar{A}')$ without iteration. The length of the reduced series $(\bar{A}')$ equals the reduced length α' of the series $(\bar{A})$. Conversely, $(\bar{A})$ may be lengthened by inserting a finite number of inessential members between any two neighbouring members $\bar{A}_\gamma$, $\bar{A}_{\gamma+1}$ or, if convenient, before (after) the first (last) member $\bar{A}_1$ $(\bar{A}_\alpha)$ of $(\bar{A})$. Every series of decompositions, generated by reducing or extending (lengthening) $(\bar{A})$, naturally, has the same reduced length as $(\bar{A})$.

If $\alpha_1 < \cdots < \alpha_\beta$ are arbitrary numbers of the set $\{1, \ldots, \alpha\}$, then even

$$\bar{A}_{\alpha_1} \geqq \cdots \geqq \bar{A}_{\alpha\beta}$$

is a series of decompositions on G, called a *partial series* or a *part of* $(\bar{A})$.

If, moreover, A is a nonempty subset of G, then the sequence

$$\bar{A}_{\alpha_1} \sqcap A \geqq \cdots \geqq \bar{A}_{\alpha\beta} \sqcap A$$

is a series of decompositions on A.

10.2. Local chains

Suppose $((\bar{A}) =)\ \bar{A}_1 \geqq \cdots \geqq \bar{A}_\alpha$ is a series of decompositions on G of length $\alpha \geqq 1$.

Let $\bar{a} \in \bar{A}_\alpha$ be an arbitrary element and $\bar{a}_\gamma \in \bar{A}_\gamma$ the element of $\bar{A}_\gamma$ containing $\bar{a}$ ($\gamma = 1, \ldots, \alpha$). There evidently holds:

$$\bar{a}_1 \supset \cdots \supset \bar{a}_\alpha \ (\bar{a}_\alpha = \bar{a}).$$

Furthermore,

$$\bar{K}_\gamma = \bar{a}_\gamma \sqcap \bar{A}_{\gamma+1} \qquad (\bar{A}_{\alpha+1} = \bar{A}_\alpha)$$

is a decomposition on $\bar{a}_\gamma$, forming a part of $\bar{A}_{\gamma+1}$ and, simultaneously, $\bar{a}_{\gamma+1} \in \bar{K}_\gamma$ ($\bar{a}_{\alpha+1} = \bar{a}_\alpha$). We observe that

$$([\bar{K}] =)\ \bar{K}_1 \to \ldots \to \bar{K}_\alpha$$

is a chain of decompositions of sets from $\bar{a}_1$ to $\bar{a}_{\alpha+1}$ $(= \bar{a})$ (2.5). It is called the *local chain of the series* $(\bar{A})$, *corresponding to the element* $\bar{a} \in \bar{A}_\alpha$, briefly: the local chain with the base $\bar{a}$. Notation as above or, more accurately: $([\bar{K}\bar{a}] =)\ \bar{K}_1\bar{a} \to \cdots \to \bar{K}_\alpha\bar{a}$. The element $\bar{a} \in \bar{A}_\alpha$ is called the *base of the chain* $[\bar{K}]$. By its base $\bar{a}$ the chain $[\bar{K}]$ is uniquely determined.

Let us remark that the final member $\bar{K}_\alpha$ of $[\bar{K}]$ is the greatest decomposition of $\bar{a}$, hence inessential. $\bar{K}_\gamma$ may, with respect to $\bar{A}_\gamma \geqq \bar{A}_{\gamma+1}$, also be defined by the formula $\bar{K}_\gamma = \bar{a}_\gamma \sqsubset \bar{A}_{\gamma+1}$.

The local chain $[\bar{K}]$ *is an elementary chain from* $\bar{a}_1$ *to* $\bar{a}_{\alpha+1}$ $(= \bar{a})$ *over* $\bar{A}_{\alpha+1}$.

Indeed, since $\bar{A}_{\gamma+1}$ is a covering of $\bar{A}_{\alpha+1}$ ($\gamma = 1, \ldots, \alpha$), $\bar{a}_\gamma \sqcap \bar{A}_{\gamma+1}$ is a covering of $\bar{a}_\gamma \sqcap \bar{A}_{\alpha+1}$.

The length of $[\bar{A}]$ is, obviously, α and therefore equal to the length of $(\bar{A})$. If a member $\bar{A}_{\gamma+1}$ of $(\bar{A})$ is inessential and so $\bar{A}_{\gamma+1} = \bar{A}_\gamma$, then there holds $\bar{a}_{\gamma+1} = \bar{a}_\gamma$; hence $\bar{K}_\gamma$ is an inessential member of $[\bar{K}]$. Consequently, for the reduced lengths α' and $\varkappa'$ of $(\bar{A})$ and the local chain $[\bar{K}]$, there holds: $\varkappa' \leqq \alpha'$. Thus, if a local chain of (A) has no iteration, except the final member which is always inessential, then $(\bar{A})$ is a series without iteration.

10.3. Refinements of series of decompositions

Suppose, again, that $((\bar{A}) =)\ \bar{A}_1 \geqq \cdots \geqq \bar{A}_\alpha$ is a series of decompositions of length $\alpha \geqq 1$ on the set G.

By a *refinement of* $(\bar{A})$ we mean a series of decompositions on G such that $(\bar{A})$ is a part of that series. Thus every refinement of $(\bar{A})$ is of the form:

$$\bar{A}_{1,1} \geqq \cdots \geqq \bar{A}_{1,\beta_1-1} \geqq \bar{A}_{1,\beta_1} \geqq \bar{A}_{2,1} \geqq \cdots \geqq \bar{A}_{2,\beta_2-1} \geqq \bar{A}_{2,\beta_2} \geqq \cdots \geqq$$
$$\geqq \bar{A}_{\alpha,\beta_\alpha} \geqq \bar{A}_{\alpha+1,1} \geqq \cdots \geqq \bar{A}_{\alpha+1,\beta_{\alpha+1}-1}.$$

In the above formulae, $\bar{A}_{\gamma,\beta_\gamma} = \bar{A}_\gamma$ holds for $\gamma = 1, \ldots, \alpha$, whereas $\beta_1, \ldots, \beta_{\alpha+1}$ are natural numbers. If $\beta_\delta = 1$, then the members $\bar{A}_{\delta,1} \geqq \ldots \geqq \bar{A}_{\delta,\beta_\delta-1}$ are not read. From the definition it is clear that any refinement of $(\bar{A})$ is obtained by way of inserting between two neighbouring members $\bar{A}_\gamma, \bar{A}_{\gamma+1}$ and, maybe, also before $\bar{A}_1$ and after $\bar{A}_\alpha$, a suitable series of decompositions. Note that every lengthening of $(\bar{A})$ is its own refinement.

Let us consider a refinement $(\mathring{A})$ of $(\bar{A})$ and use the same notation as above. In particular, $\bar{A}_{\gamma,\beta_\gamma} = \bar{A}_\gamma$ for $\gamma = 1, \ldots, \alpha$. The indices μ, ν will, in what follows, denote: for $\beta_{\alpha+1} = 1$, the numbers $\mu = 1, \ldots, \alpha$; $\nu = 1, \ldots, \beta_\mu$ and for $\beta_{\alpha+1} > 1$, even the numbers $\mu = \alpha + 1$, $\nu = 1, \ldots, \beta_{\alpha+1} - 1$.

Let $\bar{a} \in \bar{A}_\alpha$ or $\bar{a} \in \bar{A}_{\alpha+1,\beta_{\alpha+1}-1}$ stand for an element of $\bar{A}_\alpha$ or of $\bar{A}_{\alpha+1,\beta_{\alpha+1}-1}$ according as $\beta_{\alpha+1} = 1$ or $\beta_{\alpha+1} > 1$. Let, moreover, $\bar{a}_{\mu,\nu}$ and $\bar{a}_\gamma$ denote the elements of $\bar{A}_{\mu,\nu}$, $\bar{A}_\gamma$ for which $\bar{a} \subset \bar{a}_{\mu,\nu} \in \bar{A}_{\mu,\nu}$ and $\bar{a} \subset \bar{a}_\gamma \in \bar{A}_\gamma$, respectively; so we have, in particular, $\bar{a}_{\gamma,\beta_\gamma} = \bar{a}_\gamma$.

The local chain $[\mathring{K}]$ of $(\mathring{A})$, with the base $\bar{a}$, is

$$([\mathring{K}] =)\ \bar{K}_{1,1} \to \cdots \to \bar{K}_{1,\beta_1} \to \bar{K}_{2,1} \to \cdots \to \bar{K}_{2,\beta_2} \to \cdots \to \bar{K}_{\alpha,\beta_\alpha}$$
$$\to \bar{K}_{\alpha+1,1} \to \cdots \to \bar{K}_{\alpha+1,\beta_{\alpha+1}-1},$$

where $\bar{K}_{\mu,\nu} = \bar{a}_{\mu,\nu} \sqcap \bar{A}_{\mu,\nu+1}$, $\bar{A}_{\mu,\beta_\mu+1} = \bar{A}_{\mu+1,1}$ and, moreover, $\bar{A}_{\alpha+1,1} = \bar{A}_{\alpha,\beta_\alpha}$ in case of $\beta_{\alpha+1} = 1$ and $\bar{A}_{\alpha+1,\beta_{\alpha+1}} = \bar{A}_{\alpha+1,\beta_{\alpha+1}-1}$ in case of $\beta_{\alpha+1} > 1$.

We observe that the local chain $[\mathring{K}]$ is obtained by replacing each member $\bar{K}_\gamma = \bar{a}_\gamma \sqcap \bar{A}_{\gamma+1}$ of the local chain $[\bar{K}]$ of $(\bar{A})$, with the base $\bar{a}_\alpha \in \bar{A}_\alpha$, by a chain from the set $\bar{a}_\gamma$ to $\bar{a}_{\gamma+1}$:

$$\bar{K}_{\gamma,\beta_\gamma} \to \bar{K}_{\gamma+1,1} \to \cdots \to \bar{K}_{\gamma+1,\beta_{\gamma+1}-1}.$$

(if $\beta_{\gamma+1} = 1$, then we read only the initial member $\bar{K}_{\gamma,\beta_\gamma}$) and, moreover, if $\beta_1 > 1$, we add, at the beginning of $[\bar{K}]$, a chain from the set $\bar{a}_{1,1}$ to $\bar{a}_1$: $\bar{K}_{1,1} \to \cdots \to \bar{K}_{1,\beta_1-1}$. The above chains are, evidently, elementary chains from $\bar{a}_\gamma$ to $\bar{a}_{\gamma+1}$ or from $\bar{a}_{1,1}$ to $\bar{a}_1$ over the decompositions $\bar{a}_\gamma \sqcap \bar{A}_{\gamma+1}$ or $\bar{a}_{1,1} \sqcap \bar{A}_1$, respectively. Thus *the local chain of every refinement of* $(\bar{A})$, *with the base* $\bar{a} \subset \bar{a}_\alpha$ *is a refinement of the local chain of* $(\bar{A})$, *with the base* $\bar{a}_\alpha$.

10.4. Manifolds of local chains

Let us consider a series of decompositions on the set G:

$$((\bar{A}) =)\ \bar{A}_1 \geqq \cdots \geqq \bar{A}_\alpha\ (\alpha \geqq 1).$$

To every element $\bar{a} \in \bar{A}_\alpha$ there corresponds a local chain of $(\bar{A})$, with the base $\bar{a}$:

$$([\bar{K}\bar{a}] =)\ \bar{K}_1\bar{a} \to \cdots \to \bar{K}_\alpha\bar{a}.$$

The set consisting of local chains whose bases are the individual elements of $\bar{A}_\alpha$ is called the *manifold of local chains, corresponding to* $(\bar{A})$; notation: $\tilde{A}$. It is obviously an α-grade structure with regard to the sequence of decompositions $\bar{A}_2, \ldots, \bar{A}_{\alpha+1}$ $(\bar{A}_{\alpha+1} = \bar{A}_\alpha)$ in the sense of the definition introduced in 1.9.

Associating, with every point $a \in G$, the local chain $[\bar{K}\bar{a}] \in \tilde{A}$ with the base $\bar{a} = \bar{a}_\alpha \in \bar{A}_\alpha$ for which $a \in \bar{a}$, we obtain a mapping called the *natural mapping of the set G onto the manifold of the local chains* $\tilde{A}$. The decomposition of G corresponding to this mapping, naturally, coincides with $\bar{A}_\alpha$. By a *local chain of* $(\bar{A})$, *corresponding to* a, we mean the local chain $[\bar{K}\bar{a}]$.

Now let

$$((\bar{A}) =) \bar{A}_1 \geqq \cdots \geqq \bar{A}_\alpha,$$
$$((\bar{B}) =) \bar{B}_1 \geqq \cdots \geqq \bar{B}_\beta \ (\alpha, \beta \geqq 1)$$

be series of decompositions on G such that their end-members $\bar{A}_\alpha$, $\bar{B}_\beta$ coincide: $\bar{A}_\alpha = \bar{B}_\beta$.

Consider the manifolds of local chains, $\tilde{A}$ and $\tilde{B}$, corresponding to the series $(\bar{A})$ and $(\bar{B})$, respectively.

Associating, with every element $[\bar{K}\bar{a}] \in \tilde{A}$, the local chain $[\bar{L}\bar{a}] \in \tilde{B}$ with the same base $a \in \bar{A}_\alpha = \bar{B}_\beta$, we obtain a simple mapping of $\tilde{A}$ onto $\tilde{B}$, called *co-basal.*

We see that the manifolds of local chains, corresponding to two series of decompositions with coinciding end-members, are equivalent sets and that the co-basal mapping is a one-to-one mapping of one onto the other.

10.5. Chain-equivalent series of decompositions

Suppose

$$((\bar{A}) =) \bar{A}_1 \geqq \cdots \geqq \bar{A}_\alpha,$$
$$((\bar{B}) =) \bar{B}_1 \geqq \cdots \geqq \bar{B}_\alpha$$

are arbitrary chains of decompositions on G of the same length α $(\geqq 1)$.

Let again $\tilde{A}$, $\tilde{B}$ denote the manifolds of local chains corresponding to $(\bar{A}), (\bar{B})$.

$(\bar{B})$ is said to be *chain-equivalent to* $(\bar{A})$ if the manifold of the local chains, $\tilde{B}$, is strongly equivalent to the manifold $\tilde{A}$.

If $(\bar{B})$ is chain-equivalent to $(\bar{A})$, then $(\bar{A})$ is chain-equivalent to $(\bar{B})$, (6.9.1). With respect to this symmetry, we speak about chain-equivalent series $(\bar{A})$, $(\bar{B})$.

By the above definition, $(\bar{B})$ is chain-equivalent to $(\bar{A})$ if there exists a strong equivalence-mapping of the manifold of the local chains, $\tilde{A}$, onto the manifold $\tilde{B}$ (6.9.1). If, in particular, the end-members $\bar{A}_\alpha$, $\bar{B}_\alpha$ of $(\bar{A})$, $(\bar{B})$, respectively, coincide and, simultaneously, the co-basal mapping of $\tilde{A}$ onto $\tilde{B}$ is a strong equivalence, then $(\bar{B})$ is said to be *co-basally chain-equivalent to* $(\bar{A})$ and we speak about *co-basally chain-equivalent series* $(\bar{A})$, $(\bar{B})$.

Let us now assume that $(\bar{A})$, $(\bar{B})$ are chain-equivalent.

Let $\boldsymbol{f}$ be a strong equivalence-mapping of the manifold $\tilde{A}$ onto $\tilde{B}$. By 6.9.1, $\boldsymbol{f}$ is a one-to-one mapping of $\tilde{A}$ onto $\tilde{B}$, where every two associated elements of $\tilde{A}$, $\tilde{B}$ are in certain mutual relations. This situation can more accurately be described as follows:

There exists a permutation $\boldsymbol{p}$ of the set $\{1, \ldots, \alpha\}$ with the following effect:

Let $[\bar{K}] \in \tilde{A}$, $\boldsymbol{f}[\bar{K}] = [\bar{L}] \in \tilde{B}$ be two arbitrary local chains of the series $(\bar{A})$, $(\bar{B})$, respectively:

$$([\bar{K}] =)\ \bar{K}_1 \to \cdots \to \bar{K}_\alpha,$$

$$([\bar{L}] =)\ \bar{L}_1 \to \cdots \to \bar{L}_\alpha,$$

where $[\bar{L}]$ is the image of $[\bar{K}]$ under the mapping $\boldsymbol{f}$. We know that every member $\bar{K}_\gamma$ $(\bar{L}_\gamma)$ $(\gamma = 1, \ldots, \alpha)$ is a decomposition in G which is a part of $\bar{A}_{\gamma+1}$ $(\bar{B}_{\gamma+1})$ while $\bar{A}_{\alpha+1} = \bar{A}_\alpha$, $\bar{B}_{\alpha+1} = \bar{B}_\alpha$. The effect of $\boldsymbol{p}$ consists in that to every member $\bar{K}_\gamma$ of $[\bar{K}]$ there exists a one-to-one function $\boldsymbol{a}_\gamma$ mapping the member $\bar{K}_\gamma$ onto the member $\bar{L}_\delta$ of $[\bar{L}]$ while $\delta = \boldsymbol{p}\gamma$.

We observe that any two members $\bar{K}_\gamma$, $\bar{L}_\delta$ of the local chains $[\bar{K}]$, $[\bar{L}]$ with the indices $\gamma, \delta = \boldsymbol{p}\gamma$ are equivalent sets. Consequently, such members $\bar{K}_\gamma$, $\bar{L}_\gamma$ are, in the local chains $[\bar{K}]$, $[\bar{L}]$, simultaneously either essential or inessential. Hence *any two local chains corresponding to each other under* $\boldsymbol{f}$ *are of the same reduced length.*

Our object now is to show that even $(\bar{A})$, $(\bar{B})$ *are of the same reduced length.*

That is, first of all, evident if $\alpha = 1$, as the initial members $\bar{A}_1$, $\bar{B}_1$ of $(\bar{A})$, $(\bar{B})$ are aways essential.

Let $\alpha > 1$. Consider an arbitrary essential member $\bar{A}_{\gamma+1}$ $(1 \leqq \gamma < \alpha)$ of $(\bar{A})$. Then there exists an element $\bar{a}_\gamma \in \bar{A}_\gamma$ such that $\bar{a}_\gamma \sqcap \bar{A}_{\gamma+1}$ comprises more than one element. Let $\bar{a} = \bar{a}_\alpha \in \bar{A}_\alpha$ be an arbitrary element of $\bar{A}_\alpha$ such that: $\bar{a} \subset \bar{a}_\gamma$. Furthermore, let $[\bar{K}]$ be the local chain $(\bar{A})$ with the base $\bar{a}$ and $[\bar{L}] = \boldsymbol{f}[\bar{K}]$ denote the local chain of $(\bar{B})$ associated with $[\bar{K}]$ under the function $\boldsymbol{f}$. The members of $[\bar{K}]$, $[\bar{L}]$ are denoted as above. Then we have, in particular, $\bar{K}_\gamma = \bar{a}_\gamma \sqcap \bar{A}_{\gamma+1}$, $\bar{L}_\delta = \bar{b}_\delta \sqcap \bar{B}_{\delta+1}$ where $\delta = \boldsymbol{p}\gamma$ and $\bar{b}_\delta \in \bar{B}_\delta$. According to the above considerations, $\bar{L}_\delta$ is a set equivalent to $\bar{K}_\gamma$ and therefore contains more than one element. Consequently, the member $\bar{B}_{\delta+1}$ of $(\bar{B})$ is essential; in particular, we have $1 \leqq \delta < \alpha$. It is obvious that $(\bar{B})$ contains at least as many essential members as $(\bar{A})$ so that, for the reduced lengths α', β' of $(\bar{A})$, $(\bar{B})$, there holds $\alpha' \leqq \beta'$. For analogous reasons there also holds $\beta' \leqq \alpha'$ and the proof is accomplished.

10.6. Semi-joint (loosely joint) and joint series of decompositions

Let us again consider two series of decompositions $(\bar{A})$, $(\bar{B})$ on the set G, of length α $(\geqq 1)$, and use the above notation. The symbols $\tilde{A}$, $\tilde{B}$ then denote the manifolds of the local chains, corresponding to the series (A), $(\bar{B})$.

$(\bar{B})$ *is* said to be *semi-joint* or *loosely joint* (*joint*) *with* $(\bar{A})$ if the manifold of the local chains, $\tilde{B}$, is equivalent to and loosely coupled with (equivalent to and coupled with) the manifold $\tilde{A}$.

If $(\bar{B})$ is semi-joint (joint) with $(\bar{A})$, then $(\bar{A})$ is also semi-joint (joint) with $(\bar{B})$ (6.9.2). Taking account of this symmetry, we speak about the semi-joint or loosely joint (joint) series $(\bar{A})$, $(\bar{B})$.

By the above definition, $(\bar{B})$ is semi-joint (joint) with $(\bar{A})$ if there exists an equivalence connected with loose coupling (equivalence connected with coupling) of the manifold of the local chains, $\tilde{A}$, onto the manifold $\tilde{B}$ (6.9.2). If, in particular, the final members $\bar{A}_\alpha$, $\bar{B}_\beta$ of $(\bar{A})$, (B) coincide and the co-basal mapping of the manifold of the local chains, $\tilde{A}$, onto the manifold $\tilde{B}$ is an equivalence connected with loose coupling (an equivalence connected with coupling) then $(\bar{B})$ *is* said to be *co-basally semi-joint* or *co-basally loosely joint* (*co-basally joint*) *with* $(\bar{A})$; in that case we also speak of *co-basally semi-joint* or *co-basally loosely joint* (*co-basally joint*) series $(\bar{A})$, $(\bar{B})$.

Let us now assume $(\bar{A})$, $(\bar{B})$ to be loosely joint (joint).

Let $\boldsymbol{f}$ stand for an equivalence-mapping connected with loose coupling (equivalence-mapping connected with coupling) of the manifold of the local chains, $\tilde{A}$, onto the manifold $\tilde{B}$. The mapping $\boldsymbol{f}$ is therefore simple (one-to-one) (6.9) and the situation may be described as follows (6.9.2):

There exists a permutation $\boldsymbol{p}$ of the set $\{1, \ldots, \alpha\}$ with the following effect:

Let $[\bar{K}] \in \tilde{A}$, $\boldsymbol{f}[\bar{K}] = [\bar{L}] \in \tilde{B}$ be arbitrary local chains of $(\bar{A})$, $(\bar{B})$ associated with each other under the function $\boldsymbol{f}$. Then every two members $\bar{K}_\gamma$, $\bar{L}_\delta$ of $[\bar{K}]$, $[\bar{L}]$ are loosely coupled (coupled) decompositions in G; at the same time, $\delta = \boldsymbol{p}\gamma$. More accurately: each member of either of the mentioned decompositions is incident with at most one (exactly one) element of the other while there always occurs at least one incidence. The closures $\mathrm{H}\bar{K}_\gamma = \bar{L}_\delta \sqsubset \bar{K}_\gamma$, $\mathrm{H}\bar{L}_\delta = \bar{K}_\gamma \sqsubset \bar{L}_\delta$ $(= \emptyset)$ are coupled.

If $(\bar{A})$, $(\bar{B})$ are joint, then the mapping $\boldsymbol{a}_\gamma$ of $\bar{K}_\gamma$ onto $\bar{L}_\delta$, given by the incidence of the elements $\bar{K}_\gamma$, $\bar{L}_\delta$, is simple.

We see that *two joint series of decompositions are chain-equivalent*. In particular, hey are of the same reduced length.

10.7. Modular series of decompositions

Suppose that

$$((\bar{A}) =) \; \bar{A}_1 \geqq \cdots \geqq \bar{A}_\alpha,$$
$$((\bar{B}) =) \; \bar{B}_1 \geqq \cdots \geqq \bar{B}_\beta$$

are series of decompositions on G, of lengths $\alpha, \beta \geqq 1$, respectively.

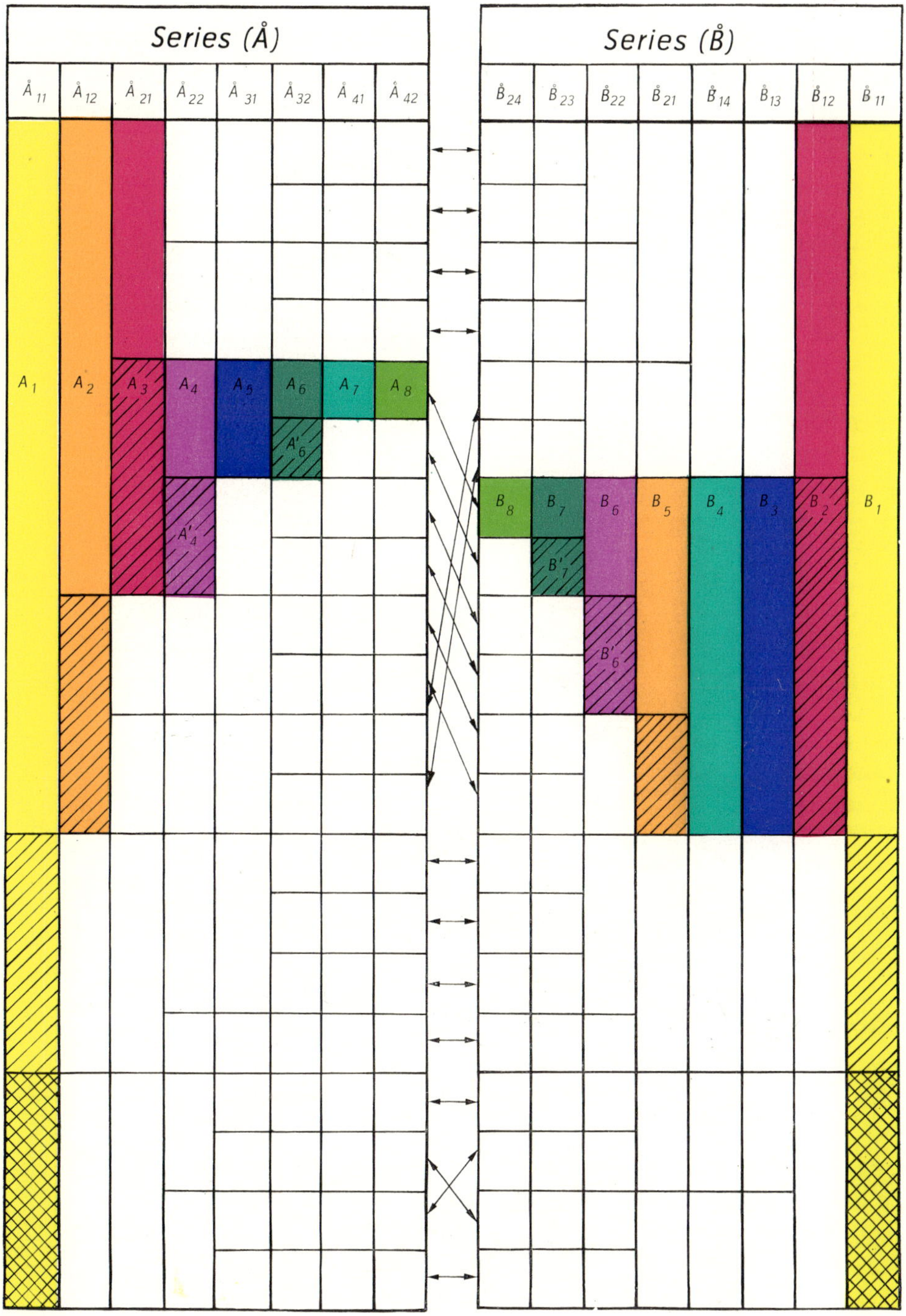
Series (Å)
Series (B̊)
Å11
Å12
Å21
Å22
Å31
Å32
Å41
Å42
B̊24
B̊23
B̊22
B̊21
B̊14
B̊13
B̊12
B̊11
A1
A2
A3
A4
A5
A6
A7
A8
A'6
A'4
B8
B7
B6
B5
B4
B3
B2
B1
B'7
B'6

$(\bar{A})$, $(\bar{B})$ are called *modular* if each member $\bar{A}_\mu$ of $(\bar{A})$ is modular with regard to every two neighbouring members $\bar{B}_{\delta-1}$, $\bar{B}_\delta$ of (B) and, simultaneously, each member $\bar{B}_\nu$ of $(\bar{B})$ is modular with regard to every two neighbouring members $\bar{A}_{\gamma-1}$, $\bar{A}_\gamma$ of $(\bar{A})$; that is to say, if there holds:

$$[\bar{A}_\gamma, (\bar{A}_{\gamma-1}, \bar{B}_\nu)] = (\bar{A}_{\gamma-1}, [\bar{A}_\gamma, \bar{B}_\nu]),$$
$$[\bar{B}_\delta, (\bar{B}_{\delta-1}, \bar{A}_\mu)] = (\bar{B}_{\delta-1}, [\bar{B}_\delta, \bar{A}_\mu]). \tag{1}$$

In what follows we shall assume the series $(\bar{A})$, $(\bar{B})$ to be modular.

Then the following theorem is true:

The series $(\bar{A})$, $(\bar{B})$ have co-basally loosely joint refinements $(\mathring{A})$, $(\mathring{B})$ with equal initial and final members. The refinements are given by the construction described in part a) *of the following proof.*

Proof. a) Let us denote:

$$[\bar{A}_1, \bar{B}_1] = \bar{U}, \quad (\bar{A}_\alpha, \bar{B}_\beta) = \bar{V},$$
$$\bar{A}_0 = \bar{B}_0 = \bar{G}_{\max}, \quad \bar{A}_{\alpha+1} = \bar{B}_{\beta+1} = \bar{V}.$$

Then the above formulae (1) are true for $\gamma, \mu = 1, \ldots, \alpha + 1; \delta, \nu = 1, \ldots, \beta + 1$.

Let us denote the decompositions on either side of the first (second) formula (1) by $\mathring{A}_{\gamma,\nu}$ and $\mathring{B}_{\delta,\mu}$, respectively, the indices γ, μ; δ, ν having the above values.

From the definition of the decompositions $\mathring{A}_{\gamma,\nu}$, $\mathring{B}_{\delta,\mu}$ there follows:

$$\bar{A}_{\gamma-1} \geqq \mathring{A}_{\gamma,\nu}, \quad \mathring{A}_{\gamma,\beta+1} = \bar{A}_\gamma,$$
$$\bar{B}_{\delta-1} \geqq \mathring{B}_{\delta,\mu}, \quad \mathring{B}_{\delta,\alpha+1} = \bar{B}_\delta.$$

For $\nu \leqq \beta$ there holds $\bar{B}_\nu \geqq \bar{B}_{\nu+1}$; hence, by 3.7.2,

$$(\bar{A}_{\gamma-1}, \bar{B}_\nu) \geqq (\bar{A}_{\gamma-1}, \bar{B}_{\nu+1})$$

and, furthermore,

$$[\bar{A}_\gamma, (\bar{A}_{\gamma-1}, \bar{B}_\nu)] \geqq [\bar{A}_\gamma, (\bar{A}_{\gamma-1}, \bar{B}_{\nu+1})].$$

In a similar way we deduce, for $\mu \leqq \alpha$, the relation

$$[\bar{B}_\delta, (\bar{B}_{\delta-1}, \bar{A}_\mu)] \geqq [\bar{B}_\delta, (\bar{B}_{\delta-1}, \bar{A}_{\mu+1})].$$

So we have, for $\nu \leqq \beta$, $\mu \leqq \alpha$, the relations:

$$\mathring{A}_{\gamma,\nu} \geqq \mathring{A}_{\gamma,\nu+1}, \quad \mathring{B}_{\delta,\mu} \geqq \mathring{B}_{\delta,\mu+1}$$

and arrive at the following series of decompositions from $\mathring{A}_{\gamma,1}$ to $\bar{A}_\gamma$ and from $\mathring{B}_{\delta,1}$ to $\bar{B}_\delta$:

$$\mathring{A}_{\gamma,1} \geqq \cdots \geqq \mathring{A}_{\gamma,\beta+1},$$
$$\mathring{B}_{\delta,1} \geqq \cdots \geqq \mathring{B}_{\delta,\alpha+1}.$$

We observe that the following series of the decompositions $(\mathring{A})$, $(\mathring{B})$ on G are refinements of $(\bar{A})$, $(\bar{B})$:

$$\big((\mathring{A}) =\big)\ \bar{U} = \mathring{A}_{1,1} \geqq \cdots \geqq \mathring{A}_{1,\beta+1} \geqq \mathring{A}_{2,1} \geqq \cdots \geqq \mathring{A}_{2,\beta+1}$$
$$\geqq \cdots \geqq \mathring{A}_{\alpha+1,1} \geqq \cdots \geqq \mathring{A}_{\alpha+1,\beta+1} = \bar{V},$$
$$\big((\mathring{B}) =\big)\ \bar{U} = \mathring{B}_{1,1} \geqq \cdots \geqq \mathring{B}_{1,\alpha+1} \geqq \mathring{B}_{2,1} \geqq \cdots \geqq \mathring{B}_{2,\alpha+1}$$
$$\geqq \cdots \geqq \mathring{B}_{\beta+1,1} \geqq \cdots \geqq \mathring{B}_{\beta+1,\alpha+1} = \bar{V}.$$

The series $(\mathring{A})$, $(\mathring{B})$ obviously have the same length $(\alpha + 1)(\beta + 1)$ and their initial and final members coincide: $(\bar{U} =)\ \mathring{A}_{1,1} = \mathring{B}_{1,1}$, $\mathring{A}_{\alpha+1,\beta+1} = \mathring{B}_{\beta+1,\alpha+1} (= \bar{V})$. *The series* $(\mathring{A})$, $(\mathring{B})$ *are the mentioned co-basally loosely joint refinements of the series* $(\bar{A})$, $(\bar{B})$, *respectively.*

b) Now let us show that $(\mathring{A})$, $(\mathring{B})$ are co-basally loosely joint.

We shall, first, define the permutation $\boldsymbol{p}$ of the set

$$\{1, \ldots, (\alpha + 1)(\beta + 1)\}$$

as follows:

$$\boldsymbol{p}[(\mu - 1)(\beta + 1) + \nu - 1] = (\nu - 1)(\alpha + 1) + \mu - 1$$
$$(\mu = 1, \ldots, \alpha + 1; \quad \nu = 1, \ldots, \beta + 1; \quad \mu + \nu > 2),$$
$$\boldsymbol{p}(\alpha + 1)(\beta + 1) = (\beta + 1)(\alpha + 1).$$

Let $\bar{a} \in \bar{V}$ be an arbitrary element and

$$([\mathring{K}\bar{a}] =) \quad \mathring{K}_1 \to \cdots \to \mathring{K}_{(\alpha+1)(\beta+1)},$$
$$([\mathring{L}\bar{a}] =) \quad \mathring{L}_1 \to \cdots \to \mathring{L}_{(\beta+1)(\alpha+1)}$$

the local chains of $(\mathring{A})$, $(\mathring{B})$ corresponding to the base $\bar{a}$.

Let, moreover, $\bar{a}_{\mu-1}, \bar{b}_{\nu-1}; \mathring{a}_{\mu,\nu}, \mathring{b}_{\nu,\mu}$ be elements given by the relations:

$$\bar{a} \subset \bar{a}_{\mu-1} \in \bar{A}_{\mu-1}, \quad \bar{a} \subset \bar{b}_{\nu-1} \in \bar{B}_{\nu-1},$$
$$\bar{a} \subset \mathring{a}_{\mu,\nu} \in \mathring{A}_{\mu,\nu}, \quad \bar{a} \subset \mathring{b}_{\nu,\mu} \in \mathring{B}_{\nu,\mu}$$
$$(\mu = 1, \ldots, \alpha + 1; \quad \nu = 1, \ldots, \beta + 1; \quad \bar{a}_0 = \bar{b}_0 = G).$$

Then we have:

$$\mathring{K}_{(\mu-1)(\beta+1)+\nu-1} = \mathring{a}_{\mu,\nu-1} \sqcap \mathring{A}_{\mu,\nu},$$
$$\mathring{L}_{(\nu-1)(\alpha+1)+\mu-1} = \mathring{b}_{\nu,\mu-1} \sqcap \mathring{B}_{\nu,\mu}$$
$$(\mu + \nu > 2; \quad \mathring{a}_{\mu,0} = \bar{a}_{\mu-1}, \quad \mathring{b}_{\nu,0} = \bar{b}_{\nu-1}),$$
$$\mathring{K}_{(\alpha+1)(\beta+1)} = \bar{a} \sqcap \bar{V} = \mathring{L}_{(\beta+1)(\alpha+1)}. \tag{1}$$

We shall show that the decompositions $\mathring{K}_{(\mu-1)(\beta+1)+\nu-1}$ and $\mathring{L}_{(\nu-1)(\alpha+1)+\mu-1}$ ($\mu + \nu > 2$) as well as $\mathring{K}_{(\alpha+1)(\beta+1)}$ and $\mathring{L}_{(\beta+1)(\alpha+1)}$ associated under the permutation $\boldsymbol{p}$ are loosely coupled.

From $\bar{a}_{\mu,\nu-1} \in (\bar{A}_{\mu-1}, [\bar{A}_\mu, \bar{B}_{\nu-1}])$ we have $\mathring{a}_{\mu,\nu-1} = \bar{a}_{\mu-1} \cap \bar{v}$ where $\bar{v} \in [\bar{A}_\mu, \bar{B}_{\nu-1}])$ is the sum of all the elements of the decomposition $\bar{B}_{\nu-1}$ that can be connected with the element $\bar{b}_{\nu-1}$ in $\bar{A}_\mu$. In particular, there holds $\bar{b}_{\nu-1} \subset \bar{v}$ and, therefore, even $\bar{a}_{\mu-1} \cap \bar{b}_{\nu-1} \subset \mathring{a}_{\mu,\nu-1}$.

Analogously, $\mathring{b}_{\nu,\mu-1} = \bar{b}_{\nu-1} \cap \bar{u}$ where $\bar{u} \in [\bar{B}_\nu, \bar{A}_{\mu-1}]$ is the sum of all the elements that can be connected with the element $\bar{a}_{\mu-1}$ in $\bar{B}_\nu$. In particular, we have $\bar{a}_{\mu-1} \subset \bar{u}$ and, therefore, even $\bar{b}_{\nu-1} \cap \bar{a}_{\mu-1} \subset \mathring{b}_{\nu,\mu-1}$.

Consequently:

$$(\bar{a}_{\mu-1} \cap \bar{b}_{\nu-1}) \subset (\mathring{a}_{\mu,\nu-1} \cap \mathring{b}_{\nu,\mu-1}) = (\bar{a}_{\mu-1} \cap \bar{v}) \cap (\bar{b}_{\nu-1} \cap \bar{u}) \subset (\bar{a}_{\mu-1} \cap \bar{b}_{\nu-1})$$

so that we have

$$\bar{a}_{\mu-1} \cap \bar{b}_{\nu-1} = \mathring{a}_{\mu,\nu-1} \cap \mathring{b}_{\nu,\mu-1}.$$

By (1), $\mathring{K}_{(\mu-1)(\beta+1)+\nu-1}$ is a decomposition on $\mathring{a}_{\mu,\nu-1}$ and $\mathring{L}_{(\nu-1)(\alpha+1)+\mu-1}$ a decomposition on $b_{\nu,\mu-1}$. To simplify the notation, let us put

$$\mathring{K}_{\mu,\nu} = \mathring{K}_{(\mu-1)(\beta+1)+\nu-1}, \quad \mathring{L}_{\nu,\mu} = \mathring{L}_{(\nu-1)(\alpha+1)+\mu-1}.$$

Then the above equality may be written in the form:

$$\bar{a}_{\mu-1} \cap \bar{b}_{\nu-1} = \mathbf{s}\mathring{K}_{\mu,\nu} \cap \mathbf{s}\mathring{L}_{\nu,\mu}.$$

Any element $\mathring{x} \in \mathring{K}_{\mu,\nu}$ is incident with an element of $\mathring{L}_{\nu,\mu}$ if and only if there holds $\mathring{x} \in (\bar{a}_{\mu-1} \cap \bar{b}_{\nu-1}) \sqsubset \mathring{K}_{\mu,\nu}$. In fact, if $\mathring{x}$ is incident with some element of $\mathring{L}_{\nu,\mu}$, then it is incident with the set $\mathbf{s}\mathring{K}_{\mu,\nu} \cap \mathbf{s}L_{\nu,\mu}$ and therefore also with $\bar{a}_{\mu-1} \cap b_{\nu-1}$ so that we have: $\mathring{x} \in (\bar{a}_{\mu-1} \cap \bar{b}_{\nu-1}) \sqsubset \mathring{K}_{\mu,\nu}$; if, conversely, the latter relation applies, then $\mathring{x}$ is incident with the set $\bar{a}_{\mu-1} \cap \bar{b}_{\nu-1}$, hence even with $\mathbf{s}\mathring{K}_{\mu,\nu} \cap \mathbf{s}\mathring{L}_{\nu,\mu}$ and, consequently, with at least one element of $\mathring{L}_{\nu,\mu}$.

In a similar way we can verify that any element $\mathring{y} \in \mathring{L}_{\nu,\mu}$ is incident with some element of $\mathring{K}_{\mu,\nu}$ if and only if there holds $\mathring{y} \in (\bar{b}_{\nu-1} \cap \bar{a}_{\mu-1}) \sqsubset \mathring{L}_{\nu,\mu}$.

It is easy to show that $\mathring{K}_{\mu,\nu}$ and $\mathring{L}_{\nu,\mu}$ are loosely coupled.

Let us, first, note that the intersection $\mathring{K}_{\mu,\nu} \sqcap \mathring{L}_{\nu,\mu}$ is not empty, for $\bar{a} \subset \subset \mathring{a}_{\mu,\nu} \cap \mathring{b}_{\nu,\mu}$. Moreover, we shall find that each element of $\mathring{K}_{\mu,\nu}$ is incident with, at most, one element of $\mathring{L}_{\nu,\mu}$. Indeed, if an element $\mathring{x} \in \mathring{K}_{\mu,\nu}$ does not lie in the closure $(\bar{a}_{\mu-1} \cap \bar{b}_{\nu-1}) \sqsubset \mathring{K}_{\mu,\nu}$, then it is not incident with any element of $\mathring{L}_{\nu,\mu}$. In the opposide case, $\mathring{x} \in \mathring{K}_{\mu,\nu}$ is incident with at least one element of $\mathring{L}_{\nu,\mu}$, and all the elements of $\mathring{L}_{\mu,\nu}$ incident with $\mathring{x}$ belong to the closure $(\bar{b}_{\nu-1} \cap \bar{a}_{\mu-1}) \sqsubset \mathring{L}_{\nu,\mu}$; by 4.3, the closures $(\bar{a}_{\mu-1} \cap \bar{b}_{\nu-1}) \sqsubset \mathring{K}_{\mu,\nu}$ and $(\bar{b}_{\nu-1} \cap \bar{a}_{\mu-1}) \sqsubset \mathring{L}_{\nu,\mu}$ are coupled and so, in $L_{\nu,\mu}$, there is exactly one element incident with $\mathring{x}$. Thus we have shown that each element of $\mathring{K}_{\mu,\nu}$ is incident with, at most, one element of $\mathring{L}_{\nu,\mu}$. In a similar way we can verify that each element of $\mathring{L}_{\nu,\mu}$, is incident with, at most, one element of $\mathring{K}_{\mu,\nu}$. It follows that the decompositions $\mathring{K}_{\mu,\nu}$, $\mathring{L}_{\nu,\mu}$ are loosely coupled.

To accomplish the proof it remains to verify that even $\mathring{K}_{(\alpha+1)(\beta+1)}$ and $\mathring{L}_{(\beta+1)(\alpha+1)}$ are loosely coupled. But that is obvious, since these decompositions consist of the single element $\bar{a}$.

10.8. Complementary series of decompositions

Let again $(\bar{A})$, $(\bar{B})$ stand for arbitrary series of decompositions on G, of lengths $\alpha, \beta \geqq 1$; notation as above.

$(\bar{A})$, $(\bar{B})$ *are* called *complementary* if any member of $(\bar{A})$ is complementary to any member of $(\bar{B})$.

Let us assume that $(\bar{A})$, $(\bar{B})$ are complementary. Then, on taking account of 5.5 and 5.4, the following theorems apply:

Every two local chains with the same ends, corresponding to the series $(\bar{A})$, $(\bar{B})$, *respectively, are adjoint.*

The series $(\bar{A})$, $(\bar{B})$ *are modular.*

Furthermore, we shall prove that

$(\bar{A})$, $(\bar{B})$ *have co-basally joint refinements* $(\mathring{A})$, $(\mathring{B})$ *with the same initial and final members.* $(\mathring{A})$, $(\mathring{B})$ *are given by the above construction of co-basally loosely joint refinements of modular series* (part a) of the above proof).

Proof. Since $(\bar{A})$, $(\bar{B})$ are not only modular but even complementary, we have to modify the part b) of the above proof so as to show that the decompositions

$$(\mathring{K}_{\mu,\nu} =)\ \mathring{K}_{(\mu-1)(\beta+1)+\nu-1} = \mathring{a}_{\mu,\nu-1} \sqcap \mathring{A}_{\mu,\nu},$$
$$(\mathring{L}_{\mu,\nu} =)\ \mathring{L}_{(\nu-1)(\alpha+1)+\mu-1} = \mathring{b}_{\nu,\mu-1} \sqcap \mathring{B}_{\nu,\mu},$$
$$(\mathring{a}_{\mu,0} = \bar{a}_{\mu-1},\quad \mathring{b}_{\nu,0} = \bar{b}_{\nu-1};\quad \mu+\nu > 2)$$

are coupled.

As we know from 5.3, the decompositions $\bar{A}_{\mu}$, $(\bar{A}_{\mu-1}, \bar{B}_{\nu-1})$ are complementary; hence, on taking account of the first theorem in 5.3, we observe that the element $\mathring{a}_{\mu,\nu-1} \in [\bar{A}_{\mu}, (\bar{A}_{\mu-1}, \bar{B}_{\nu-1})]$ is the sum of all the elements of the decomposition $\bar{A}_{\mu}$ that are incident with the element $\bar{a}_{\mu-1} \cap \bar{b}_{\nu-1} \in (\bar{A}_{\mu-1}, \bar{B}_{\nu-1})$. Even an arbitrary element $\mathring{x} \in \mathring{A}_{\mu,\nu}$ is the sum of certain elements of $\bar{A}_{\mu}$; we observe that $\mathring{x} \in \mathring{A}_{\mu,\nu}$ is incident with $\mathring{a}_{\mu,\nu-1}$ if and only if it is incident with the set $\bar{a}_{\mu-1} \cap \bar{b}_{\nu-1}$. It follows:

$$\mathring{K}_{\mu,\nu} = (\bar{a}_{\mu-1} \cap \bar{b}_{\nu-1}) \sqsubset \mathring{A}_{\mu,\nu}.$$

In a similar way we obtain:

$$\mathring{L}_{\nu,\mu} = (\bar{b}_{\nu-1} \cap \bar{a}_{\mu-1}) \sqsubset \mathring{B}_{\nu,\mu}.$$

As the decompositions on both sides of the above equalities are coupled (5.5), the proof is complete.

10.9. Example of co-basally joint series of decompositions

In the figure behind p. 80 we find an example of co-basally joint series of decompositions $(\mathring{A})$, $(\mathring{B})$ on the set G consisting of 20 elements (cf. p. 205, N°39). The elements of G, or the one-point sets formed by these elements, are in the inner columns, denoted by A_8, B_8, ...; the arrows show which of the elements are the same. The individual members of the co-basally joint series

$$((\mathring{A}) =)\ \mathring{A}_{11} \geqq \mathring{A}_{12} \geqq \mathring{A}_{21} \geqq \mathring{A}_{22} \geqq \mathring{A}_{31} \geqq \mathring{A}_{32} \geqq \mathring{A}_{41} \geqq \mathring{A}_{42},$$
$$((\mathring{B}) =)\ \mathring{B}_{11} \geqq \mathring{B}_{12} \geqq \mathring{B}_{13} \geqq \mathring{B}_{14} \geqq \mathring{B}_{21} \geqq \mathring{B}_{22} \geqq \mathring{B}_{23} \geqq \mathring{B}_{24}$$

are in the appropriate columns.

The starting-point for the construction of the series $(\mathring{A})$, $(\mathring{B})$ are the complementary series of the decompositions of G:

$$((\bar{A}) =)\ \bar{A}_1 \geqq \bar{A}_2 \geqq \bar{A}_3,$$
$$((\bar{B}) =)\ \bar{B}_1 \geqq \bar{B}_2\ (=\bar{A}_3),$$

the individual members of which are, in $(\mathring{A})$, $(\mathring{B})$, denoted by $\mathring{A}_{12}$, $\mathring{A}_{22}$, $\mathring{A}_{32}$ and $\mathring{B}_{14}$, $\mathring{B}_{24}$, respectively. From the figure it is clear that each member of $(\bar{B})$ is complementary to each member of $(\bar{A})$.

The coupled members contained in two local chains of $(\mathring{A})$, $(\mathring{B})$, with the same base, are found in the columns marked by $\mathring{A}_{\gamma\delta}$, $\mathring{B}_{\delta\gamma}$.

The local chains of $(\mathring{A})$, $(\mathring{B})$, with the base $A_8 = B_8$, are marked in colours. We observe that the members of these local chains, introduced in the columns $\mathring{A}_{\gamma\delta}$, $\mathring{B}_{\delta\gamma}$, are coupled decompositions. Incident elements of two coupled decompositions are marked in the same colour. For example, to the decomposition consisting of the elements A_4, A_4' there corresponds the decomposition formed by the elements B_6, B_6'; A_4 (B_6) is incident with the single element B_6 (A_4) and A_4' (B_6') with the single element B_6' (A_4').

10.10. Connection with the theory of mappings of sets onto sets of finite sequences

The above theory of the series of decompositions of sets is closely connected with a study of mappings of sets onto sets formed by finite sequences of the same length.

Consider a nonempty set $\mathcal{A}$ consisting of finite α-membered sequences $(\alpha \geqq 1)$ and a mapping $\boldsymbol{a}$ of the set G onto $\mathcal{A}$.

To the set $\mathcal{A}$ there belongs, as we know from 1.7, a number α of sets of the main parts, $\mathcal{A}_1, \ldots, \mathcal{A}_\alpha$ $(= \mathcal{A})$.

Choosing an arbitrary γ $(= 1, \ldots, \alpha)$, we first define the mapping $\boldsymbol{a}_\gamma$ of G onto $\mathcal{A}_\gamma$ by associating, with each point $a \in G$, the γ-th main part $a^{(\gamma)} \in \mathcal{A}_\gamma$ of the sequence $\boldsymbol{a}a$. The mapping $\boldsymbol{a}_\alpha$ is, of course, the same as $\boldsymbol{a}$.

To the mapping $\boldsymbol{a}_\gamma$ there belongs a certain decomposition of G, denoted by $\bar{A}_\gamma$. $\bar{A}_\alpha$, naturally, coincides with the decomposition belonging to $\boldsymbol{a}$.

Let $a \in G$ be an arbitrary point of G.

To the element $\boldsymbol{a}_\gamma a = a^{(\gamma)} \in \mathcal{A}_\gamma$ there corresponds the set of its successors (1.7) $M(a^{(\gamma)}) \subset \mathcal{A}_{\gamma+1}$ $(1 \leqq \gamma < \alpha)$. It is useful to employ the notation $M(a^{(\alpha)}) = \{a^{(\alpha)}\}$. The sequence of the sets

$$([Ma] =)\ M(a^{(1)}) \to \cdots \to M(a^{(\alpha)})$$

is called the *chain of successor-sets* that belongs to a.

Consider the element $a^{(\gamma)} \in \mathcal{A}_\gamma$ and the element $\bar{a}_\gamma \in \bar{A}_\gamma$ consisting of $\boldsymbol{a}_\gamma$-inverse images of $a^{(\gamma)}$. Under the mapping $\boldsymbol{a}_{\gamma+1}$ $(1 \leqq \gamma < \alpha)$ every point lying in $\bar{a}_\gamma$ is mapped onto a certain successor of $a^{(\gamma)}$; at the same time, each successor of $a^{(\gamma)}$ has, under the mapping $\boldsymbol{a}_{\gamma+1}$, one or more inverse images lying in G; all of them are contained in $\bar{a}_\gamma$. It is obvious that the sets of the $\boldsymbol{a}_{\gamma+1}$-inverse images of the individual successors of $a^{(\gamma)}$, i.e., the sets of the $\boldsymbol{a}_{\gamma+1}$-inverse images of the individual elements of the set $M(a^{(\gamma)})$, form a decomposition of the element $\bar{a}_\gamma \in \bar{A}_\gamma$; it is the decomposition $(\bar{K}_\gamma a =)\ \bar{a}_\gamma \sqcap \bar{A}_{\gamma+1}$ belonging to the partial mapping $\boldsymbol{a}_{\gamma+1}$ of $\bar{a}_\gamma$ onto the successor-set $M(a^{(\gamma)})$. The latter is, by the first equivalence theorem (6.8), equivalent to the decomposition $\bar{K}_\gamma a$. The set $M(a^{(\alpha)})$ is, of course, equivalent to $\bar{K}_\alpha a$.

Thus we arrive at the following description of the situation:

The set of sequences, $\mathcal{A}$, and the mapping $\boldsymbol{a}$ of G onto $\mathcal{A}$ determine, on G, a series of decompositions, of length α, the so-called *model series*

$$((\bar{A}) =)\ \bar{A}_1 \geqq \cdots \geqq \bar{A}_\alpha$$

whose members are the decompositions belonging to the individual mappings $\boldsymbol{a}_1, \ldots, \boldsymbol{a}_\alpha$.

To each point $a \in G$ there corresponds a chain of successor-sets

$$([Ma] =)\ M(a^{(1)}) \to \cdots \to M(a^{(\alpha)})$$

and a local chain of the series $(\bar{A})$

$$([\bar{K}a] =)\ \bar{K}_1 a \to \cdots \to \bar{K}_\alpha a.$$

Every two members $M(a^{(\gamma)})$, $\bar{K}_\gamma a$ of these chains, with the same index γ, are equivalent sets.

Let us now consider two nonempty sets $\mathcal{A}$, $\mathcal{B}$ consisting of finite $\alpha(\geqq 1)$-membered sequences and arbitrary mappings $\boldsymbol{a}, \boldsymbol{b}$ of G onto $\mathcal{A}, \mathcal{B}$, respectively. Then we have the corresponding sets of the main parts, $\mathcal{A}_1, \ldots, \mathcal{A}_\alpha$ $(= \mathcal{A})$; $\mathcal{B}_1, \ldots, \mathcal{B}_\alpha$ $(= \mathcal{B})$, furthermore, the mappings $\boldsymbol{a}_1, \ldots, \boldsymbol{a}_\alpha$ $(= \boldsymbol{a})$; $\boldsymbol{b}_1, \ldots, \boldsymbol{b}_\alpha$ $(= \boldsymbol{b})$ of G onto the corresponding sets of the main parts and, finally, the model-series

$$((\bar{A}) =)\ \bar{A}_1 \geqq \cdots \geqq \bar{A}_\alpha,$$
$$((\bar{B}) =)\ \bar{B}_1 \geqq \cdots \geqq \bar{B}_\alpha.$$

To each point $a \in G$ there correspond two chains of successor-sets:

$$([Ma] =)\ M(a^{(1)}) \to \cdots \to M(a^{(\alpha)}),$$

$$([Na] =)\ N(b^{(1)}) \to \cdots \to N(b^{(\alpha)})$$

and, furthermore, the local chains of the series $(\bar{A})$, $(\bar{B})$:

$$([\bar{K}a] =)\ \bar{K}_1 a \to \cdots \to \bar{K}_\alpha a,$$

$$([\bar{L}a] =)\ \bar{L}_1 a \to \cdots \to \bar{L}_\alpha a.$$

Every two members $M(a^{(\gamma)})$, $\bar{K}_\gamma a$ or $N(b^{(\gamma)})$, $\bar{L}_\gamma a$ of these chains, respectively, with the same index γ, are equivalent sets.

Let us now assume that the model-series $(\bar{A})$, $(\bar{B})$ are co-basally chain-equivalent.

In that case, first, the final members $\bar{A}_\alpha$, $\bar{B}_\alpha$ of the model series $(\bar{A})$, $(\bar{B})$ coincide, hence $\bar{A}_\alpha = \bar{B}_\alpha$. Moreover, we can show that:

There exists a permutation $\boldsymbol{p}$ of the set $\{1, \ldots, \alpha\}$, such that the member $M(a^{(\gamma)})$, with an arbitrary index γ, of the chain of successor-sets, $[Ma]$, corresponding to an arbitrary point $a \in G$ and the member $N(b^{(\delta)})$, with the index $\delta = \boldsymbol{p}\gamma$, of the chain of successor-sets, $[Na]$, corresponding to the same point a, are equivalent sets.

Proof. The co-basal mapping of the manifold of the local chains, $\tilde{A}$, of $(\bar{A})$ onto the manifold of the local chains, $\tilde{B}$, of $(\bar{B})$ is, on our assumption, a strong equivalence. That means that there exists a permutation $\boldsymbol{p}$ of the set $\{1, \ldots, \alpha\}$ with the following effect:

Let $a \in G$ stand for an arbitrary point and $\bar{a}$ for that element of the decomposition $\bar{A}_\alpha = \bar{B}_\alpha$ which comprises it.

Let, moreover, $[\bar{K}\bar{a}]$, $[\bar{L}\bar{a}]$ be the local chains of $(\bar{A})$, $(\bar{B})$ with the base $\bar{a}$. Then every two members $\bar{K}_\gamma \bar{a}$, $\bar{L}_\delta \bar{a}$ of $[\bar{K}\bar{a}]$, $[\bar{L}\bar{a}]$ for which $\delta = \boldsymbol{p}\gamma$ are equivalent sets.

Consider the member $M(a^{(\gamma)})$ of $[Ma]$, with an arbitrary index γ, corresponding to the point a and the member $N(b^{(\delta)})$ of $[Na]$, with index $\delta = \boldsymbol{p}\gamma$, corresponding to the same point a. Then we have $\bar{K}_\gamma a = \bar{K}_\gamma \bar{a}$, $\bar{L}_\delta a = \bar{L}_\delta \bar{a}$. Since $M(a^{(\gamma)})$, $\bar{K}_\gamma a$ and $\bar{L}_\delta a$ are equivalent to $\bar{K}_\gamma a\ (= \bar{K}_\gamma \bar{a})$, $\bar{L}_\delta a\ (= \bar{L}_\delta \bar{a})$ and $N(b^{(\delta)})$, respectively, it is obvious (6.10.7) that $M(a^{(\gamma)})$ is equivalent to $N(b^{(\delta)})$ and the proof is accomplished.

The above theorem leads to the following observation: If an arbitrary point $a \in G$ is mapped, under the functions $\boldsymbol{a}_\gamma$, $\boldsymbol{b}_\delta$, into the sets of the main parts, $\mathcal{A}_\gamma$, $\mathcal{B}_\delta$, γ and δ being in the above relation, then the successor-sets of both images are equivalent.

10.11. Some remarks on the use of the preceding theory in scientific classifications

The theory of the series of decompositions of sets is of interesting use in scientific classifications. In this respect, however, we shall content ourselves with a few remarks, for a more detailed study would exceed the limits of this book.

A *scientific classification* $(\mathcal{A})$ *of the set* G, briefly, *a classification of* G is a non-empty set $\mathcal{A}$ formed by finite α-membered sequences ($\alpha \geqq 1$) and a mapping $\boldsymbol{a}$ of G onto $\mathcal{A}$. The γ-th member of the sequence $\boldsymbol{a}a$ is called the *γ-th characteristic* or the *characteristic of order* γ of the element a. The elements of $\mathcal{A}$ are therefore sometimes called *sequences of the characteristics*. The above notions concerning mappings onto sets of sequences may, of course, be directly applied to classifications.

In case of scientific classifications, the elements of G are called *individuals*, the sets of the main parts are the *characteristic-sets* and the model-series is the so-called *classification-series*.

In an actual construction of a classification, the choice of the characteristics is restricted by special conditions which influence, in particular, the properties of the classification-series. In natural sciences, for example, the chosen characteristics of the individuals are particular properties of the latter, given by nature herself.

Any individual a in the classification $(\mathcal{A})$ is determined by finding the corresponding sequence of the characteristics, $\boldsymbol{a}a$. In actual cases, however, it sometimes happens that some of the characteristics cannot be ascertained, e.g., for deficiency of adequate means to do so or if the individual is damaged or pathological. In such cases the given individual cannot be determined by means of $(\mathcal{A})$.

Hence there arises the following problem:

We are to describe the principle of constructing two so-called *harmonious classifications* of the set G in convenient mutual relations. It is required that: 1) both classifications lead to the same result, i.e., that the individuals which are not considered to be different from one another be in both classifications the same; 2) that the characteristics missing in one classification may, for each individual, be replaced by adequate characteristics in the other.

Our results concerning the functions whose values are sequences point out the way of solving this, rather difficult, problem. Let us start with two suitably chosen complementary series of decompositions of the classified set G and choose, according to the construction introduced in 10.7, the characteristics in both classifications in a way that the corresponding classification-series be co-basally joint (10.8). If we have succeeded, then we are able to determine, for each individual, the $(\gamma + 1)$-th characteristic in one of these classifications from the knowledge of its first γ characteristics in the latter and of its $\delta + 1$ characteristics in the other classification; we can do so by means of the simple mappings existing between the corresponding successor-sets. But the possibility of constructing such harmonious classifications is, in actual cases, rarely available, as the choice of the characteristics depends on the postulates imposed on them. In this respect, however, the latter grants a certain freedom because the complementary series of decompositions from which it starts may be arbitrarily chosen.

10.12. Exercises

1. The manifold of local chains corresponding to a series of decompositions $((\bar{A}) =) \bar{A}_1 \geqq \cdots \geqq \bar{A}_\alpha$ on G is a set of sequences, $\mathscr{A}$. Associating, with every point $a \in G$, the corresponding local chain $[\bar{K}a]$, we obtain a mapping $\boldsymbol{a}$ of G onto $\mathscr{A}$. The corresponding model-series is $(\bar{A})$. The γ-th main part ($\gamma = 1, \ldots, a$) of the sequence $\boldsymbol{a}a$ associated with an arbitrary point $a \in G$ is the chain $\bar{K}_1 a \to \cdots \to \bar{K}_\gamma a$. For $1 \leqq \gamma < \alpha$, all the successors of the latter are obtained by adding, at its end, always one decomposition $\bar{x}_{\gamma+1} \sqcap \bar{A}_{\gamma+2}$ while $\bar{x}_{\gamma+1}$ runs over all the elements of $\bar{a}_\gamma \sqcap \bar{A}_{\gamma+1}$ ($a \in \bar{a}_\gamma \in \bar{A}_\gamma$; $\bar{A}_{\alpha+1} = \bar{A}_\alpha$). There exist mappings of G onto sets of sequences with arbitrarily given model-series.
2. The figure behind p. 80 may be regarded as a scheme of two harmonious classifications (with co-basally joint classification-series). The sequences of characteristics corresponding to the single individuals or classes of individuals that are not distinguished from one another are introduced in the single rows; the arrows point to both sequences of characteristics belonging to the same individual. The corresponding equivalent sets of successors are introduced in two columns denoted by $\mathring{A}_{\gamma\delta}$ and $\mathring{B}_{\delta\gamma}$. If, for example, a certain individual has, in the classification $(\mathring{A})$, the characteristics A_1, A_2, A_3, A_4, A_5 and, in the classification $(\mathring{B})$, the characteristics $B_1, B_2, B_3, B_4, B_5, B_6, B_7$ (or B_7'), then it has, in $(\mathring{A})$, also the characteristic A_6 (or A_6'). A detailed study of this problem may be left to the reader.

II. GROUPOIDS

11. Multiplication in sets

11.1. Basic concepts

By a *multiplication* or a *binary operation in the set* G we mean a relationship between the elements of G by which there corresponds, to every two-membered sequence of the elements $a, b \in G$, exactly one element $c \in G$; in other words, a relationship by which every two-membered sequence of the elements a, b of G is mapped onto an element c of the same set G. The element c is called the *product of a and b* and is denoted by $a \,.\, b$ or ab; so we have $c = ab$, where a, b is the *first, second factor of the product* c, respectively.

From these definitions it is obvious that the word "multiplication" only expresses a relationship between the elements of G which need not have, in actual cases, anything in common with arithmetic multiplication; the same applies to the product and the symbols $a \,.\, b$, ab.

In what sense the concept of multiplication in G generalizes that of a mapping of G into itself can easily be understood by comparing the two definitions: Every mapping of G into itself associates, with each element of G, again an element of G; every multiplication in G associates, with every two-membered sequence of elements of G, again an element of G.

It is obvious that a multiplication in G may also be defined as a mapping of the Cartesian square $G \times G$ (1.8) into G. Then the products are images of the individual elements of this Cartesian square. The theory of groupoids which, as we shall see, is based on the concept of multiplication in a set can, in this way, be included in the general theory of mappings of sets. The following considerations are, however, based on the above concept of multiplication, since the theory of groupoids, developed in this way, without a detailed study of the properties of Cartesian squares, is simpler and better suited to our purpose. The reader is, nevertheless, advised to follow our study even from the aspect of mappings of sets; it will help him to gain an independent view of the single situations.

If the multiplication in G is given, then, in particular, the product of each element $a \in G$ and a itself is uniquely determined; instead of aa we sometimes write, briefly, a^2.

11.2. Commutative (Abelian) multiplication

A multiplication in G may have particular properties. The above definition does not exclude, e.g., that a multiplication associates, with two inversely arranged pairs of elements of G, two different elements; thus it may happen that the product of some elements a and b is different from the product of the elements b and a, i.e., $ab \neq ba$.

If, for two elements $a, b \in G$, there holds $ab = ba$, then a, b are called *interchangeable*; if every two elements of G are interchangeable, then the multiplication is called *commutative* or *Abelian*.

Multiplication in a set may, of course, have other remarkable properties. In the following paragraph we shall give examples of multiplications that will often be referred to later.

11.3. Examples of multiplication in a set

a) Let G be the set of all integers and let the multiplication be defined as follows: The product of an element $a \in G$ and an element $b \in G$ is the number $a + b$. The multiplication is, in this case, addition in the usual sense. From the equality $a + b = b + a$, true for every two elements $a, b \in G$, there follows that it is an Abelian multiplication.

b) Let n be an arbitrary positive integer and G a set consisting of non-negative integers and containing all the numbers $0, \ldots, n - 1$. The multiplication in G is defined as follows: The product ab of an element $a \in G$ and an element $b \in G$ is the remainder of the division of $a + b$ by n. The product ab is therefore always one of the numbers $0, \ldots, n - 1$. This multiplication is called *addition modulo* n; evidently, it is also Abelian.

c) Assuming G to be the set of all the permutations of a finite set of order n ($\geqq 1$), let the multiplication be defined as follows: The product $\boldsymbol{p} \,.\, \boldsymbol{q}$ of two arbitrary elements $\boldsymbol{p} \in G$ and $\boldsymbol{q} \in G$ is the composite permutation $\boldsymbol{qp}$. Hence the multiplication is the composition of permutations. We know, from 8.7.2, that it need not be Abelian.

d) Suppose G is the set of all decompositions of a certain set and let the multiplication be defined as follows: The product of two arbitrary elements $A \in G$ and $B \in G$ is the decomposition $[A, B]$ or (A, B), respectively. By 3.4 and 3.5, both the multiplications are Abelian.

11.4. Multiplication tables

1. *Description of a multiplication table.* If the set G is finite and consists, for example, of the elements $a, b, \ldots, m$, then any multiplication in G may be described by means of a *multiplication table* constructed as follows:

The first row and the first column, usually separated from the others by a horizontal and a vertical line, contains all the letters $a, b, \ldots, m$, generally in the same order: in the first row from left to right, in the first column from top to bottom. On the right-hand side of each letter x in the first column, under the single letters a, $b, \ldots, m$ in the first row, there are the letters denoting the single products $xa, xb, \ldots$, xm.

The first row and the first column are the headings of the table. Every multiplication table contains, moreover, exactly as many rows and columns as the number of elements of G. If the letters $a, b, \ldots, m$ in both headings are written in the same order, then an Abelian multiplication is apparent from the symmetry of the table with regard to the main diagonal; that is to say, in any j-th row and any k-th column beyond both headings there is the same element as in the k-th row and the j-th column.

2. *Examples of multiplication tables.* Let us introduce, e.g., tables for the multiplication in G of all the permutations of a set H consisting of $n = 1, 2, 3$ elements, the multiplication being the composition of permutations described in 11.3c. Since the number of all the permutations in H and, therefore, of all the elements of G is $n! = 1, 2, 6$, these multiplication tables contain, besides the two headings, $n! = 1, 2, 6$ rows and the same number of columns.

For $n = 1$. The set G consists of the identical permutation $\boldsymbol{e}$. If the unique element of H is denoted by the letter a, then the symbol of the permutation is $\begin{pmatrix} a \\ a \end{pmatrix}$ and the multiplication table is:

	$\boldsymbol{e}$
$\boldsymbol{e}$	$\boldsymbol{e}$

For $n = 2$. The set G consists of two permutations. If the elements of the set H are a and b, then the symbols of the permutations are $\begin{pmatrix} a\,b \\ a\,b \end{pmatrix}, \begin{pmatrix} a\,b \\ b\,a \end{pmatrix}$. The former permutation is the identical permutation $\boldsymbol{e}$, the latter is denoted, e.g., by $\boldsymbol{a}$. The composite permutations are: $\boldsymbol{ee} = \boldsymbol{e}$, $\boldsymbol{ae} = \boldsymbol{a}$, $\boldsymbol{ea} = \boldsymbol{a}$, $\boldsymbol{aa} = \boldsymbol{e}$, whence we have the following multiplication table:

	$\boldsymbol{e}$	$\boldsymbol{a}$
$\boldsymbol{e}$	$\boldsymbol{e}$	$\boldsymbol{a}$
$\boldsymbol{a}$	$\boldsymbol{a}$	$\boldsymbol{e}$

For $n = 3$. The set G consists of six permutations. If the elements of H are a, b, c, then the symbols of the permutations are:

$$\begin{pmatrix} a & b & c \\ a & b & c \end{pmatrix}, \begin{pmatrix} a & b & c \\ b & c & a \end{pmatrix}, \begin{pmatrix} a & b & c \\ c & a & b \end{pmatrix}, \begin{pmatrix} a & b & c \\ a & c & b \end{pmatrix}, \begin{pmatrix} a & b & c \\ c & b & a \end{pmatrix}, \begin{pmatrix} a & b & c \\ b & a & c \end{pmatrix}.$$

The first symbol expresses the identical permutation $\boldsymbol{e}$, the others are denoted by $\boldsymbol{a}, \boldsymbol{b}, \boldsymbol{c}, \boldsymbol{d}, \boldsymbol{f}$, respectively. The composite permutations are:

$$
\begin{array}{llllll}
ee = e, & ae = a, & be = b, & ce = c, & de = d, & fe = f,\\
ea = a, & aa = b, & ba = e, & ca = d, & da = f, & fa = c,\\
eb = b, & ab = e, & bb = a, & cb = f, & db = c, & fb = d,\\
ec = c, & ac = f, & bc = d, & cc = e, & dc = b, & fc = a,\\
ed = d, & ad = c, & bd = f, & cd = a, & dd = e, & fd = b,\\
ef = f, & af = d, & bf = c, & cf = b, & df = a, & ff = e,
\end{array}
$$

and we have the following multiplication table:

	e	a	b	c	d	f
e	e	a	b	c	d	f
a	a	b	e	d	f	c
b	b	e	a	f	c	d
c	c	f	d	e	b	a
d	d	c	f	a	e	b
f	f	d	c	b	a	e

All the above multiplication tables contain, in both headings, the symbols $\boldsymbol{e}, \boldsymbol{a}, \ldots, \boldsymbol{f}$ of the individual elements of G in the same order and we observe that, for $n = 1, 2$, the tables are symmetric with regard to the main diagonal, whereas, for $n = 3$, the table is not symmetric. Consequently, the above multiplication in G is Abelian for $n = 1, 2$, whereas, for $n = 3$, it is not.

We could give as many examples of multiplication in sets as we wished just by taking an arbitrary abstract nonempty set G and uniquely associating, with every two-membered sequence of elements $a, b \in G$, an arbitrarily chosen element of G. If G is finite, then the correspondence may be defined in a table where the symbols of the chosen elements are written in the single places under the horizontal and to the right of the vertical headings. Each choice of these elements determines a certain multiplication to which the resulting multiplication table applies.

11.5. Exercises

1. In the set of all Euclidian motions on an straight line, $\boldsymbol{f}[a]$, as well as in the set of all Euclidean motions on a straight line, $\boldsymbol{f}[a]$, $\boldsymbol{g}[a]$, (6.10.4), the multiplication may be defined by means of composing the motions in a similar way as in Example 11.3c). An analogous result applies to the set of all Euclidean motions in a plane, $\boldsymbol{f}[\alpha; a, b]$, and to the set of all Euclidean motions in a plane, $\boldsymbol{f}[\alpha; a, b]$, $\boldsymbol{g}[\alpha; a, b]$ (6.10.5).
2. In the set of $2n$ permutations of the vertices of a regular n-gon in a plane ($n \geqq 3$), described in Exercise 8.8.4, the multiplication may be defined by composing the permutations similarly as in Example 11.3c). Construct the appropriate multiplication tables for $n = 4, 5, 6$.

3. In 11.3 b, the set G may consist only of the numbers $0, \ldots, n-1$. Construct the appropriate multiplication tables if $n = 1, 2, 3, 4, 5$.

4. If the positive integers a, b are less than or equal to a positive integer $n \geqq 5$, then the number of the prime factors of the number $10a + b$ is $\leqq n$. Hence a multiplication in the set G, consisting of the numbers $1, 2, \ldots, n$, can be defined as follows: The product $a \,.\, b$ of an element $a \in G$ and an element $b \in G$ is the number of the prime factors of $10a + b$. The reader may verify that, for $n = 6$, the corresponding table is

	1	2	3	4	5	6
1	1	3	1	2	2	4
2	2	2	1	4	2	2
3	1	5	2	2	2	4
4	1	3	1	3	3	2
5	2	3	1	4	2	4
6	1	2	3	6	2	3

5. In the system of all the subsets of a nonempty set the multiplication can be defined by associating, with each ordered pair of subsets, their sum. May the multiplication be similarly defined by means of intersection?

6. Find some other examples of multiplication in sets.

12. Basic notions relative to groupoids

12.1. Definition

A nonempty set G together with a multiplication $\boldsymbol{M}$ in G is called a *groupoid*. G is the *field* and $\boldsymbol{M}$ the *multiplication of* or *in the groupoid*. The groupoids will generally be denoted by German capitals corresponding to the Latin capitals used for their fields. Thus, for a groupoid whose field is denoted by G, we use the notation $\mathfrak{G}$; if a groupoid is denoted by $\mathfrak{G}$, then G generally stands for its field.

12.2. Further notions. The groupoids $\mathfrak{Z}$, $\mathfrak{Z}_n$, $\mathfrak{S}_n$

To groupoids we may apply the notions and symbols defined for their fields. So we speak, for example, about elements of a groupoid instead of elements of the field of a groupoid and write $a \in \mathfrak{G}$ instead of $a \in G$; we speak about subsets of a groupoid and write, e.g., $A \subset \mathfrak{G}$ or $\mathfrak{G} \supset A$, we speak about decompositions in a groupoid and on a groupoid, about the order of a groupoid, a mapping of a groupoid into a certain set, into a certain groupoid or onto a groupoid, etc. A nonempty subset of a groupoid is also called a *complex*. If G is an abstract set, then the groupoid $\mathfrak{G}$ is called *abctract.*

The notions and symbols defined for multiplication apply to groupoids as well. Thus, in particular, every two-membered sequence of elements $a, b \in \mathfrak{G}$ has a well determined product $a \,.\, b$, briefly ab; if for each $a, b \in \mathfrak{G}$ there holds $ab = ba$, then the groupoid is called *commutative* or *Abelian.* With every finite groupoid we can also associate a multiplication table describing the multiplication in $\mathfrak{G}$. In 11.3 we have given several examples of multiplication; each of them simultaneously applies to a groupoid.

In what follows we shall often refer to three groupoids denoted by $\mathfrak{Z}, \mathfrak{Z}_n, \mathfrak{S}_n$: $\mathfrak{Z}$ consists of the set Z of all integers and its multiplication is defined by the usual addition (11.3a). $\mathfrak{Z}_n$ consists of the set $Z = \{0, \ldots, n - 1\}$ where n is a positive integer and the multiplication is defined by addition modulo n (11.3b). The groupoid $\mathfrak{S}_n$ consists of the set S_n formed by all permutations of a finite set H of order n ($\geqq 1$) and the multiplication is defined by the composition of permutations. Any groupoid whose elements are permutations of a (finite or infinite) set and the multiplication is defined by composing permutations is called a *permutation groupoid*, e.g., the groupoid $\mathfrak{S}_n$.

12.3. Interchangeable subsets

Let $\mathfrak{G}$ denote (throughout the book) a groupoid.

Suppose A, B are subsets of $\mathfrak{G}$. The subset of $\mathfrak{G}$ consisting of the products ab of each element $a \in A$ and each element $b \in B$ is called the *product of the subsets A and B*; notation: $A \,.\, B$ or AB. If any of the subsets A, B is empty, then the symbols $A \,.\, B$, AB denote the empty set. For $a \in \mathfrak{G}$ we generally write aA instead of $\{a\}A$ and, similarly, Aa instead of $A\{a\}$; for example, aA denotes the set of all the products of a and each element of A or, if $A = \emptyset$, the empty set. Instead of AA we sometimes write, briefly, A^2.

If $AB = BA$, then the subsets A, B are called *interchangeable.* In that case the product of any element $a \in A$ and any element $b \in B$ is the product of an element $b' \in B$ and an element $a' \in A$; simultaneously, the product of any element $b \in B$ and any element $a \in A$ is the product of an element $a' \in A$ and an element $b' \in B$. If $\mathfrak{G}$ is Abelian, then, of course, every two subsets of $\mathfrak{G}$ are interchangeable. In the opposite case there holds, for some elements $a, b \in \mathfrak{G}$, the inequality $ab \neq ba$, hence every two subsets $A, B \subset \mathfrak{G}$ need not be interchangeable, for example, if $A = \{a\}$, $B = \{b\}$. The product AB of the subset $A = \{1\}$ and the subset $B = \{\ldots, -2, 0, 2, \ldots\}$ of the groupoid $\mathfrak{Z}$ is $\{\ldots, -1, 1, 3, \ldots\}$ and, evidently, equals the product BA; if $A = \{0, 1\}$, $B = \{\ldots, -2, 0, 2, \ldots\}$, then we have $AB = BA = Z$. Note that for every groupoid $\mathfrak{G}$ the relation $GG \subset \mathfrak{G}$ is true.

12.4. Subgroupoids, supergroupoids, ideals

Suppose A stands for a certain complex in $\mathfrak{G}$. If $AA \subset A$, that is to say, if the product of any $a \in A$ and $b \in A$ is again an element of A, then A is said to be a *groupoidal subset of* $\mathfrak{G}$. In that case the multiplication M in $\mathfrak{G}$ determines a, so-called *partial multiplication* M_A *in* A, defined as follows: M_A associates, with any two-membered sequence of elements $a, b \in A$, the same product $ab \in A$ as the multiplication M. The set A together with the partial multiplication M_A is a groupoid $\mathfrak{A}$. We say that $\mathfrak{A}$ is a *subgroupoid of* $\mathfrak{G}$ and $\mathfrak{G}$ *a supergroupoid of* $\mathfrak{A}$ and we write: $\mathfrak{A} \subset \mathfrak{G}$ or $\mathfrak{G} \supset \mathfrak{A}$. If A is a proper subset of $\mathfrak{G}$, then $\mathfrak{A}$ is said to be a *proper subgroupoid of* $\mathfrak{G}$ and $\mathfrak{G}$ a *proper supergroupoid of* $\mathfrak{A}$. $\mathfrak{G}$ always contains the *greatest subgroupoid,* identical with itself.

If even $GA \subset A$ (or $AG \subset A$ or, simultaneously, $GA \subset A \supset AG$), then $\mathfrak{A}$ is called a *left* (or a *right* or a *bilateral*) *ideal of* $\mathfrak{G}$. The case of $A \neq G$ is again characterized by the attribute: *proper.*

For example, the complex of $\mathfrak{Z}$, consisting of all integer multiples of a given positive integer m, is groupoidal because the product (i.e., the sum in the usual sense) of any two integer multiples of m is again an integer multiple of m; this complex together with addition in the usual sense is therefore a subgroupoid of $\mathfrak{Z}$; in case of $m > 1$ it is obviously a proper subgroupoid of $\mathfrak{Z}$. Another example: The subset of all elements of $\mathfrak{S}_n$ that leave a given element $a \in H$ invariant is groupoidal because, if any two permutations $\boldsymbol{p}, \boldsymbol{q} \in \mathfrak{S}_n$ do not change the element a, then the same, naturally, holds for their product $\boldsymbol{p} \cdot \boldsymbol{q}$ (i.e., for the composite permutation $\boldsymbol{qp}$); this subset, together with the composition of permutations in the usual sense, is therefore a subgroupoid of $\mathfrak{S}_n$.

It is easy to see that for any groupoids $\mathfrak{A}, \mathfrak{B}, \mathfrak{G}$ there evidently hold the following statements:

If $\mathfrak{B}$ is a subgroupoid of $\mathfrak{A}$ and $\mathfrak{A}$ a subgroupoid of $\mathfrak{G}$, then $\mathfrak{B}$ is a subgroupoid of $\mathfrak{G}$.

If $\mathfrak{A}, \mathfrak{B}$ are subgroupoids of $\mathfrak{G}$ and for their fields A, B there holds $B \subset A$, then $\mathfrak{B}$ is a subgroupoid of $\mathfrak{A}$.

12.5. Further notions

Since we apply to groupoids the notions and symbols we have defined for their fields, we sometimes speak, e.g., about the intersection of a subset $B \subset \mathfrak{G}$ and a subgroupoid $\mathfrak{A} \subset \mathfrak{G}$ in the sense of the intersection of the subset B and the field A of $\mathfrak{A}$; analogously, we speak about the product of a subset B and a subgroupoid $\mathfrak{A}$, about the product of a subgroupoid $\mathfrak{A}$ and a subset B, about the closure of a subgroupoid $\mathfrak{A}$ in a certain decomposition $\bar{A}$, about the intersection of $\bar{A}$ and a subgroupoid $\mathfrak{A}$, etc; notation, e.g., $B \cap \mathfrak{A}$ or $\mathfrak{A} \cap B$, $B\mathfrak{A}$, $\mathfrak{A}B$, $\mathfrak{A} \sqsubset \bar{A}$ or $\bar{A} \sqsupset \mathfrak{A}$, $\bar{A} \sqcap \mathfrak{A}$ or $\mathfrak{A} \sqcap \bar{A}$, etc.

12.6. The intersection of groupoids

Let us now consider two subgroupoids $\mathfrak{A}, \mathfrak{B} \subset \mathfrak{G}$ and suppose the intersection $A \cap B$ of their fields A, B is not empty, $A \cap B \neq \emptyset$. For any elements $a, b \in A \cap B$ there holds $ab \in AA \subset A$, on the one hand, and $ab \in BB \subset B$, on the other hand, and so $ab \in A \cap B$; hence $A \cap B$ is a groupoidal subset of $\mathfrak{G}$. The corresponding subgroupoid of $\mathfrak{G}$ is called the *intersection of* $\mathfrak{A}$ *and* $\mathfrak{B}$ and denoted by $\mathfrak{A} \cap \mathfrak{B}$ or $\mathfrak{B} \cap \mathfrak{A}$. We observe that any two subgroupoids of $\mathfrak{G}$ whose fields are incident have an intersection which is a subgroupoid of $\mathfrak{G}$. This intersection is, of course, a subgroupoid of either of the two subgroupoids. Note that the concept of the intersection of two subgroupoids of $\mathfrak{G}$ is defined only if the fields of both subgroupoids have common elements. There exists, for example, the intersection of the subgroupoids $\mathfrak{A}, \mathfrak{B} \subset \mathfrak{S}_n$ where the field A of $\mathfrak{A}$ consists of all the elements of $\mathfrak{S}_n$ that do not change a certain element $a \in H$, whereas the field B of $\mathfrak{B}$ consists of all the elements of $\mathfrak{S}_n$ that do not change a certain element $b \in H$, as both A and B have at least one common element, namely, the identical permutation of H which does not change any of the elements of H.

The concept of the intersection of two subgroupoids of $\mathfrak{G}$ may easily be extended to the intersection of a system of subgroupoids of $\mathfrak{G}$: If we have a system $\{\mathfrak{a}_1, \mathfrak{a}_2, \ldots\}$ of subgroupoids of $\mathfrak{G}$ and the intersection of their fields is not empty, then this intersection is a groupoidal subset of $\mathfrak{G}$; the corresponding groupoid of $\mathfrak{G}$ is called the intersection of the system of subgroupoids $\{\mathfrak{a}_1, \mathfrak{a}_2, \ldots\}$ and denoted $\mathfrak{a}_1 \cap \mathfrak{a}_2 \cap \ldots$, briefly, $\bigcap \mathfrak{a}$ or similarly.

12.7. Product of a finite sequence of elements

1. *Definition.* Consider an n-membered sequence of elements $a_1, \ldots, a_n \in \mathfrak{G}$, where $n \geqq 2$. What do we mean by the product of this sequence? The product of a two-membered sequence a_1, a_2 ($n = 2$) has already been defined and denoted $a_1 \,.\, a_2$ or a_1a_2. The product of a three-membered sequence a_1, a_2, a_3 ($n = 3$) is defined as follows: It is the set consisting of the so-called product-elements: $a_1(a_2a_3)$, $(a_1a_2)a_3$. This product is denoted by $\{a_1 \,.\, a_2 \,.\, a_3\}$ or $\{a_1a_2a_3\}$; the symbol $a_1 \,.\, a_2 \,.\, a_3$ or $a_1a_2a_3$ stands for any of the product-elements so that it denotes the product of a_1 and a_2a_3 as well as the product of a_1a_2 and a_3. The product of a four-membered sequence a_1, a_2, a_3, a_4 ($n = 4$) is the set consisting of the three product-elements $a_1(a_2a_3a_4)$, $(a_1a_2)(a_3a_4)$, $(a_1a_2a_3)a_4$. It is denoted by $\{a_1 \,.\, a_2 \,.\, a_3 \,.\, a_4\}$ or $\{a_1a_2a_3a_4\}$; the symbol $a_1 \,.\, a_2 \,.\, a_3 \,.\, a_4$ or $a_1a_2a_3a_4$ stands for any of the product-elements so that it denotes any of the elements: $a_1\big(a_2(a_3a_4)\big)$, $a_1\big((a_2a_3)a_4\big)$, $(a_1a_2)\,(a_3a_4)$, $\big(a_1(a_2a_3)\big)a_4$, $\big((a_1a_2)a_3\big)a_4$. From these examples we can understand the following definition:

The *product of an n-membered sequence of elements* $a_1, a_2, \ldots, a_n$ is the set $\{a_1, a_2, \ldots, a_n\}$ defined as follows: If $n = 2$, then the set $\{a_1, a_2\}$ consists of one single element a_1a_2; if $n > 2$, then it is defined by the formula

$$\{a_1a_2 \dots a_n\} = \{a_1\}\{a_2 \dots a_n\} \cup \{a_1a_2\}\{a_3 \dots a_n\} \cup \dots \cup \{a_1 \dots a_{n-1}\}\{a_n\}.$$

Sometimes we also use the notation $\{a_1 . a_2 \dots a_n\}$. The individual elements of this set, the so-called *product-elements*, are denoted by the symbol $a_1 . a_2 \dots a_n$ or $a_1a_2 \dots a_n$. Naturally, there exists only a finite number of product-elements. If $n = 2$, then we generally do not draw any difference between the product and the corresponding product-element.

2. *Associative groupoids.* From what we have said in the preceding paragraph it follows that every three-membered sequence of elements $a_1, a_2, a_3 \in \mathfrak{G}$ has, at most, two different product-elements: $a_1(a_2a_3)$, $(a_1a_2)a_3$. If they always coincide, i.e., if for any three elements $a_1, a_2, a_3 \in \mathfrak{G}$ there holds $a_1(a_2a_3) = (a_1a_2)a_3$, then the multiplication in $\mathfrak{G}$ as well as the groupoid itself is called *associative.*

The groupoids that have most been studied in mathematics have the property that every finite sequence of their elements has only one product-element; as we shall show later (18.1), it is exactly the associative groupoids that have this remarkable property.

The groupoid $\mathfrak{Z}$, for example, is associative because, by the definition of its multiplication, the product-elements $a(bc)$, $(ab)c$ of any three-membered sequence of the elements $a, b, c \in \mathfrak{Z}$ are sums in the usual sense $a + (b + c)$, $(a + b) + c$ and therefore equal.

Analogously, even the groupoid $\mathfrak{Z}_n$ $(n \geqq 1)$ is associative. Indeed, by the definition of its multiplication, the product-elements $a(bc)$, $(ab)c$ of any three-membered sequence of elements $a, b, c \in \mathfrak{Z}_n$ are the remainders of the division of the numbers $a + r$, $s + c$ by n, r (s) denoting the remainder of the division of $b + c$ $(a + b)$ by n. Since $a + r$ and $a + (b + c)$ differ only by an integer multiple of n, $a(bc)$ is the remainder of the division of $a + (b + c)$ by n; analogously, $(ab)c$ is the remainder of the division of $(a + b) + c$ by n. From $a + (b + c) = (a + b) + c$ there follows $a(bc) = (ab)c$.

The groupoid $\mathfrak{S}_n$ $(n \geqq 1)$ is associative as well because, if $\boldsymbol{p}, \boldsymbol{q}, \boldsymbol{r}$ are arbitrary elements of $\mathfrak{S}_n$, then, by the definition of the multiplication in $\mathfrak{S}_n$, the product-elements $\boldsymbol{p} . (\boldsymbol{q} . \boldsymbol{r})$, $(\boldsymbol{p} . \boldsymbol{q}) . \boldsymbol{r}$ are composite permutations $(\boldsymbol{rq})\boldsymbol{p}$, $\boldsymbol{r}(\boldsymbol{qp})$ and, with respect to the results in 8.7.3, equal.

3. *Example.* To illustrate the process of determining a product, let us find the product $\{1 . 2 . 3 . 4\}$ in the groupoid described in 11.5.4. By the appropriate multiplication table we have

$$\begin{aligned}\{1 . 2 . 3\} &= \{1\} . \{2 . 3\} \cup \{1 . 2\} . \{3\} = \{1\} . \{1\} \cup \{3\} . \{3\} \\ &= \{1\} \cup \{2\} = \{1, 2\};\\ \{2 . 3 . 4\} &= \{2\} . \{3 . 4\} \cup \{2 . 3\} . \{4\} = \{2\} . \{2\} \cup \{1\} . \{4\} \\ &= \{2\} \cup \{2\} = \{2\};\end{aligned}$$

$$\{1\,.\,2\,.\,3\,.\,4\} = \{1\}\,.\,\{2\,.\,3\,.\,4\} \cup \{1\,.\,2\}\,.\,\{3\,.\,4\} \cup \{1\,.\,2\,.\,3\}\,.\,\{4\}$$
$$= \{1\}\,.\,\{2\} \cup \{3\}\,.\,\{2\} \cup \{1, 2\}\,.\,\{4\} = \{3\} \cup \{5\} \cup \{2, 4\}$$
$$= \{2, 3, 4, 5\}.$$

All the product-elements $1\,.\,2\,.\,3\,.\,4$ are therefore 2, 3, 4, 5.

12.8. The product of a finite sequence of subsets

1. *Definition.* Now let $A_1, \ldots, A_n$ $(n \geqq 2)$ stand for arbitrary subsets of $\mathfrak{G}$.

The *product of the n-membered sequence of subsets* $A_1, A_2, \ldots, A_n$ is the sum of all the products $\{a_1 a_2 \ldots a_n\}$, the elements $a_1 \in A_1, a_2 \in A_2, \ldots, a_n \in A_n$ running over all the elements of the corresponding subsets $A_1, A_2, \ldots, A_n$. We denote it by $A_1 \,.\, A_2 \ldots A_n$ or $A_1 A_2 \ldots A_n$. If any of the subsets $A_1, \ldots, A_n$ is empty, then the product in question is defined as the empty set. By the above definition and the meaning of the symbol $\{a_1 \ldots a_n\}$, each element $a \in A_1 A_2 \ldots A_n$ is the result of the multiplication of a product-element $a_1 \ldots a_k$ and one of the elements $a_{k+1} \ldots a_n$ where $1 \leqq k \leqq n - 1$; hence

$$a \in (A_1 \ldots A_k)\,(A_{k+1} \ldots A_n).$$

Conversely, the product of any element of the set $A_1 \ldots A_k$ and any element $A_{k+1} \ldots A_n$ is an element $a \in A_1 \ldots A_n$. So we have

$$A_1 \ldots A_n = A_1(A_2 \ldots A_n) \cup (A_1 A_2)\,(A_3 \ldots A_n) \cup \ldots \cup (A_1 \ldots A_{n-1}) A_n.$$

If A denotes a subset of $\mathfrak{G}$, then we write A^n instead of $\underbrace{A \ldots A}_{n}$ so that, for $n \geqq 2$, we have

$$A^n = A A^{n-1} \cup A^2 A^{n-2} \cup \cdots \cup A^{n-1} A.$$

The above definitions of the product of a finite sequence of elements or sets obviously generalize the definitions of the product of a two-membered sequence of elements or sets, respectively.

2. *Example.* Let A denote the subset $\{1, 2, 4\}$ of the groupoid described in 11.5.4. Then:

$$A^2 = \{1, 2, 4\}\,.\,\{1, 2, 4\}$$
$$= \{1\,.\,1, 1\,.\,2, 1\,.\,4, 2\,.\,1, 2\,.\,2, 2\,.\,4, 4\,.\,1, 4\,.\,2, 4\,.\,4\}$$
$$= \{1, 2, 3, 4\};$$

$$A^3 = \{1, 2, 4\}\,.\,\{1, 2, 3, 4\} \cup \{1, 2, 3, 4\}\,.\,\{1, 2, 4\}$$
$$= \{1, 2, 3, 4, 5\};$$

$$A^4 = \{1, 2, 4\}\,.\,\{1, 2, 3, 4, 5\} \cup \{1, 2, 3, 4\}\,.\,\{1, 2, 3, 4\}$$
$$\cup \{1, 2, 3, 4, 5\}\,.\,\{1, 2, 4\}$$
$$= \{1, 2, 3, 4, 5\}.$$

12.9. Exercises

1. If $A \subset \mathfrak{G}$ and $B \subset \mathfrak{G}$ are the sums of some subsets $\bar{a}_1, \bar{a}_2, \ldots$ and $\bar{b}_1, \bar{b}_2, \ldots$, respectively, then AB is the sum of the products of each subset $\bar{a}_1, \bar{a}_2, \ldots$ and each $\bar{b}_1, \bar{b}_2, \ldots$.

2. If the subsets $A \subset \mathfrak{G}$ and $B \subset \mathfrak{G}$ are the intersections of some subsets $\bar{a}_1, \bar{a}_2, \ldots$ and $\bar{b}_1, \bar{b}_2, \ldots$, respectively, then AB is a part of the intersection of the products of each subset $\bar{a}_1, \bar{a}_2, \ldots$ and each $\bar{b}_1, \bar{b}_2, \ldots$. Thus for any subsets $A, B, C \subset \mathfrak{G}$ there hold, in particular, the relations: a) $(A \cap B)C \subset AC \cap BC$; b) $C(A \cap B) \subset CA \cap CB$. Give suitable examples to show that the symbol $\subset$ can, in these relations, not always be replaced by $=$.

3. Show that the number N_n of the product-elements of an n-membered sequence of elements of $\mathfrak{G}$ ($n \geqq 2$) is expressed, in general, by the formula $N_n = (2n-2)!/(n-1)!n!$

4. Let A stand for a subset of $\mathfrak{G}$ and m, n denote arbitrary positive integers. Then the following relations are true: a) $A^m A^n \subset A^{m+n}$; b) $(A^m)^n \subset A^{mn}$.

5. Suppose $A \subset B$ are subsets of $\mathfrak{G}$ and n denotes an arbitrary positive integer. There holds $A^n \subset B^n$.

6. Let n be an arbitrary positive integer. For the field G of $\mathfrak{G}$ there holds the relation $G^n \supset G^{n+1}$ so that $G \supset G^2 \supset G^3 \supset \cdots$.

7. Let G, n be the same as in Exercise 6. G^n is a groupoidal subset of $\mathfrak{G}$ and the corresponding subgroupoid of $\mathfrak{G}$ is a bilateral ideal. — Remark. The latter is denoted by $\mathfrak{G}^n$.

8. If $\mathfrak{G}$ is an associative groupoid, then: a) every subgroupoid of $\mathfrak{G}$ is associative; b) for any subsets $A, B, C \subset \mathfrak{G}$ there holds $A(BC) = (AB)C$.

9. If $\mathfrak{G}$ is an associative groupoid and A, B are groupoidal and interchangeable subsets of $\mathfrak{G}$, then the subset AB is groupoidal as well. — Remark. If $\mathfrak{A}, \mathfrak{B}$ are interchangeable subgroupoids of $\mathfrak{G}$, then the subgroupoid of $\mathfrak{G}$, corresponding to the product of their fields, is called the *product of the subgroupoids* $\mathfrak{A}, \mathfrak{B}$ and denoted by $\mathfrak{A}\mathfrak{B}$ or $\mathfrak{B}\mathfrak{A}$.

10. If $\mathfrak{G}$ is an associative groupoid, then the set of all the elements of $\mathfrak{G}$ that are interchangeable with each element of $\mathfrak{G}$ is groupoidal unless it is empty. — Remark. The corresponding subgroupoid of $\mathfrak{G}$ is called the *center of* $\mathfrak{G}$.

11. Suppose $\mathfrak{G}$ is a groupoid whose field consists of all positive integers, and the multiplication is defined as follows: The product of any element $a \in \mathfrak{G}$ and any element $b \in \mathfrak{G}$ is the least common multiple or the greatest common divisor of the numbers a and b. Show that in both cases $\mathfrak{G}$ is Abelian and associative.

13. Homomorphic mappings (deformations) of groupoids

13.1. Definition

Let $\mathfrak{G}$, $\mathfrak{G}^*$ be arbitrary groupoids. As we have already said (in 12.2), a mapping of $\mathfrak{G}$ into $\mathfrak{G}^*$ is a mapping of the field G of $\mathfrak{G}$ into the field G^* of $\mathfrak{G}^*$. In a similar way we apply to groupoids all the other concepts and symbols we have described (in Chapter 6) while studying the mappings of sets. By the above definition, the concept of a mapping of $\mathfrak{G}$ into $\mathfrak{G}^*$ concerns only the fields and does in no way depend on the multiplications in the groupoids. Some mappings may, however, be in certain relations with the multiplications in $\mathfrak{G}$ and $\mathfrak{G}^*$. Of great importance to the theory of groupoids are the so-called homomorphic mappings characterized by preserving the multiplications of both groupoids. A detailed definition:

A mapping $\boldsymbol{d}$ of the groupoid $\mathfrak{G}$ into $\mathfrak{G}^*$ is called *homomorphic* if the product ab of an arbitrary element $a \in \mathfrak{G}$ and an element $b \in \mathfrak{G}$ is mapped onto the product of the $\boldsymbol{d}$-image of a and the $\boldsymbol{d}$-image of b, i.e., if, for $a, b \in \mathfrak{G}$, there holds $\boldsymbol{d}ab = \boldsymbol{d}a \,.\, \boldsymbol{d}b$.

For convenience, a homomorphic mapping of the groupoid $\mathfrak{G}$ into $\mathfrak{G}^*$ is called a *deformation of the groupoid $\mathfrak{G}$ into $\mathfrak{G}^*$*. A deformation of $\mathfrak{G}$ *onto* $\mathfrak{G}^*$ is sometimes called a *homomorphism*.

While studying the mapping of sets, we have realized that there need not always exist a mapping of a given set onto another set; consequently, a mapping of $\mathfrak{G}$ onto $\mathfrak{G}^*$ and, of course, a deformation of $\mathfrak{G}$ onto $\mathfrak{G}^*$ need not exist at all. If it exists, then *the groupoid $\mathfrak{G}^*$ is* said to be *homomorphic with* $\mathfrak{G}$.

13.2. Example of a deformation

Let n denote a positive integer and $\boldsymbol{d}$ the mapping of the groupoid $\mathfrak{Z}$ onto $\mathfrak{Z}_n$, defined as follows: $\boldsymbol{d}a \in \mathfrak{Z}_n$ is, for $a \in \mathfrak{Z}$, the remainder of the division of a by n. It is easy to verify that $\boldsymbol{d}$ is a deformation and therefore a homomorphism of $\mathfrak{Z}$ onto $\mathfrak{Z}_n$. Indeed, let a, b stand for arbitrary elements of $\mathfrak{Z}$. The product ab of a and b is, by the definition of the multiplication in $\mathfrak{Z}$, the sum $a + b$ and $\boldsymbol{d}a, \boldsymbol{d}b, \boldsymbol{d}ab$ are, by the definition of the mapping $\boldsymbol{d}$, the remainders of the division of $a, b, a + b$ by n, respectively. The product $\boldsymbol{d}a\boldsymbol{d}b$ of $\boldsymbol{d}a$ and $\boldsymbol{d}b$ is, by the definition, the remainder of the division of $\boldsymbol{d}a + \boldsymbol{d}b$ by n and, since the numbers $\boldsymbol{d}a + \boldsymbol{d}b$ and $a + b$ differ only by an integral multiple of n, the product $\boldsymbol{d}a\boldsymbol{d}b$ is the remainder of the division $a + b$ by n. Hence we have $\boldsymbol{d}a\boldsymbol{d}b = \boldsymbol{d}ab$ and see that $\boldsymbol{d}$ is a deformation. In the following study of groupoids we shall often meet with cases of deformation, so we shall, meanwhile, be satisfied with this single example.

13.3. Properties of deformations

Let $\boldsymbol{d}$ be an arbitrary deformation of $\mathfrak{G}$ into $\mathfrak{G}^*$.

Suppose A, B, C are nonempty subsets of $\mathfrak{G}$.

1. The symbol $\boldsymbol{d}A$ denotes, as we know, the image of the set A under the extended mapping $\boldsymbol{d}$, i.e., the subset of $\mathfrak{G}^*$ consisting of the $\boldsymbol{d}$-images of the individual elements of A.

It is easy to show that *there holds*

$$\boldsymbol{d}(AB) = \boldsymbol{d}A \,.\, \boldsymbol{d}B.$$

Every element $c^* \in \boldsymbol{d}(AB)$ is, on the one hand, the $\boldsymbol{d}$-image of the product ab of an element $a \in A$ and an element $b \in B$ so that $c^* = \boldsymbol{d}ab = \boldsymbol{d}a \,.\, \boldsymbol{d}b \in \boldsymbol{d}A \,.\, \boldsymbol{d}B$; consequently, there holds $\boldsymbol{d}(AB) \subset \boldsymbol{d}A \,.\, \boldsymbol{d}B$. On the other hand, every element $c^* \in \boldsymbol{d}A \,.\, \boldsymbol{d}B$ is the product of an element $a^* \in \boldsymbol{d}A$ and an element $b^* \in \boldsymbol{d}B$ so that there exist elements $a \in A$, $b \in B$ such that $a^* = \boldsymbol{d}a$, $b^* = \boldsymbol{d}b$ and we have: $c^* = a^*b^* = \boldsymbol{d}a \,.\, \boldsymbol{d}b = \boldsymbol{d}ab \in \boldsymbol{d}(AB)$; consequently: $\boldsymbol{d}A \,.\, \boldsymbol{d}B \subset \boldsymbol{d}(AB)$ and the proof is complete.

2. With respect to this result we conclude that *if the set AB is a part of C, then the set $\boldsymbol{d}A \,.\, \boldsymbol{d}B$ is a part of $\boldsymbol{d}C$*; that is to say, $AB \subset C$ yields $\boldsymbol{d}A \,.\, \boldsymbol{d}B \subset \boldsymbol{d}C$.

3. If A is the field of a subgroupoid $\mathfrak{A} \subset \mathfrak{G}$ so that it is groupoidal, then we have $AA \subset A$ whence $\boldsymbol{d}A \,.\, \boldsymbol{d}A \subset \boldsymbol{d}A$ and we see that *the $\boldsymbol{d}$-image of the field of the subgroupoid $\mathfrak{A}$ is a groupoidal subset of $\mathfrak{G}^*$*. The subgroupoid of $\mathfrak{G}^*$ whose field is $\boldsymbol{d}A$ is called the *image of the subgroupoid $\mathfrak{A}$ under the deformation $\boldsymbol{d}$* and is denoted $\boldsymbol{d}\mathfrak{A}$; the subgroupoid $\mathfrak{A}$ is called an *inverse image of $\boldsymbol{d}\mathfrak{A}$ under the deformation $\boldsymbol{d}$*. It is obvious that $\boldsymbol{d}$ is a deformation of $\mathfrak{A}$ onto $\boldsymbol{d}\mathfrak{A}$ so that $\boldsymbol{d}\mathfrak{A}$ is homomorphic with $\mathfrak{A}$.

The above notions and results apply, in particular, in case of the field G of $\mathfrak{G}$. We observe that *the $\boldsymbol{d}$-image $\boldsymbol{d}\mathfrak{G}$ of $\mathfrak{G}$ is a subgroupoid of $\mathfrak{G}^*$, homomorphic with $\mathfrak{G}$*. If $\boldsymbol{d}$ is a deformation of $\mathfrak{G}$ onto $\mathfrak{G}^*$, then we, naturally, have $\mathfrak{G}^* = \boldsymbol{d}\mathfrak{G}$.

4. If $\boldsymbol{d}$ is a deformation of $\mathfrak{G}$ into $\mathfrak{G}^*$ and $\boldsymbol{f}$ a deformation of $\mathfrak{G}^*$ into a groupoid $\mathfrak{F}$, then $\boldsymbol{fd}$ is a deformation of $\mathfrak{G}$ into $\mathfrak{F}$. Indeed, in accordance with the definition of the composite mapping $\boldsymbol{fd}$, and $\boldsymbol{d}, \boldsymbol{f}$ being deformations, there holds, for $a, b \in \mathfrak{G}$:

$$\boldsymbol{fd}(ab) = \boldsymbol{f}(\boldsymbol{d}ab) = \boldsymbol{f}(\boldsymbol{d}a \,.\, \boldsymbol{d}b) = \boldsymbol{f}(\boldsymbol{d}a) \,.\, \boldsymbol{f}(\boldsymbol{d}b) = \boldsymbol{fd}a \,.\, \boldsymbol{fd}b\,,$$

and therefore, in fact, $\boldsymbol{fd}(ab) = \boldsymbol{fd}a \,.\, \boldsymbol{fd}b$.

13.4. Isomorphic mappings

1. The concept of a deformation includes other important notions, first of all, the notion of a simple deformation of the groupoid $\mathfrak{G}$ into $\mathfrak{G}^*$, i.e., a deformation in which each element of $\mathfrak{G}^*$ has, at most, one inverse image. A simple deformation of $\mathfrak{G}$ into (onto) $\mathfrak{G}^*$ is called *isomorphic mapping* of $\mathfrak{G}$ into (onto) $\mathfrak{G}^*$.

From the results in 6.7 and 13.3.4 there follows that *if* $\boldsymbol{d}$ *is an isomorphic mapping of* $\mathfrak{G}$ *into* $\mathfrak{G}^*$ *and* $\boldsymbol{f}$ *an isomorphic mapping of* $\mathfrak{G}^*$ *into* $\mathfrak{F}$, *then the composite mapping* $\boldsymbol{fd}$ *of* $\mathfrak{G}$ *into* $\mathfrak{F}$ *is also isomorphic.*

2. An isomorphic mapping of $\mathfrak{G}$ *onto* $\mathfrak{G}^*$ is called *isomorphism.* To every simple deformation $\boldsymbol{d}$ of $\mathfrak{G}$ onto $\mathfrak{G}^*$ there, naturally, exists an inverse mapping $\boldsymbol{d}^{-1}$ of $\mathfrak{G}^*$ onto $\mathfrak{G}$ which is simple and, as we shall easily verify, a deformation. Assuming a^*, b^* to be arbitrary elements of $\mathfrak{G}^*$, let $a, b \in \mathfrak{G}$ be their inverse images under $\boldsymbol{d}$ so that $\boldsymbol{d}a = a^*$, $\boldsymbol{d}b = b^*$, $\boldsymbol{d}ab = a^*b^*$. Hence we have, by the definition of the inverse mapping $\boldsymbol{d}^{-1}$, the equalities: $a = \boldsymbol{d}^{-1}a^*$, $b = \boldsymbol{d}^{-1}b^*$, $ab = \boldsymbol{d}^{-1}a^*b^*$ which, in fact, yield $\boldsymbol{d}^{-1}a^*b^* = \boldsymbol{d}^{-1}a^* \,.\, \boldsymbol{d}^{-1}b^*$. Thus, if there exists an isomorphism $\boldsymbol{d}$ of $\mathfrak{G}$ onto $\mathfrak{G}^*$, then there exists an isomorphism $\boldsymbol{d}^{-1}$ of $\mathfrak{G}^*$ onto $\mathfrak{G}$; in that case we say that $\mathfrak{G}$ ($\mathfrak{G}^*$) is *isomorphic* with $\mathfrak{G}^*$ ($\mathfrak{G}$) or that $\mathfrak{G}$, $\mathfrak{G}^*$ are isomorphic and write $\mathfrak{G} \simeq \mathfrak{G}^*$ or $\mathfrak{G}^* \simeq \mathfrak{G}$. It is obvious that the fields of any two isomorphic groupoids are equivalent sets.

A mapping composite of two isomorphisms is again an isomorphism.

3. *Examples.* The abstract groupoid with the field $\{\boldsymbol{e}\}$ and the multiplication described in the first multiplication table in 11.4.2 is isomorphic with the groupoid $\mathfrak{S}_1$. The abstract groupoid with the field $\{\boldsymbol{e}, \boldsymbol{a}\}$ and the multiplication described in the second multiplication table in 11.4.2 is isomorphic with the groupoid $\mathfrak{S}_2$; the abstract groupoid with the field $\{\boldsymbol{e}, \boldsymbol{a}, \boldsymbol{b}, \boldsymbol{c}, \boldsymbol{d}, \boldsymbol{f}\}$ and the multiplication described in the third multiplication table in 11.4.2 is isomorphic with the groupoid $\mathfrak{S}_3$.

13.5. Operators, meromorphic and automorphic mappings

1. Further notions included in the concept of a deformation concern the case of a deformation of $\mathfrak{G}$ into or onto itself.

A deformation of $\mathfrak{G}$ into itself is also called an *operator on* (or *of*) *the groupoid* $\mathfrak{G}$ or an *endomorphic mapping of* $\mathfrak{G}$.

A simple operator on $\mathfrak{G}$, i.e., an isomorphic mapping of $\mathfrak{G}$ into itself is sometimes called a *meromorphic mapping of* $\mathfrak{G}$. If the image of $\mathfrak{G}$ is a proper subgroupoid of $\mathfrak{G}$, then the meromorphic mapping of $\mathfrak{G}$ is said to be *proper.*

2. An isomorphic mapping of $\mathfrak{G}$ onto itself is also called an *automorphic mapping of* $\mathfrak{G}$, briefly, an *automorphism of* $\mathfrak{G}$.

3. *Examples.* The mapping of the groupoid $\mathfrak{Z}$ into itself where each element $a \in \mathfrak{Z}$ is mapped onto the product (in arithmetic sense) $ka \in \mathfrak{Z}$, k denoting a non-negative integer, is an operator on $\mathfrak{Z}$. For $k \geqq 1$ it is a meromorphic mapping of $\mathfrak{Z}$, for

$k = 1$ it is an automorphism of $\mathfrak{Z}$ and for $k = 0$ an operator but not a meromorphic mapping of $\mathfrak{Z}$.

The simplest example of an automorphism of any groupoid $\mathfrak{G}$ is the identical mapping of $\mathfrak{G}$, the so-called *identical automorphism of* $\mathfrak{G}$.

13.6. Exercises

1. If any two elements of $\mathfrak{G}$ are interchangeable, then their images under every deformation of $\mathfrak{G}$ into $\mathfrak{G}^*$ are also interchangeable. The image of every Abelian groupoid is also Abelian.
2. If the product of a three-membered sequence of elements $a, b, c \in \mathfrak{G}$ consists of a single element, then the same holds for the sequence of images $\boldsymbol{d}a, \boldsymbol{d}b, \boldsymbol{d}c \in \mathfrak{G}^*$ under any deformation $\boldsymbol{d}$ of $\mathfrak{G}$ into $\mathfrak{G}^*$. The image of every associative groupoid under any deformation is also associative.
3. If $\mathfrak{G}$ is associative and has a center, then the image of the center under any deformation of $\mathfrak{G}$ onto $\mathfrak{G}^*$ lies in the center of $\mathfrak{G}^*$.
4. The inverse image of a groupoidal subset of $\mathfrak{G}^*$ under a deformation of $\mathfrak{G}$ onto $\mathfrak{G}^*$ need not be groupoidal.
5. Every meromorphic mapping of a finite groupoid $\mathfrak{G}$ is an automorphism of $\mathfrak{G}$.
6. For isomorphisms of the groupoids $\mathfrak{A}, \mathfrak{B}, \mathfrak{C}$ the following statements are true: a) $\mathfrak{A} \simeq \mathfrak{A}$ (reflexivity); b) $\mathfrak{A} \simeq \mathfrak{B}$ yields $\mathfrak{B} \simeq \mathfrak{A}$ (symmetry); c) from $\mathfrak{A} \simeq \mathfrak{B}$, $\mathfrak{B} \simeq \mathfrak{C}$ there follows $\mathfrak{A} \simeq \mathfrak{C}$ (transitivity).
7. It is left to the reader to give some examples of deformation himself.

14. Generating decompositions

14.1. Basic concepts

Suppose $\mathfrak{G}$ is an arbitrary groupoid.

Definition. Any decomposition $\bar{A}$ in $\mathfrak{G}$ is called *generating* if there exists, to any two-membered sequence of the elements $\bar{a}, \bar{b} \in \bar{A}$, an element $\bar{c} \in \bar{A}$ such that $\bar{a}\bar{b} \subseteq \bar{c}$.

As to the generating decompositions on the groupoid $\mathfrak{G}$, note that the greatest decomposition $\bar{G}_{max}$ and the least decomposition $\bar{G}_{min}$ are generating. On every groupoid there exist at least these two extreme generating decompositions.

The equivalence belonging to a generating decomposition (9.3) is usually called a *congruence*.

14.2. Deformation decompositions

Let $\mathfrak{G}$, $\mathfrak{G}^*$ denote arbitrary groupoids.

Suppose there exists a deformation $\boldsymbol{d}$ of $\mathfrak{G}$ onto $\mathfrak{G}^*$. Since $\boldsymbol{d}$ is a mapping of G onto G^*, it determines a decomposition $\bar{D}$ on $\mathfrak{G}$, corresponding to $\boldsymbol{d}$; each element $\bar{a}$ of $\bar{D}$ consists of all the inverse $\boldsymbol{d}$-images of an element $a^* \in \mathfrak{G}^*$. $\bar{D}$ is called the *deformation decomposition with regard to* $\boldsymbol{d}$ or the *decomposition corresponding (belonging) to the deformation* $\boldsymbol{d}$. Since $\boldsymbol{d}$ preserves the multiplications in both groupoids, it may be expected that $\bar{D}$ is in a certain relationship with the multiplication in $\mathfrak{G}$. Consider any two elements $\bar{a}, \bar{b} \in \bar{D}$. By the definition of $\bar{D}$, there exist elements $a^*, b^* \in \mathfrak{G}^*$ such that $\bar{a}$ ($\bar{b}$) is the set of all inverse $\boldsymbol{d}$-images of a^* (b^*). Consider the product $\bar{a}\bar{b}$ of $\bar{a}$ and $\bar{b}$. Each element $c \in \bar{a}\bar{b}$ is the product of an element $a \in \bar{a}$ and an element $b \in \bar{b}$ and is, with respect to $\boldsymbol{d}c = \boldsymbol{d}ab = \boldsymbol{d}a \,.\, \boldsymbol{d}b = a^*b^*$, an $\boldsymbol{d}$-inverse image of a^*b^*. Hence c is contained in that element $\bar{c} \in \bar{D}$ which consists of the inverse images of a^*b^*. Thus we have verified that the relation $\bar{a}\bar{b} \subset \bar{c}$ is true, hence $\bar{D}$ is generating. Consequently, *the decomposition of the groupoid* $\mathfrak{G}$, *corresponding to any deformation of* $\mathfrak{G}$ *onto another groupoid is generating.*

14.3. Generating decompositions in groupoids

Let us now study the properties of generating decompositions in groupoids.

1. *The sum of the elements of a generating decomposition.* Let $\bar{A}$ denote a generating decomposition in $\mathfrak{G}$.

The subset $\mathbf{s}\bar{A} \subset \mathfrak{G}$, that is to say, the subset of $\mathfrak{G}$, consisting of all the elements contained in some element of $\bar{A}$, *is groupoidal.* Indeed, to any elements $a, b \in \mathbf{s}\bar{A}$ there correspond elements $\bar{a}, \bar{b}, \bar{c} \in \bar{A}$ such that $a \in \bar{a}, b \in \bar{b}, \bar{a}\bar{b} \subset \bar{c}$ whence $ab \in \bar{a}\bar{b} \subset \bar{c} \subset \mathbf{s}\bar{A}$; thus ab is an element of $\mathbf{s}\bar{A}$. The corresponding subgroupoid of $\mathfrak{G}$ is denoted by $\mathbf{s}\mathfrak{A}$. It is evident that $\bar{A}$ is a generating decomposition on $\mathbf{s}\mathfrak{A}$.

2. *Closures and intersections.* Let B denote a groupoidal subset and $\bar{A}, \bar{C}$ be generating decompositions in $\mathfrak{G}$.

If $B \cap \mathbf{s}\bar{C} \neq \emptyset$, *then the closure* $B \sqsubset \bar{C}$ *and the intersection* $B \sqcap \bar{C}$ *are generating decompositions in* $\mathfrak{G}$. More generally: *if* $\mathbf{s}\bar{A} \cap \mathbf{s}\bar{C} \neq \emptyset$, *then the closure* $\bar{A} \sqsubset \bar{C}$ *and the intersection* $\bar{A} \sqcap \bar{C}$ *are generating decompositions in* $\mathfrak{G}$.

Proof. The decomposition $\bar{B}_{\max}$ consisting of a single element B is obviously a generating decomposition in $\mathfrak{G}$. If $B \cap \mathbf{s}\bar{C} \neq \emptyset$, then $\mathbf{s}\bar{B}_{\max} \cap \mathbf{s}\bar{C} \neq \emptyset$ and, furthermore, $B \sqsubset \bar{C} = \bar{B}_{\max} \sqsubset \bar{C}$, $B \sqcap \bar{C} = \bar{B}_{\max} \sqcap \bar{C}$. Consequently, the second part of the above statement is, in fact, a generalization of the first part and so it is only the latter we have to prove.

a) As there holds $\bar{A} \sqsubset \bar{C} = \mathbf{s}\bar{A} \sqsubset \bar{C}$, it is sufficient to show that the decomposition $\mathbf{s}\bar{A} \sqsubset \bar{C}$ is generating. Consider any two elements $\bar{c}_1, \bar{c}_2 \in \mathbf{s}\bar{A} \sqsubset \bar{C}$. Since the decomposition $\bar{C}$ is generating, there exists an element $\bar{c} \in \bar{C}$ such that $\bar{c}_1\, \bar{c}_2 \subset \bar{c}$. Choose two arbitrary points $x \in \mathbf{s}\bar{A} \cap \bar{c}_1$, $y \in \mathbf{s}\bar{A} \cap \bar{c}_2$. Then we have $xy \in \mathbf{s}\bar{A} \,.\, \mathbf{s}\bar{A} \cap \bar{c}_1\bar{c}_2 \subset \mathbf{s}\bar{A} \cap \bar{c}$ whence $\mathbf{s}\bar{A} \cap \bar{c} \neq \emptyset$. There follows $\bar{c} \in \mathbf{s}\bar{A} \sqsubset \bar{C}$.

b) Let $\bar{x}, \bar{y} \in \bar{A} \sqcap \bar{C}$ be arbitrary elements. By the definition of $\bar{A} \sqcap \bar{C}$ there exist elements $\bar{a}_1, \bar{a}_2 \in \bar{A}$; $\bar{c}_1, \bar{c}_2 \in \bar{C}$ such that $\bar{x} = \bar{a}_1 \cap \bar{c}_1$, $\bar{y} = \bar{a}_2 \cap \bar{c}_2$. Since the decomposition $\bar{A}$ ($\bar{C}$) is generating, there exists an element $\bar{a} \in \bar{A}$ ($\bar{c} \in \bar{C}$) such that $\bar{a}_1\bar{a}_2 \subset \bar{a}$ ($\bar{c}_1\bar{c}_2 \subset \bar{c}$). So we have

$$\bar{x}\bar{y} \subset \bar{a}_1\bar{a}_2 \cap \bar{c}_1\bar{c}_2 \subset \bar{a} \cap \bar{c} \in \bar{A} \sqcap \bar{C}$$

and the proof is accomplished.

Now let us add the following remarks:

If $\bar{C}$ lies on $\mathfrak{G}$, then the above assumption: $B \cap \mathbf{s}\bar{C} \neq \emptyset$ is satisfied because $\mathbf{s}\bar{C} = G \supset B$ and we have $B \cap \mathbf{s}\bar{C} = B \neq \emptyset$; the decomposition $B \sqcap \bar{C}$ then lies on B. Hence every generating decomposition $\bar{C}$ on $\mathfrak{G}$ and a groupoidal subset B of $\mathfrak{G}$ uniquely determine two generating decompositions in $\mathfrak{G}$: $B \sqsubset \bar{C}$, $\bar{C} \sqcap B$; the former is a subset of $\bar{C}$, the latter a decomposition on B.

In a similar way, every pair of generating decompositions $\bar{A}$, $\bar{C}$ in $\mathfrak{G}$ of which, e.g., $\bar{C}$ lies on $\mathfrak{G}$ determines two generating decompositions in $\mathfrak{G}$: $\bar{A} \sqsubset \bar{C}$, $\bar{A} \sqcap \bar{C}$; the former is a part of $\bar{C}$, the latter a decomposition on $\mathbf{s}\bar{A}$.

Finally, if both $\bar{A}$ and $\bar{C}$ lie on $\mathfrak{G}$, then $\bar{A} \sqcap \bar{C} = (\bar{A}, \bar{C})$ (3.5). We see that *the greatest common refinement of two generating decompositions lying on $\mathfrak{G}$ is again generating* (14.4.3).

3. *Enforced coverings.* Let again $\bar{A}$, $\bar{C}$ stand for generating decompositions in $\mathfrak{G}$. Suppose $\bar{A} = \bar{C} \sqsubset \bar{A}$, $\bar{C} = \bar{A} \sqsubset \bar{C}$ and let $\bar{B}$ denote a common covering of $\bar{A} \sqcap \mathbf{s}\bar{C}$, and $\bar{C} \sqcap \mathbf{s}\bar{A}$; these decompositions obviously lie on the set $\mathbf{s}\bar{A} \cap \mathbf{s}\bar{C}$. Let us, moreover, consider the coverings $\mathring{A}$, $\mathring{C}$ of $\bar{A}$, $\bar{C}$, enforced by $\bar{B}$ (4.1). $\mathring{A}$ and $\mathring{C}$ are coupled and $\bar{B}$ is their intersection: $\mathring{A} \sqcap \mathring{C} = \bar{B}$.

We shall prove that *if $\bar{B}$ is generating, then $\mathring{A}$ and $\mathring{C}$ are generating as well.*

Proof. Suppose $\bar{B}$ is generating and show that, e.g., $\mathring{A}$ has the same property. To simplify the notation, put $A = \mathbf{s}\bar{A}$, $C = \mathbf{s}\bar{C}$.

Let $\bigcup_1\bar{a}_1$, $\bigcup_2\bar{a}_2 \in \mathring{A}$ so that $\bar{a}_1$, $\bar{a}_2$ are elements of $\bar{A}$ and $\bigcup_1(\bar{a}_1 \cap C)$, $\bigcup_2(\bar{a}_2 \cap C)$ elements of $\bar{B}$. Since $\bar{A}$ is generating, there exists, to every product $\bar{a}_1\bar{a}_2$, an element $\bar{a}_{12} \in \bar{A}$ such that $\bar{a}_1\bar{a}_2 \subset \bar{a}_{12}$ whence even $(\bar{a}_1 \cap C)\,(\bar{a}_2 \cap C) \subset \bar{a}_{12} \cap C$. As $\bar{B}$ is generating as well, there exists an element $\bigcup_3(\bar{a}_3 \cap C) \in \bar{B}$ such that

$$\bigcup_1(\bar{a}_1 \cap C) \,.\, \bigcup_2(\bar{a}_2 \cap C) = \bigcup_1\bigcup_2(\bar{a}_1 \cap C)\,(\bar{a}_2 \cap C) \subset \bigcup_3(\bar{a}_3 \cap C),$$

where $\bar{a}_3$ denotes elements of $\bar{A}$ characterized by $\bigcup_3\bar{a}_3 \in \mathring{A}$. For each element $\bar{a}_1$ ($\bar{a}_2$) to which the symbol $\bigcup_1$ ($\bigcup_2$) applies we then have:

$$(\bar{a}_1 \cap C)\,(\bar{a}_2 \cap C) \subset (\bar{a}_{12} \cap C) \in \bigcup_3(\bar{a}_3 \cap C).$$

But the intersections $\bar{a}_{12} \cap C$, $\bar{a}_3 \cap C$ are elements of $\bar{A} \sqcap C$ lying on $A \cap C$. Consequently, among the elements $\bar{a}_3$ to which $\cup_3$ applies there exists an element $\bar{a}_3$ such that $\bar{a}_{12} \cap C = \bar{a}_3 \cap C$ and we have $\bar{a}_{12} = \bar{a}_3$. Hence there holds $\cup_1\bar{a}_1\cup_2\bar{a}_2 \subset \cup_1\cup_2\bar{a}_{12} \subset \cup_3\bar{a}_3 \in \mathring{A}$ and the proof is complete.

14.4. Generating decompositions on groupoids

Now we shall deal with generating decompositions on groupoids. The results will be useful even in case of generating decompositions in groupoids because every generating decomposition $\bar{A}$ in the groupoid $\mathfrak{G}$ is simultaneously a generating decomposition on the subgroupoid $\mathbf{s}\mathfrak{A}$.

1. *Local properties of coverings and refinements.* Let $\bar{A} \geqq \bar{B}$ denote any two generating decompositions on $\mathfrak{G}$.

Consider two arbitrary elements $\bar{a}_1, \bar{a}_2 \in \bar{A}$. Since $\bar{A}$ is generating, there exists an element $\bar{a}_3 \in \bar{A}$ such that $\bar{a}_1\bar{a}_2 \subset \bar{a}_3$. Next, consider the decompositions in $\mathfrak{G}$: $\bar{a}_1 \sqcap \bar{B}$, $\bar{a}_2 \sqcap \bar{B}$, $\bar{a}_3 \sqcap \bar{B}$. The latter represent, with regard to $\bar{A} \geqq \bar{B}$, nonempty parts of $\bar{B}$. As $\bar{B}$ is generating, there exists, to any pair of elements $\bar{x} \in \bar{a}_1 \sqcap \bar{B}, \bar{y} \in \bar{a}_2 \sqcap \bar{B}$, an element $\bar{z} \in \bar{B}$ such that $\bar{x}\bar{y} \subset \bar{z}$.

We shall show that $\bar{z}$ *is an element of* $\bar{a}_3 \sqcap \bar{B}$, hence $\bar{z} \in \bar{a}_3 \sqcap \bar{B}$.

Indeed, from $\bar{x} \subset \bar{a}_1$, $\bar{y} \subset \bar{a}_2$, $\bar{a}_1\bar{a}_2 \subset \bar{a}_3$ there follows $\bar{x}\bar{y} \subset \bar{a}_3$. So we have $\bar{x}\bar{y} \subset \bar{z} \cap \bar{a}_3$ whence, with respect to $\bar{B} \leqq \bar{A}$, there follows $\bar{z} \subset \bar{a}_3$ (3.2) and, consequently, $\bar{z} \in \bar{a}_3 \sqcap \bar{B}$.

We observe, in particular, that if the subset $\bar{a}_1 \subset \mathfrak{G}$ is groupoidal, then $\bar{a}_1 \sqcap \bar{B}$ is a generating decomposition (14.3.2).

2. *The least common covering.* Let $\bar{A}$, $\bar{B}$ stand for arbitrary generating decompositions on $\mathfrak{G}$.

We shall show that *their least common covering* $[\bar{A}, \bar{B}]$ *is generating as well.*

To that purpose we shall consider an arbitrary ordered pair of elements $\bar{u}, \bar{v} \in [\bar{A}, \bar{B}]$. We are to verify that there exists an element $\bar{w} \in [\bar{A}, \bar{B}]$ such that $\bar{u}\bar{v} \subset \bar{w}$.

Suppose $\bar{a} \in \bar{A}$ and $\bar{b} \in \bar{A}$ are arbitrary elements lying in $\bar{u}$ and $\bar{v}$, respectively, and so $\bar{a} \subset \bar{u}$, $\bar{b} \subset \bar{v}$. Since $\bar{A}$ is generating, there exists an element $\bar{c} \in \bar{A}$ such that $\bar{a}\bar{b} \subset \bar{c}$. The element $\bar{c}$ lies in a certain element $\bar{w} \in [\bar{A}, \bar{B}]$ and we have $\bar{c} \subset \bar{w}$.

Every element $p \in \bar{u}$ lies in a certain element $\bar{p} \in \bar{A}$ which is a part of $\bar{u}$; similarly, every element $q \in \bar{v}$ lies in a certain element $\bar{q} \in \bar{A}$ which is a part of $\bar{v}$. Moreover, the set $\bar{p}\bar{q}$ is a part of a certain element $\bar{r} \in \bar{A}$ and so $pq \in \bar{p}\bar{q} \subset \bar{r}$. From this we see that all we need to prove that $\bar{u}\bar{v} \subset \bar{w}$ applies is to verify that the element $\bar{r} \in \bar{A}$ comprising the set $\bar{p}\bar{q}$ is, for any two elements $\bar{p}, \bar{q} \in \bar{A}$, $\bar{p} \subset \bar{u}, \bar{q} \subset \bar{v}$, a part of $\bar{w}$, i.e., $\bar{r} \subset \bar{w}$.

Now, let $\bar{p}, \bar{q} \in \bar{A}$, $\bar{p} \subset \bar{u}$, $\bar{q} \subset \bar{v}$ denote arbitrary elements.

Taking account of the definition of the decomposition $\bar{A}$, $\bar{B}$ and of the fact that the elements $\bar{a}$ and $\bar{b}$ lie in $\bar{u}$ and $\bar{v}$, respectively, we conclude that there exists a binding $\{\bar{A}, \bar{B}\}$ from $\bar{a}$ to $\bar{p}$,

$$\bar{a}_1, \ldots, \bar{a}_\alpha \qquad (\text{where } \bar{a}_1 = \bar{a}, \bar{a}_\alpha = \bar{p}), \tag{1}$$

and, similarly, a binding $\{\bar{A}, \bar{B}\}$ from $\bar{b}$ to $\bar{q}$,

$$\bar{b}_1, \ldots, \bar{b}_\beta \qquad (\text{where } \bar{b}_1 = \bar{b}, \bar{b}_\beta = \bar{q}). \tag{2}$$

We may assume that $\beta = \alpha$ because if, for example, $\beta < \alpha$, then it is sufficient to denote the element $\bar{b}_\beta$ by the further symbols: $\bar{b}_{\beta+1}, \ldots, \bar{b}_\alpha$.

Since $\bar{A}$ is generating, there exist elements of $\bar{A}$

$$\bar{c}_1, \ldots, \bar{c}_\alpha \qquad (\text{where } \bar{c}_1 = \bar{c}, \bar{c}_\alpha = \bar{r}) \tag{3}$$

such that $\bar{a}_1\bar{b}_1 \subset \bar{c}_1, \ldots, \bar{a}_\alpha\bar{b}_\alpha \subset \bar{c}_\alpha$. With respect to the definition of $[\bar{A}, \bar{B}]$ and to the fact that the element $\bar{c}$ lies in $\bar{w}$, the relation $\bar{r} \subset \bar{w}$ will be proved by verifying that the sequence (3) is a binding $\{\bar{A}, \bar{B}\}$ from $\bar{c}$ to $\bar{r}$.

Since (1) and (2) are bindings $\{\bar{A}, \bar{B}\}$, there exists to every two elements $\bar{a}_\nu, \bar{a}_{\nu+1}$ and, similarly, to every two elements $\bar{b}_\nu, \bar{b}_{\nu+1}$ an element $\bar{x}_\nu \in \bar{B}$ and an element $\bar{y}_\nu \in \bar{B}$ $(\nu = 1, \ldots, \alpha - 1)$, respectively, incident with both. As $\bar{B}$ is generating, there exists a certain element $\bar{z}_\nu \in \bar{B}$ for which $\bar{x}_\nu\bar{y}_\nu \subset \bar{z}_\nu$. Since $\bar{x}_\nu$ and $\bar{y}_\nu$ are incident with $\bar{a}_\nu$ and $\bar{b}_\nu$, respectively, the set $\bar{x}_\nu\bar{y}_\nu$ is incident with $\bar{a}_\nu\bar{b}_\nu$; consequently, $\bar{z}_\nu$ is incident with $\bar{a}_\nu\bar{b}_\nu$ and therefore also with $\bar{c}_\nu$. Analogously, we observe that $\bar{z}_\nu$ is incident with $\bar{c}_{\nu+1}$. Hence every two elements $\bar{c}_\nu, \bar{c}_{\nu+1}$ are incident with a certain element $\bar{z}_\nu \in \bar{B}$ and, consequently, the sequence (3) is a binding $\{\bar{A}, \bar{B}\}$ from $\bar{c}$ to $\bar{r}$.

3. *The greatest common refinement.* Let again $\bar{A}$, $\bar{B}$ denote arbitrary generating decompositions on $\mathfrak{G}$.

Theorem. *The greatest common refinement* $(\bar{A}, \bar{B})$ *of the decompositions* $\bar{A}$, $\bar{B}$ *is also generating.*

This theorem has already been proved (in 14.3.2) on the ground of $(\bar{A}, \bar{B}) = \bar{A} \sqcap \bar{B}$ by verifying that the intersection $\bar{A} \sqcap \bar{B}$ of the generating decompositions $\bar{A}$, $\bar{B}$ is also generating.

14.5. Exercises

1. If an element $\bar{a} \in \bar{A}$ of a generating decomposition $\bar{A}$ in the groupoid $\mathfrak{G}$ contains a groupoidal subset $X \subset \mathfrak{G}$ so that $X \subset \bar{a}$, then the element $\bar{a}$ is groupoidal as well.
2. Let $\mathfrak{G}$ denote the groupoid whose field consists of all positive integers and whose multiplication is defined as follows: the product ab $(a, b \in \mathfrak{G})$ is the number $a_1 \ldots a_\alpha b_1 \ldots b_\beta$, where the numbers $a_1, \ldots, a_\alpha$ and $b_1, \ldots, b_\beta$ are the digits of a and b, respectively, in the decimal system. Thus, for example, $14.23 = 1423$. Show that: a) the groupoid $\mathfrak{G}$ is asso-

ciative; b) the decomposition of $\mathfrak{G}$, the elements of which are the sets of all the numbers in $\mathfrak{G}$ expressed, in the decimal system, by symbols containing the same number of digits, is generating.

3. The groupoid $\mathfrak{G}$, whose field is an arbitrary set and the multiplication given by $ab = a$ ($ab = b$) for $a, b \in \mathfrak{G}$, is associative and all its decompositions are generating.

15. Factoroids

The notion of a factoroid we shall now be concerned with plays an important part throughout the following theory.

15.1. Basic concepts

Let again $\bar{A}$ denote an arbitrary generating decomposition in $\mathfrak{G}$. With $\bar{A}$ we can uniquely associate a groupoid denoted $\bar{\mathfrak{A}}$ and defined as follows: The field of $\bar{\mathfrak{A}}$ is the decomposition $\bar{A}$ and the multiplication is defined in the following way: the product of any element $\bar{a} \in \bar{A}$ and any element $\bar{b} \in \bar{A}$ is the element $\bar{c} \in \bar{A}$ for which $\bar{a}\bar{b} \subset \bar{c}$. Then we generally write

$$\bar{a} \circ \bar{b} = \bar{c},$$

and we have $\bar{a}\bar{b} \subset \bar{a} \circ \bar{b} \in \bar{\mathfrak{A}}$. We employ the symbol $\circ$ to denote the products in $\bar{\mathfrak{A}}$ in the same way as we use the symbol . to denote the products in $\mathfrak{G}$.

$\bar{\mathfrak{A}}$ is called a *factoroid in* $\mathfrak{G}$; if $\bar{A}$ is *on* $\mathfrak{G}$, then it is a *factoroid on* $\mathfrak{G}$ or a *factoroid of* $\mathfrak{G}$. Every generating decomposition in $\mathfrak{G}$ uniquely determines a certain factoroid in $\mathfrak{G}$, namely the one whose field it is; we say that to every generating decomposition in $\mathfrak{G}$ there *corresponds* or *belongs* a certain factoroid in $\mathfrak{G}$.

Note that on $\mathfrak{G}$ there exist at least two factoroids, namely the so-called *greatest factoroid*, $\bar{\mathfrak{G}}_{\max}$, belonging to the greatest generating decomposition $\bar{G}_{\max}$ and the *least factoroid*, $\bar{\mathfrak{G}}_{\min}$, belonging to the least generating decomposition $\bar{G}_{\min}$ of the groupoid $\mathfrak{G}$. These extreme factoroids on $\mathfrak{G}$ are either different from each other or coincide according as $\mathfrak{G}$ contains more than one or precisely one element.

15.2. Example of a factoroid

Consider, for example, the groupoid $\mathfrak{Z}$. Let n be an arbitrary positive integer and $\bar{a}_i$, where i runs over the numbers $0, \ldots, n-1$, stand for the set of all the elements of $\mathfrak{Z}$ that, in the division by n, leave the remainder i. The sets $\bar{a}_0, \ldots, \bar{a}_{n-1}$ are:

$$\begin{array}{llllll}
\bar{a}_0 = \{\ldots, -2n, & -n, & 0, & n, & 2n, & \ldots\}, \\
\bar{a}_1 = \{\ldots, -2n+1, & -n+1, & 1, & n+1, & 2n+1, & \ldots\}, \\
\bar{a}_2 = \{\ldots, -2n+2, & -n+2, & 2, & n+2, & 2n+2, & \ldots\}, \\
\hline
\multicolumn{6}{l}{\bar{a}_{n-1} = \{\ldots, -2n+(n-1), -n+(n-1), n-1, n+(n-1), 2n+(n-1), \ldots\}.}
\end{array}$$

We see that the system $\{\bar{a}_0, \ldots, \bar{a}_{n-1}\}$ is a decomposition of $\mathfrak{Z}$; let us denote it $\bar{Z}_n$ and show that it is generating. To that purpose we shall verify that the product $\bar{a}_i\bar{a}_j$ of an element $\bar{a}_i \in \bar{Z}_n$ and an element $\bar{a}_j \in \bar{Z}_n$ is a part of an element $\bar{a}_k \in \bar{Z}_n$. By its definition, the set $\bar{a}_i \,.\, \bar{a}_j$ consists of the products $a \,.\, b$ where a and b run over all the elements of $\bar{a}_i$ and $\bar{a}_j$, respectively. Now let a be an element of $\bar{a}_i$ so that the remainder in the division of a by n is i, and let b denote an element of $\bar{a}_j$ so that the remainder in the division of b by n is j. By the definition of the multiplication in $\mathfrak{Z}$, we have $a \,.\, b = a + b \in \bar{a}_k$ where k is the remainder in the division of $i + j$ by n because both $a + b$ and $i + j$ leave, in the division by n, the same remainder. So we have $\bar{a}_i\bar{a}_j \subset \bar{a}_k$, hence $\bar{Z}_n$ is generating. The corresponding factoroid $\bar{\mathfrak{Z}}_n$ therefore consists of n elements: $\bar{a}_0, \ldots, \bar{a}_{n-1}$ and its multiplication is defined by the rule that the product $\bar{a}_i \,.\, \bar{a}_j$ is the element $\bar{a}_k$ where k is the remainder in the division of $i + j$ by n. Obviously $\bar{\mathfrak{Z}}_1$ is the greatest factoroid on $\mathfrak{Z}$.

15.3. Factoroids in groupoids

Before proceeding with our study, let us remember that we apply, to groupoids, all the concepts, symbols and results defined for their fields and multiplication. The same holds for factoroids. The most important concepts, symbols and results arrived at in this way are:

1. *Coverings and refinements.* Let $\bar{\mathfrak{A}}, \bar{\mathfrak{B}}$ stand for factoroids in $\mathfrak{G}$.

$\bar{\mathfrak{A}}$ ($\bar{\mathfrak{B}}$) is called a *covering* (*refinement*) *of* $\bar{\mathfrak{B}}$ ($\bar{\mathfrak{A}}$) if, for the fields $\bar{A}$, $\bar{B}$ of $\bar{\mathfrak{A}}, \bar{\mathfrak{B}}$, there holds $\bar{A} \geqq \bar{B}$. We write $\bar{\mathfrak{A}} \geqq \bar{\mathfrak{B}}$ or $\bar{\mathfrak{B}} \leqq \bar{\mathfrak{A}}$. The meaning of a *normal* and a *pure covering* (*refinement*) *of* $\bar{\mathfrak{B}}$ ($\bar{\mathfrak{A}}$) is obvious (2.4). The relation $\bar{\mathfrak{A}} \geqq \bar{\mathfrak{B}}$ yields $\mathbf{s}\bar{\mathfrak{A}} \supset \mathbf{s}\bar{\mathfrak{B}}$ and, in case of a pure covering (refinement): $\mathbf{s}\bar{\mathfrak{A}} = \mathbf{s}\bar{\mathfrak{B}}$. If $\bar{\mathfrak{A}} \geqq \bar{\mathfrak{B}}$ and, at the same time, $\bar{\mathfrak{A}} \neq \bar{\mathfrak{B}}$, then $\bar{\mathfrak{A}}$ ($\bar{\mathfrak{B}}$) is a *proper covering* (*proper refinement*) *of* $\bar{\mathfrak{B}}$ ($\bar{\mathfrak{A}}$); then we sometimes write $\bar{\mathfrak{A}} > \bar{\mathfrak{B}}$ or $\bar{\mathfrak{B}} < \bar{\mathfrak{A}}$.

2. *Closures and intersections.* Let $\mathfrak{B} \subset \mathfrak{G}$ stand for a subgroupoid and $\bar{\mathfrak{A}}, \bar{\mathfrak{C}}$ for factoroids in $\mathfrak{G}$.

If $B \cap \mathbf{s}\bar{C} \neq \emptyset$, then (14.3.2) $B \sqsubset \bar{C}$ and $B \sqcap \bar{C}$ are generating decompositions in $\mathfrak{G}$. The corresponding factoroids in $\mathfrak{G}$ are called the *closure of the subgroupoid* $\mathfrak{B}$ *in the factoroid* $\bar{\mathfrak{C}}$ and the *intersection of the subgroupoid* $\mathfrak{B}$ (*factoroid* $\bar{\mathfrak{C}}$) *and the factoroid* $\bar{\mathfrak{C}}$ (*subgroupoid* $\mathfrak{B}$); notation for closure: $\mathfrak{B} \sqsubset \bar{\mathfrak{C}}$ or $\bar{\mathfrak{C}} \sqsupset \mathfrak{B}$, for intersection: $\mathfrak{B} \sqcap \bar{\mathfrak{C}}$ or $\bar{\mathfrak{C}} \sqcap \mathfrak{B}$.

The meaning of the concepts defined for $\mathbf{s}\bar{A} \cap \mathbf{s}\bar{C} \neq \emptyset$ and denoted $\bar{\mathfrak{A}} \sqsubset \bar{\mathfrak{C}}$ or $\bar{\mathfrak{C}} \sqsupset \bar{\mathfrak{A}}$ and $\bar{\mathfrak{A}} \sqcap \bar{\mathfrak{C}}$ is obvious as well; the former is called the *closure* of $\bar{\mathfrak{A}}$ *in* $\bar{\mathfrak{C}}$, the latter is the *intersection of* $\bar{\mathfrak{A}}$ *and* $\bar{\mathfrak{C}}$. Evidently: $\bar{\mathfrak{A}} \sqcap \bar{\mathfrak{C}} = \bar{\mathfrak{C}} \sqcap \bar{\mathfrak{A}}$.

Note that $\mathfrak{B} \sqsubset \bar{\mathfrak{C}}$ is a subgroupoid in $\bar{\mathfrak{C}}$ and $\mathfrak{B} \sqcap \bar{\mathfrak{C}}$ a factoroid in $\mathfrak{B}$.

If, in particular, $\bar{\mathfrak{C}}$ lies on $\mathfrak{G}$, then the above assumption $B \cap \mathbf{s}\bar{C} \neq \emptyset$ is satisfied and $\mathfrak{B} \sqcap \bar{\mathfrak{C}}$ is a factoroid on $\mathfrak{B}$. Every factoroid $\bar{\mathfrak{C}}$ on $\mathfrak{G}$ and a subgroupoid $\mathfrak{B}$ of $\mathfrak{G}$ thus uniquely determine a subgroupoid $\mathfrak{B} \sqsubset \bar{\mathfrak{C}}$ in $\bar{\mathfrak{C}}$ and a factoroid $\mathfrak{B} \sqcap \bar{\mathfrak{C}}$ on $\mathfrak{B}$.

Similarly, a factoroid $\bar{\mathfrak{A}}$ in $\mathfrak{G}$ and a factoroid $\bar{\mathfrak{C}}$ on $\mathfrak{G}$ determine a factoroid $\bar{\mathfrak{A}} \sqsubset \bar{\mathfrak{C}}$ and a factoroid $\bar{\mathfrak{A}} \sqcap \bar{\mathfrak{C}}$; the former is a subgroupoid of $\bar{\mathfrak{C}}$ and the latter a factoroid on $\mathbf{s}\bar{\mathfrak{A}}$.

Finally, let us remark that *if* $\bar{\mathfrak{A}}$ *and* $\bar{\mathfrak{C}}$ *cover* $\mathfrak{G}$, *then their intersection coincides with the greatest common refinement* $(\bar{\mathfrak{A}}, \bar{\mathfrak{C}})$ *of* $\bar{\mathfrak{A}}, \bar{\mathfrak{C}}$ and so $\bar{\mathfrak{A}} \sqcap \bar{\mathfrak{C}} = (\bar{\mathfrak{A}}, \bar{\mathfrak{C}})$ (15.4.5).

Example. In order to illustrate the above notions by an example, let us again consider the factoroid $\bar{\mathfrak{Z}}_n$ on the groupoid $\mathfrak{Z}$ ($n \geqq 1$). Let $\mathfrak{A}_m$ denote the subgroupoid of $\mathfrak{Z}$, with the field consisting of all multiples of a given positive integer m and suppose (to simplify our example) that the greatest common divisor of m and n is 1.

Which elements do the factoroids $\mathfrak{A}_m \sqsubset \bar{\mathfrak{Z}}_n$, $\bar{\mathfrak{Z}}_n \sqcap \mathfrak{A}_m$ consist of?

Consider which of the elements $\bar{a}_0, \ldots, \bar{a}_{n-1} \in \bar{\mathfrak{Z}}_n$ are incident with the subgroupoid $\mathfrak{A}_m$. Any element $\bar{a}_i \in \bar{\mathfrak{Z}}_n$ is incident with $\mathfrak{A}_m$ if and only if it comprises a multiple xm of m (x integer). Since each element of $\bar{a}_i$ is of the form $yn + i$ where y also denotes an integer, we see that $\bar{a}_i$ is incident with $\mathfrak{A}_m$ if and only if the equation $xm = yn + i$ and therefore even $xm - yn = i$ has an integral solution. Since the greatest common divisor of m and n is 1, there exist integers a, b satisfying $am - bn = 1$. Consequently, $xm - yn = i$ has, for every number $i = 0, \ldots, n - 1$, an integral solution, namely $x = ai, y = bi$, hence every element $\bar{a}_i \in \bar{\mathfrak{Z}}_n$ is incident with $\mathfrak{A}_m$. Thus the factoroid $\mathfrak{A}_m \sqsubset \bar{\mathfrak{Z}}_n$ is identical with $\bar{\mathfrak{Z}}_n$ and the elements of $\bar{\mathfrak{Z}}_n \sqcap \mathfrak{A}_m$ are sets consisting of all the multiples of m contained in the individual elements $\bar{a}_0, \ldots, \bar{a}_{n-1}$ of the factoroid $\bar{\mathfrak{Z}}_n$.

3. *Semi-coupled or loosely coupled and coupled factoroids.* Let $\bar{\mathfrak{A}}, \bar{\mathfrak{C}}$ be factoroids in $\mathfrak{G}$. The factoroids $\bar{\mathfrak{A}}, \bar{\mathfrak{C}}$ are said to be *semi-coupled* or *loosely coupled* (*coupled*) if their fields $\bar{A}, \bar{C}$ have the same property (4.1).

For example, the closure $\mathfrak{X} \sqsubset \bar{\mathfrak{Y}}$ of an arbitrary subgroupoid $\mathfrak{X} \subset \mathfrak{G}$ in the factoroid $\bar{\mathfrak{Y}}$ in $\mathfrak{G}$ and the intersection $\bar{\mathfrak{Y}} \sqcap \mathfrak{X}$ ($X \cap \mathbf{s}\bar{Y} \neq \emptyset$) are coupled factoroids.

In what follows we shall assume that $\bar{\mathfrak{A}} = \bar{\mathfrak{C}} \sqsubset \bar{\mathfrak{A}}, \bar{\mathfrak{C}} = \bar{\mathfrak{A}} \sqsubset \bar{\mathfrak{C}}$.

In that case there lies, in $\mathfrak{G}$, the subgroupoid $\mathbf{s}\bar{\mathfrak{A}} \cap \mathbf{s}\bar{\mathfrak{C}}$ and, on the latter, the factoroids $\bar{\mathfrak{A}} \sqcap \mathbf{s}\bar{\mathfrak{C}}, \bar{\mathfrak{C}} \sqcap \mathbf{s}\bar{\mathfrak{A}}$. From the theorem in 14.3.3 we conclude that *every common covering* $\overline{\mathfrak{B}}$ *of* $\overline{\mathfrak{A}} \sqcap \mathbf{s}\overline{\mathfrak{C}}$ *and* $\overline{\mathfrak{C}} \sqcap \mathbf{s}\overline{\mathfrak{A}}$ *enforces coupled coverings* $\mathring{\mathfrak{A}} \geqq \overline{\mathfrak{A}}$, $\mathring{\mathfrak{C}} \geqq \overline{\mathfrak{C}}$ *of* $\overline{\mathfrak{A}}, \overline{\mathfrak{C}}$ *intersecting each other in the factoroid* $\overline{\mathfrak{B}}$: $\mathring{\mathfrak{A}} \sqcap \mathring{\mathfrak{C}} = \overline{\mathfrak{B}}$.

4. *Adjoint factoroids.* Let $\overline{\mathfrak{A}}, \overline{\mathfrak{C}}$ be factoroids and $\mathfrak{B}, \mathfrak{D}$ subgroupoids of $\mathfrak{G}$. Denote $\mathfrak{A} = \mathbf{s}\overline{\mathfrak{A}}$, $\mathfrak{C} = \mathbf{s}\overline{\mathfrak{C}}$ and let B, D stand for the fields of the subgroupoids $\mathfrak{B}, \mathfrak{D}$.

Suppose there holds: $\mathfrak{B} \in \overline{\mathfrak{A}}$, $\mathfrak{D} \in \overline{\mathfrak{C}}$; $\mathfrak{B} \cap \mathfrak{D} \neq \emptyset$, the first formulae expressing the relations $B \in \overline{\mathfrak{A}}$, $D \in \overline{\mathfrak{C}}$.

The factoroids $\overline{\mathfrak{A}}, \overline{\mathfrak{C}}$ are said to be *adjoint with regard to* $\mathfrak{B}, \mathfrak{D}$ if the decompositions $\bar{A}, \bar{C}$ have the same property with regard to B, D (4.2). This may be expressed by the formula:

$$\mathbf{s}(\mathfrak{D} \sqsubset \overline{\mathfrak{A}} \sqcap \mathfrak{C}) = \mathbf{s}(\mathfrak{B} \sqsubset \overline{\mathfrak{C}} \sqcap \mathfrak{A}).$$

Suppose $\overline{\mathfrak{A}}, \overline{\mathfrak{C}}$ are adjoint with regard to $\mathfrak{B}, \mathfrak{D}$. Then

$$\overline{\mathfrak{A}}_1 = \mathfrak{C} \sqsubset \overline{\mathfrak{A}}, \quad \overline{\mathfrak{A}}_2 = \mathfrak{D} \sqsubset \overline{\mathfrak{A}},$$

$$\overline{\mathfrak{C}}_1 = \mathfrak{A} \sqsubset \overline{\mathfrak{C}}, \quad \overline{\mathfrak{C}}_2 = \mathfrak{B} \sqsubset \overline{\mathfrak{C}}$$

are factoroids in $\mathfrak{G}$. Denote $\mathfrak{A}_1 = \mathbf{s}\overline{\mathfrak{A}}_1$, $\mathfrak{A}_2 = \mathbf{s}\overline{\mathfrak{A}}_2$; $\mathfrak{C}_1 = \mathbf{s}\overline{\mathfrak{C}}_1$, $\mathfrak{C}_2 = \mathbf{s}\overline{\mathfrak{C}}_2$. From the result in 4.2 there follows, with respect to 14.4.2 and 14.3.3, the following theorem:

The factoroids $\overline{\mathfrak{A}}_1, \overline{\mathfrak{C}}_1$ *have coupled coverings* $\mathring{\mathfrak{A}}, \mathring{\mathfrak{C}}$ *such that* $\mathfrak{A}_2 \in \mathring{\mathfrak{A}}$, $\mathfrak{C}_2 \in \mathring{\mathfrak{C}}$; the coverings $\mathring{\mathfrak{A}}, \mathring{\mathfrak{C}}$ are given by the construction described in 4.2a. *The subgroupoids* $\mathfrak{A}_2, \mathfrak{C}_2$ *are incident.*

5. *Chains of factoroids.* Let $\mathfrak{A} \supset \mathfrak{B}$ denote subgroupoids of $\mathfrak{G}$.

A *chain of factoroids from* $\mathfrak{A}$ *to* $\mathfrak{B}$, briefly, a *chain from* $\mathfrak{A}$ *to* $\mathfrak{B}$, is a finite sequence consisting of α ($\geqq 1$) factoroids $\overline{\mathfrak{K}}_1, \ldots, \overline{\mathfrak{K}}_\alpha$ in $\mathfrak{G}$ with the following properties: a) the factoroid $\overline{\mathfrak{K}}_1$ lies on $\mathfrak{A}$; b) for $1 \leqq \gamma \leqq \alpha - 1$ the factoroid $\overline{\mathfrak{K}}_{\gamma+1}$ lies on an element of $\overline{\mathfrak{K}}_\gamma$; c) $\mathfrak{B} \in \overline{\mathfrak{K}}_\alpha$. Such a chain is denoted

$$\overline{\mathfrak{K}}_1 \to \cdots \to \overline{\mathfrak{K}}_\alpha,$$

briefly: $[\overline{\mathfrak{K}}]$.

The notions relative to chains of decompositions, defined in 2.5 and 4.2, can be directly applied to chains of factoroids. In particular, the concept of adjoint chains of factoroids is defined as follows:

Let $\mathfrak{A} \supset \mathfrak{B}$, $\mathfrak{C} \supset \mathfrak{D}$ stand for subgroupoids of $\mathfrak{G}$ and let

$$([\overline{\mathfrak{K}}] =)\ \overline{\mathfrak{K}}_1 \to \cdots \to \overline{\mathfrak{K}}_\alpha,$$

$$([\overline{\mathfrak{L}}] =)\ \overline{\mathfrak{L}}_1 \to \cdots \to \overline{\mathfrak{L}}_\beta$$

be chains of factoroids from $\mathfrak{A}$ to $\mathfrak{B}$ and from $\mathfrak{C}$ to $\mathfrak{D}$.

The chains $[\overline{\mathfrak{K}}], [\overline{\mathfrak{L}}]$ are called *adjoint* if a) their ends coincide, i.e., $\mathfrak{A} = \mathfrak{C}$, $\mathfrak{B} = \mathfrak{D}$; b) every two members $\overline{\mathfrak{K}}_\gamma, \overline{\mathfrak{L}}_\delta$ are adjoint with regard to $\mathbf{s}\overline{\mathfrak{K}}_{\gamma+1}$, $\mathbf{s}\overline{\mathfrak{L}}_{\delta+1}$ for $\gamma = 1, \ldots, \alpha$; $\delta = 1, \ldots, \beta$ while $\mathbf{s}\overline{\mathfrak{K}}_{\alpha+1} = \mathfrak{B}$, $\mathbf{s}\overline{\mathfrak{L}}_{\beta+1} = \mathfrak{D}$.

15.4. Factoroids on groupoids

Let us now deal with factoroids on groupoids. The results can often be applied even to factoroids in groupoids, since every factoroid $\bar{\mathfrak{A}}$ in $\mathfrak{G}$ lies on the subgroupoid $\mathbf{s}\bar{\mathfrak{A}}$.

1. *Coverings and refinements.* We shall start from the notions of a covering and a refinement of a factoroid in $\mathfrak{G}$, described in 15.3.1 and proceed to the case of factoroids on $\mathfrak{G}$.

Let $\bar{\mathfrak{A}}$, $\bar{\mathfrak{B}}$ denote factoroids on $\mathfrak{G}$.

We know that $\bar{\mathfrak{A}}$ ($\bar{\mathfrak{B}}$) is called a covering (refinement) of $\bar{\mathfrak{B}}$ ($\bar{\mathfrak{A}}$) and that we write $\bar{\mathfrak{A}} \geqq \bar{\mathfrak{B}}$ or $\bar{\mathfrak{B}} \leqq \bar{\mathfrak{A}}$ if, for the fields $\bar{A}$ and $\bar{B}$ of $\bar{\mathfrak{A}}$ and $\bar{\mathfrak{B}}$, respectively, there holds $\bar{A} \geqq \bar{B}$.

For example, $\bar{\mathfrak{G}}_{\max}$ ($\bar{\mathfrak{B}}$) is the greatest (least) covering of $\bar{\mathfrak{B}}$ in the sense that every covering of $\bar{\mathfrak{B}}$ is a refinement of $\bar{\mathfrak{G}}_{\max}$ and, of course, a covering of $\bar{\mathfrak{B}}$; analogously, $\bar{\mathfrak{A}}$ ($\bar{\mathfrak{G}}_{\min}$) is the greatest (least) refinement of $\bar{\mathfrak{A}}$.

If $\bar{\mathfrak{A}} \geqq \bar{\mathfrak{B}}$, then $\bar{A}$ is a covering of $\bar{B}$ so that $\bar{A}$ is enforced by a certain decomposition $\bar{\bar{B}}$ lying on $\bar{B}$ and, naturally, also on $\bar{\mathfrak{B}}$ (2.4). Note that every element $\bar{\bar{b}} \in \bar{\bar{B}}$ is a system of subsets in $\mathfrak{G}$ which are elements of $\bar{\mathfrak{B}}$ and that $\bar{A}$ is obtained by summing all the elements of $\bar{\mathfrak{B}}$ lying in the individual elements $\bar{\bar{b}}$.

Conversely, every decomposition $\bar{\bar{B}}$ on the factoroid $\bar{\mathfrak{B}}$ enforces a certain covering of $\bar{B}$ which, however, is not necessarily generating. We observe that the covering enforced by $\bar{\bar{B}}$ need not be the field of a factoroid.

Let us now prove the following theorem:

Let $\bar{\mathfrak{B}}$ stand for a factoroid on $\mathfrak{G}$, $\bar{\bar{B}}$ for a decomposition of $\bar{\mathfrak{B}}$ and $\bar{A}$ for the covering of the field $\bar{B}$ of $\bar{\mathfrak{B}}$ enforced by $\bar{\bar{B}}$. The decomposition $\bar{A}$ is generating if and only if $\bar{\bar{B}}$ is generating.

Proof. a) Suppose $\bar{\bar{B}}$ is generating. Consider arbitrary elements $\bar{a}_1, \bar{a}_2 \in \bar{A}$. We are to show that there exists an element $\bar{a}_3 \in \bar{A}$ such that $\bar{a}_1\bar{a}_2 \subset \bar{a}_3$. Now, with regard to the definition of $\bar{A}$, there holds $\bar{a}_1 = \bigcup_1 \bar{b}_1$, $\bar{a}_2 = \bigcup_2 \bar{b}_2$, the symbol $\bigcup_1$ ($\bigcup_2$) relating to all the elements of the factoroid $\bar{\mathfrak{B}}$ contained in a certain element $\bar{\bar{b}}_1$ ($\bar{\bar{b}}_2$) of $\bar{\bar{B}}$. Since $\bar{\bar{B}}$ is generating, there exists an element $\bar{\bar{b}}_3 \in \bar{\bar{B}}$ such that $\bar{\bar{b}}_1 \circ \bar{\bar{b}}_2 \subset \bar{\bar{b}}_3$. Let $\bar{a}_3$ be the sum of all the elements of $\bar{\mathfrak{B}}$ contained in $\bar{\bar{b}}_3$ so that $\bar{a}_3 \in \bar{A}$. For every element $\bar{b}_1$ ($\bar{b}_2$) to which the symbol $\bigcup_1$ ($\bigcup_2$) applies we evidently have $\bar{b}_1 \circ \bar{b}_2 \in \bar{\bar{b}}_1 \circ \bar{\bar{b}}_2 \subset \bar{\bar{b}}_3$. Hence the relations:

$$\bar{a}_1\bar{a}_2 = \bigcup_1\bigcup_2 \bar{b}_1\bar{b}_2 \subset \bigcup_1\bigcup_2 \bar{b}_1 \circ \bar{b}_2 \subset \bar{a}_3$$

which prove the first part of the theorem.

b) Suppose $\bar{A}$ is generating. Consider arbitrary elements $\bar{\bar{b}}_1, \bar{\bar{b}}_2 \in \bar{\bar{B}}$ and let $\bar{a}_1, \bar{a}_2$, $\bar{b}_1, \bar{b}_2$ have the above meaning. Since $\bar{A}$ is generating, there exists an element $\bar{a}_3 \in \bar{A}$ such that $\bar{a}_1\bar{a}_2 \subset \bar{a}_3$. By the definition of $\bar{A}$ there exist elements $\bar{b}_3 \in \bar{\mathfrak{B}}$ such that

$\bar{a}_3 = \bigcup \bar{b}_3$ and the set of these elements is an element $\bar{\bar{b}}_3 \in \bar{\bar{B}}$. For any elements $\bar{b}_1 \in \bar{\bar{b}}_1$, $\bar{b}_2 \in \bar{\bar{b}}_2$ there holds $\bar{b}_1\bar{b}_2 \subset \bar{a}_3$ and we see that there exists an element $\bar{b}_3 \in \bar{\bar{b}}_3$ such that $\bar{b}_1\bar{b}_2 \subset \bar{b}_3$, Hence $\bar{b}_1 \circ \bar{b}_2 = \bar{b}_3 \in \bar{\bar{b}}_3$ so that $\bar{\bar{b}}_1 \circ \bar{\bar{b}}_2 \subset \bar{\bar{b}}_3$ and the proof is complete.

Thus *both decompositions $\bar{A}$ and $\bar{\bar{B}}$ are simultaneously generating*, i.e., if one is generating, then the other is generating as well. If they are generating, then there corresponds to $\bar{A}$ a certain factoroid $\bar{\mathfrak{A}}$ on $\mathfrak{G}$ and there holds $\bar{\mathfrak{A}} \geqq \bar{\mathfrak{B}}$; similarly, there corresponds to $\bar{\bar{B}}$ a certain factoroid $\bar{\bar{\mathfrak{B}}}$ on $\bar{\mathfrak{B}}$. $\bar{\mathfrak{A}}$ is called the *covering of* $\bar{\mathfrak{B}}$ *enforced by* $\bar{\bar{\mathfrak{B}}}$. Every factoroid on an arbitrary factoroid $\bar{\mathfrak{B}}$ of $\mathfrak{G}$ therefore enforces a certain covering of $\bar{\mathfrak{B}}$ and, conversely, every covering of $\bar{\mathfrak{B}}$ is enforced by a factoroid on $\bar{\mathfrak{B}}$.

Example. To illustrate the above notions by an example, let us again consider the factoroid $\bar{\mathfrak{Z}}_n$ on the groupoid $\mathfrak{Z}$ (15.2). Suppose the number n is greater than 1 and is not a prime number. Then there exists a divisor $(1 <)\, d\, (< n)$ of the number n and we have $n = qd$ where q is a positive integer $1 < q < n$. We shall now be concerned with the decomposition $\bar{\bar{Z}}_d$ of $\bar{\mathfrak{Z}}_n$ whose elements are:

$$\begin{aligned} \bar{\bar{a}}_0 &= \{\bar{a}_0, \ \bar{a}_d, \ \bar{a}_{2d}, \ \ldots, \bar{a}_{(q-1)d}\}, \\ \bar{\bar{a}}_1 &= \{\bar{a}_1, \ \bar{a}_{d+1}, \ \bar{a}_{2d+1}, \ \ldots, \bar{a}_{(q-1)d+1}\}, \\ &\cdots\cdots\cdots\cdots\cdots\cdots\cdots\cdots\cdots\cdots \\ \bar{\bar{a}}_{d-1} &= \{\bar{a}_{d-1}, \bar{a}_{d+d-1}, \bar{a}_{2d+d-1}, \ldots, \bar{a}_{(q-1)d+d-1}\}, \end{aligned}$$

and so any element $\bar{\bar{a}}_i$ $(i = 0, \ldots, d-1)$ of $\bar{\bar{Z}}_d$ consists of those elements of $\bar{\mathfrak{Z}}_n$ whose indices are congruent to i modulo d. Let us prove that $\bar{\bar{Z}}_d$ is generating. Consider arbitrary elements $\bar{\bar{a}}_i, \bar{\bar{a}}_j$ of the decomposition $\bar{\bar{Z}}_d$. We shall show that there holds $\bar{\bar{a}}_i \,.\, \bar{\bar{a}}_j \subset \bar{\bar{a}}_k$ where k is the remainder of $i + j$ divided by d. Let $\bar{a}_\alpha$ and $\bar{a}_\beta$ be arbitrary elements of $\bar{\bar{a}}_i$ and $\bar{\bar{a}}_j$, respectively, so that divided by d, α leaves the remainder i and β the remainder j; consequently, $\alpha + \beta$, $i + j$ differ only by an integer multiple of d. In accordance with the definition of the multiplication in $\bar{\mathfrak{Z}}_n$, there holds $\bar{a}_\alpha \circ \bar{a}_\beta = \bar{a}_\gamma$ where γ is the remainder of $\alpha + \beta$ divided by n. Since d is a divisor of n, the numbers $\alpha + \beta$, γ and hence even $i + j$, γ differ by an integer multiple of d; consequently, γ divided by d leaves the remainder k. So we have $\bar{a}_\alpha \circ \bar{a}_\beta = \bar{a}_\gamma \in \bar{\bar{a}}_k$ which yields $\bar{\bar{a}}_i \,.\, \bar{\bar{a}}_j \subset \bar{\bar{a}}_k$. The covering of $\bar{\mathfrak{Z}}_n$, enforced by the factoroid $\bar{\bar{\mathfrak{Z}}}_d$ belonging to the generating decomposition $\bar{\bar{Z}}_d$, consists of d elements

$$\{\ldots, -n + i, -n + d + i, \ldots, -n + (q-1)d + i, i, d + i, \ldots, \\ (q-1)d + i, n + i, n + d + i, \ldots, n + (q-1)d + i, \ldots\},$$

where i denotes one of the numbers $0, \ldots, d-1$.

2. *Local properties of coverings and refinements.* Let $\bar{\mathfrak{A}} \geqq \bar{\mathfrak{B}}$ stand for arbitrary factoroids on $\mathfrak{G}$.

Consider arbitrary elements $\bar{a}_1, \bar{a}_2 \in \overline{\mathfrak{A}}$ and $\bar{b}_1, \bar{b}_2 \in \overline{\mathfrak{B}}$ such that $\bar{a}_1 \supset \bar{b}_1, \bar{a}_2 \supset \bar{b}_2$ and, furthermore, the following decompositions in $\mathfrak{G}$: $\bar{a}_1 \sqcap \overline{\mathfrak{B}}, \bar{a}_2 \sqcap \overline{\mathfrak{B}}$. With regard to the relation $\overline{\mathfrak{A}} \geqq \overline{\mathfrak{B}}$, the mentioned decompositions are complexes in $\overline{\mathfrak{B}}$.

We shall show that *there holds*:

$$\bar{a}_1 \circ \bar{a}_2 \supset \bar{b}_1 \circ \bar{b}_2, \tag{1}$$

$$(\bar{a}_1 \sqcap \overline{\mathfrak{B}}) \circ (\bar{a}_2 \sqcap \overline{\mathfrak{B}}) \subset \bar{a}_1 \circ \bar{a}_2 \sqcap \overline{\mathfrak{B}}. \tag{2}$$

Proof. a) From $\bar{b}_1\bar{b}_2 \subset \bar{b}_1 \circ \bar{b}_2 \cap \bar{a}_1\bar{a}_2 \subset \bar{b}_1 \circ \bar{b}_2 \cap \bar{a}_1 \circ \bar{a}_2$ there follows that the elements $\bar{b}_1 \circ \bar{b}_2 \in \overline{\mathfrak{B}}$, $\bar{a}_1 \circ \bar{a}_2 \in \overline{\mathfrak{A}}$ are incident. Hence, with regard to $\overline{\mathfrak{A}} \geqq \overline{\mathfrak{B}}$ (3.2), we have the formula (1).

b) The product $\bar{x} \circ \bar{y}$ with arbitrary factors $\bar{x} \in \bar{a}_1 \sqcap \overline{\mathfrak{B}}$, $\bar{y} \in \bar{a}_2 \sqcap \overline{\mathfrak{B}}$ is the element $\bar{z} \in \overline{\mathfrak{B}}$ for which $\bar{x}\bar{y} \subset \bar{z}$; $\bar{z}$ is an element of the decomposition $\bar{a}_1 \circ \bar{a}_2 \sqcap \overline{\mathfrak{B}}$ (14.4.1).

We observe, in particular, that if any element $\bar{a} \in \overline{\mathfrak{A}}$ is a groupoidal subset of $\mathfrak{G}$ and so $\bar{a} \circ \bar{a} = \bar{a}$, then the formula (2) yields (for $\bar{a}_1 = \bar{a}_2 = \bar{a}$): $(\bar{a} \sqcap \overline{\mathfrak{B}}) \circ (\bar{a} \sqcap \overline{\mathfrak{B}}) \subset \bar{a} \sqcap \overline{\mathfrak{B}}$. In that case the decomposition $\bar{a} \sqcap \overline{\mathfrak{B}}$ is a groupoidal complex in the factoroid $\overline{\mathfrak{B}}$.

If any element $\bar{a} \in \overline{\mathfrak{A}}$ is a groupoidal subset of $\mathfrak{G}$, then the decomposition $\bar{a} \sqcap \overline{\mathfrak{B}}$ generates, on the corresponding subgroupoid $\bar{\mathfrak{a}} \subset \mathfrak{G}$, the factoroid $\bar{\mathfrak{a}} \sqcap \overline{\mathfrak{B}}$.

In particular, every element $\bar{a} \in \overline{\mathfrak{A}}$ comprising an idempotent point $a \in \bar{a}$ (i.e., such that $aa = a \in \bar{a}$) is a groupoidal subset of $\mathfrak{G}$ (15.6.4).

It is easy to see that, if $a \in \mathfrak{G}$ is idempotent, then the element $\bar{a} \in \overline{\mathfrak{A}}$ containing it is a groupoidal subset of $\mathfrak{G}$ and that the decomposition $\bar{a} \sqcap \overline{\mathfrak{B}}$ generates, on the corresponding subgroupoid $\bar{\mathfrak{a}} \subset \mathfrak{G}$, the factoroid $\bar{\mathfrak{a}} \sqcap \overline{\mathfrak{B}}$.

3. *Common covering and common refinement of two factoroids.* Let $\overline{\mathfrak{A}}, \overline{\mathfrak{B}}$ denote arbitrary factoroids on $\mathfrak{G}$.

A *common covering*, briefly, a *covering of* $\overline{\mathfrak{A}}, \overline{\mathfrak{B}}$ is any factoroid on $\mathfrak{G}$ that is a covering of either of the factoroids $\overline{\mathfrak{A}}, \overline{\mathfrak{B}}$.

Analogously, by a *common refinement*, briefly, a *refinement of the factoroids* $\overline{\mathfrak{A}}, \overline{\mathfrak{B}}$ we mean any factoroid on $\mathfrak{G}$ that is a refinement of either of the factoroids $\overline{\mathfrak{A}}, \overline{\mathfrak{B}}$.

For example, the greatest factoroid $\overline{\mathfrak{G}}_{\max}$ is a common covering and the least factoroid $\overline{\mathfrak{G}}_{\min}$ a common refinement of the factoroids $\overline{\mathfrak{A}}, \overline{\mathfrak{B}}$.

It is obvious that *every covering of any common covering of* $\overline{\mathfrak{A}}, \overline{\mathfrak{B}}$ *is again a covering of* $\overline{\mathfrak{A}}, \overline{\mathfrak{B}}$; analogously, *every refinement of any common refinement of* $\overline{\mathfrak{A}}, \overline{\mathfrak{B}}$ *is again their refinement.*

4. *The least common covering of two factoroids.* From 14.4.2 we know that the least common covering of the fields of $\overline{\mathfrak{A}}, \overline{\mathfrak{B}}$ is a generating decomposition of $\mathfrak{G}$. The factoroid corresponding to the least common covering of the fields of $\overline{\mathfrak{A}}, \overline{\mathfrak{B}}$ is called the *least common covering*, briefly, the *least covering of* $\overline{\mathfrak{A}}, \overline{\mathfrak{B}}$ and is denoted by $[\overline{\mathfrak{A}}, \overline{\mathfrak{B}}]$ or $[\overline{\mathfrak{B}}, \overline{\mathfrak{A}}]$.

From the definition of the factoroid $[\overline{\mathfrak{A}}, \overline{\mathfrak{B}}]$ it follows that its field is a refinement of any common covering of the fields of $\overline{\mathfrak{A}}, \overline{\mathfrak{B}}$ and therefore also of any generating common covering of the fields of $\overline{\mathfrak{A}}, \overline{\mathfrak{B}}$. Hence the factoroid $[\overline{\mathfrak{A}}, \overline{\mathfrak{B}}]$ is the least common covering of $\overline{\mathfrak{A}}, \overline{\mathfrak{B}}$, least in the sense that any common covering of both factoroids is a covering of $[\overline{\mathfrak{A}}, \overline{\mathfrak{B}}]$.

5. *The greatest common refinement of two factoroids.* From 14.4.3 we know that the greatest common refinement of the fields of $\overline{\mathfrak{A}}, \overline{\mathfrak{B}}$ is a generating decomposition of $\mathfrak{G}$. The factoroid corresponding to the greatest common refinement of the fields of $\overline{\mathfrak{A}}, \overline{\mathfrak{B}}$ is called the *greatest common refinement*, briefly, the *greatest refinement of* $\overline{\mathfrak{A}}, \overline{\mathfrak{B}}$ and is denoted by $(\overline{\mathfrak{A}}, \overline{\mathfrak{B}})$ or $(\overline{\mathfrak{B}}, \overline{\mathfrak{A}})$.

From the definition of the factoroid $(\overline{\mathfrak{A}}, \overline{\mathfrak{B}})$ it follows that its field is a covering of any common refinement of the fields of $\overline{\mathfrak{A}}, \overline{\mathfrak{B}}$ and therefore also of any generating common refinement of the fields of $\overline{\mathfrak{A}}, \overline{\mathfrak{B}}$. Hence $(\overline{\mathfrak{A}}, \overline{\mathfrak{B}})$ is the greatest common refinement of $\overline{\mathfrak{A}}, \overline{\mathfrak{B}}$, greatest in the sense that any common refinement of both factoroids is a refinement of $(\overline{\mathfrak{A}}, \overline{\mathfrak{B}})$.

On this occasion, let us note the formula: $(\overline{\mathfrak{A}}, \overline{\mathfrak{B}}) = \overline{\mathfrak{A}} \sqcap \overline{\mathfrak{B}}$ (15.3.2).

6. *Modular factoroids.* Let $\overline{\mathfrak{X}}, \overline{\mathfrak{A}}, \overline{\mathfrak{B}}$ be factoroids on $\mathfrak{G}$ such that $\overline{\mathfrak{X}} \geqq \overline{\mathfrak{A}}$.

The factoroid $\overline{\mathfrak{B}}$ is said to be *modular with regard to* $\overline{\mathfrak{X}}, \overline{\mathfrak{A}}$ (in this order) if there holds:

$$[\overline{\mathfrak{A}}, (\overline{\mathfrak{X}}, \overline{\mathfrak{B}})] = (\overline{\mathfrak{X}}, [\overline{\mathfrak{A}}, \overline{\mathfrak{B}}]).$$

If, for example, $\overline{\mathfrak{X}} = \overline{\mathfrak{A}}$ or $\overline{\mathfrak{A}} = \overline{\mathfrak{G}}_{\max}$, then $\overline{\mathfrak{B}}$ is modular with regard to $\overline{\mathfrak{X}}, \overline{\mathfrak{A}}$.

Let $\overline{\mathfrak{X}}, \overline{\mathfrak{Y}}$ and $\overline{\mathfrak{A}}, \overline{\mathfrak{B}}$ denote arbitrary factoroids on $\mathfrak{G}$ such that $\overline{\mathfrak{X}} \geqq \overline{\mathfrak{A}}, \overline{\mathfrak{Y}} \geqq \overline{\mathfrak{B}}$ and suppose that $\overline{\mathfrak{B}}$ is modular with regard to $\overline{\mathfrak{X}}, \overline{\mathfrak{A}}$ and $\overline{\mathfrak{A}}$ is modular with regard to $\overline{\mathfrak{Y}}, \overline{\mathfrak{B}}$.

Then we have:

$$(\mathring{\mathfrak{A}} =)\ [\overline{\mathfrak{A}}, (\overline{\mathfrak{X}}, \overline{\mathfrak{B}})] = (\overline{\mathfrak{X}}, [\overline{\mathfrak{A}}, \overline{\mathfrak{B}}]),$$

$$(\mathring{\mathfrak{B}} =)\ [\overline{\mathfrak{B}}, (\overline{\mathfrak{Y}}, \overline{\mathfrak{A}})] = (\overline{\mathfrak{Y}}, [\overline{\mathfrak{B}}, \overline{\mathfrak{A}}]),$$

$\mathring{\mathfrak{A}}$ and $\mathring{\mathfrak{B}}$ denoting factoroids defined by the first and the second formula, respectively.

In this situation there hold the interpolation formulae

$$\overline{\mathfrak{X}} \geqq \mathring{\mathfrak{A}} \geqq \overline{\mathfrak{A}}, \quad \overline{\mathfrak{Y}} \geqq \mathring{\mathfrak{B}} \geqq \overline{\mathfrak{B}}$$

and, furthermore, the equalities (4.3)

$$[\mathring{\mathfrak{A}}, \mathring{\mathfrak{B}}] = [\overline{\mathfrak{A}}, \overline{\mathfrak{B}}], \quad [\overline{\mathfrak{X}}, \mathring{\mathfrak{B}}] = [\overline{\mathfrak{X}}, \overline{\mathfrak{B}}], \quad [\overline{\mathfrak{Y}}, \mathring{\mathfrak{A}}] = [\overline{\mathfrak{Y}}, \overline{\mathfrak{A}}], \tag{1}$$

$$(\mathring{\mathfrak{A}}, \mathring{\mathfrak{B}}) = (\overline{\mathfrak{X}}, \mathring{\mathfrak{B}}) = (\overline{\mathfrak{Y}}, \mathring{\mathfrak{A}}) = \big((\overline{\mathfrak{X}}, \overline{\mathfrak{Y}}), [\overline{\mathfrak{A}}, \overline{\mathfrak{B}}]\big). \tag{2}$$

7. *Complementary* (*commuting*) *factoroids.* Let $\overline{\mathfrak{A}}, \overline{\mathfrak{B}}$ stand for arbitrary factoroids on $\mathfrak{G}$. $\overline{\mathfrak{A}}, \overline{\mathfrak{B}}$ are called *complementary* (*commuting*) if their fields are complementary, i.e., if any two elements $\bar{a} \in \overline{\mathfrak{A}}, b \in \overline{\mathfrak{B}}$ lying in the same element $\bar{u} \in [\overline{\mathfrak{A}}, \overline{\mathfrak{B}}]$ are incident.

If, for example, one of the two factoroids is a covering of the other, then both factoroids are complementary.

If there holds, for a certain factoroid $\overline{\mathfrak{X}}$ on $\mathfrak{G}$, the relation $\overline{\mathfrak{X}} \geqq \overline{\mathfrak{A}}$ and $\overline{\mathfrak{A}}, \overline{\mathfrak{B}}$ are complementary, then $\overline{\mathfrak{B}}$ is modular with respect to $\overline{\mathfrak{X}}, \overline{\mathfrak{A}}$ (5.4).

Later (25.3) we shall see that there exist groupoids on which any two factoroids are complementary. Generally, however, two factoroids of a given groupoid are not complementary. For example, on the groupoid whose field consists of four elements a, b, c, d and the multiplication is given by $xy = y$, all the decompositions are generating (14.5.3); factoroids whose fields are, e.g., the two decompositions $\{a, b\}, \{c, d\}$ and $\{a\}, \{b, c, d\}$ are not complementary (5.6.2).

15.5. α-grade groupoidal structures

Let us now proceed to the definition of a more complicated notion based on the concept of an α-grade set structure, which plays an important part in the following considerations.

Let $\alpha(\geqq 1)$ be an arbitrary natural number, $([\mathfrak{A}] =) (\mathfrak{A}_1, \ldots, \mathfrak{A}_\alpha)$ be an α-membered sequence of groupoids and the symbol A_γ denote the field of $\mathfrak{A}_\gamma$, $\gamma = 1, 2, \ldots, \alpha$.

By an *α-grade groupoidal structure,* briefly, a *groupoidal structure* or a *structure with regard to the sequence* $[\mathfrak{A}]$ we mean a groupoid $\tilde{\mathfrak{A}}$ of the following form:

The field of the groupoid $\tilde{\mathfrak{A}}$ is an α-grade set structure with regard to the sequence $(A_1, \ldots, A_\alpha)$; each element

$$\bar{\bar{a}} = (\bar{a}_1, \ldots, \bar{a}_\alpha) \in \tilde{\mathfrak{A}}$$

is, consequently, an α-membered sequence every member $\bar{a}_\gamma$, $(\gamma = 1, \ldots, \alpha)$ of which is a complex in $\mathfrak{A}_\gamma$. The multiplication in $\tilde{\mathfrak{A}}$ is such that for any two elements

$$\bar{\bar{a}} = (\bar{a}_1, \ldots, \bar{a}_\alpha), \quad \bar{\bar{b}} = (\bar{b}_1, \ldots, \bar{b}_\alpha)$$

and their product

$$\bar{\bar{a}}\bar{\bar{b}} = \bar{\bar{c}} = (\bar{c}_1, \ldots, \bar{c}_\alpha) \in \tilde{\mathfrak{A}}$$

there holds

$$\bar{a}_1\bar{b}_1 \subset \bar{c}_1, \ldots, \bar{a}_\alpha\bar{b}_\alpha \subset \bar{c}_\alpha.$$

In what follows we shall be particularly concerned with the case when the groupoids $\mathfrak{A}_1, \ldots, \mathfrak{A}_\alpha$ are factoroids $\overline{\mathfrak{A}}_1, \ldots, \overline{\mathfrak{A}}_\alpha$ on $\mathfrak{G}$. Such α-grade groupoidal structures are, consequently, formed in the following way: Every element $\bar{\bar{a}} = (\bar{a}_1, \ldots, \bar{a}_\alpha) \in \tilde{\mathfrak{A}}$ is an α-membered sequence each member $\bar{a}_\gamma$, $(\gamma = 1, \ldots, \alpha)$ of which is a decomposition in $\mathfrak{G}$ and, in fact, a complex in $\overline{\mathfrak{A}}_\gamma$. The multiplication in $\tilde{\mathfrak{A}}$ is such that, for any two elements

$$\bar{\bar{a}} = (\bar{a}_1, \cdots, \bar{a}_\alpha),\ \bar{\bar{b}} = (\bar{b}_1, \ldots, \bar{b}_\alpha) \in \tilde{\mathfrak{A}}$$

and their product

$$\bar{\bar{a}}\bar{\bar{b}} = \bar{\bar{c}} = (\bar{c}_1, \ldots, \bar{c}_\alpha) \in \tilde{\mathfrak{A}},$$

there holds:

$$\bar{a}_1 \circ \bar{b}_1 \subset \bar{c}_1, \ldots, \bar{a}_\alpha \circ \bar{b}_\alpha \subset \bar{c}_\alpha.$$

15.6. Exercises

1. Show that the groupoids $\mathfrak{Z}_n$, $\overline{\mathfrak{Z}}_n$ $(n \geqq 1)$ are isomorphic.
2. Let $\mathfrak{A}_m$ stand for the subgroupoid of $\mathfrak{Z}$ whose field consists of all the integer multiples of a certain natural number $m > 1$. Of which elements do the factoroids $\mathfrak{A}_m \sqsubset \overline{\mathfrak{Z}}_n$ and $\overline{\mathfrak{Z}}_n \sqcap \mathfrak{A}_m$ $(n > 1)$ consist if m, n are not relatively prime?
3. Every factoroid on an Abelian (associative) groupoid is Abelian (associative).
4. If a groupoid $\mathfrak{G}$ contains an element a such that $aa = a$, i.e., a so-called *idempotent element* (15.4.2), then the element of any factoroid in $\mathfrak{G}$ comprising a is idempotent as well.

16. Deformations of factoroids

16.1. The isomorphism theorems for groupoids

Let us now proceed to the isomorphism theorems for groupoids. These theorems describe situations occurring under homomorphic mappings of groupoids or factoroids and connected with the concept of isomorphism. The set structure of these theorems is expressed by the equivalence theorems dealt with in 6.8.

1. *The first theorem.* Let $\mathfrak{G}$, $\mathfrak{G}^*$ be groupoids and suppose there exists a deformation $\boldsymbol{d}$ of $\mathfrak{G}$ onto $\mathfrak{G}^*$. In 14.2 we have shown that the decomposition $\overline{D}$ of $\mathfrak{G}$ corresponding to d is generating. Let $\overline{\mathfrak{D}}$ stand for the factoroid corresponding to $\overline{D}$. Associating with each element $\bar{a} \in \overline{\mathfrak{D}}$ that element $a^* \in \mathfrak{G}^*$ of whose $\boldsymbol{d}$-inverse

images the element $\bar{a}$ consists, we obtain a simple mapping of $\bar{\mathfrak{D}}$ onto $\mathfrak{G}^*$; let us denote it $\boldsymbol{i}$. By the definition of $\boldsymbol{i}$, there holds $\boldsymbol{i}\bar{a} = \boldsymbol{d}a$ for every $\bar{a} \in \bar{\mathfrak{D}}$ and $a \in \bar{a}$. Let $\bar{a}, \bar{b}$ stand for arbitrary elements of $\bar{\mathfrak{D}}$, a for an element of $\bar{a}$ and b for an element of $\bar{b}$. Then there holds: $ab \subset \bar{a}\bar{b} \subset \bar{a} \circ \bar{b} \in \bar{\mathfrak{D}}$ and hence: $\boldsymbol{i}(\bar{a} \circ \bar{b}) = \boldsymbol{d}ab = \boldsymbol{d}a \,.\, \boldsymbol{d}b = \boldsymbol{i}\bar{a} \,.\, \boldsymbol{i}\bar{b}$. So we have the equality $\boldsymbol{i}(\bar{a} \circ \bar{b}) = \boldsymbol{i}\bar{a} \,.\, \boldsymbol{i}\bar{b}$ by which $\boldsymbol{i}$ is a deformation and therefore (since it is simple) an isomorphism of $\bar{\mathfrak{D}}$ onto $\mathfrak{G}^*$. Thus we have shown that if there exists a deformation $\boldsymbol{d}$ of $\mathfrak{G}$ onto $\mathfrak{G}^*$, then there is on $\mathfrak{G}$ a factoroid isomorphic with $\mathfrak{G}^*$, namely, the factoroid $\bar{\mathfrak{D}}$ corresponding to the generating decomposition belonging to $\boldsymbol{d}$ while the mapping $\boldsymbol{i}$ is an isomorphism. $\bar{\mathfrak{D}}$ is said to *belong* or *correspond* to the deformation $\boldsymbol{d}$.

Let now, conversely, $\bar{\mathfrak{D}}$ be an arbitrary factoroid on $\mathfrak{G}$ and $\boldsymbol{d}$ a mapping of $\mathfrak{G}$ onto $\bar{\mathfrak{D}}$ defined as follows: The $\boldsymbol{d}$-image of any element $a \in \mathfrak{G}$ is that element $\bar{a} \in \bar{\mathfrak{D}}$ for which $a \in \bar{a}$. It is easy to show that $\boldsymbol{d}$ is a deformation of $\mathfrak{G}$ onto $\bar{\mathfrak{D}}$. Let us consider any elements $a, b \in \mathfrak{G}$ and the elements $\bar{a}, \bar{b} \in \bar{\mathfrak{D}}$ containing a, b, hence $\bar{a} = \boldsymbol{d}a, \bar{b} = \boldsymbol{d}b$. The relations $ab \in \bar{a}\bar{b} \subset \bar{a} \circ \bar{b} \in \bar{\mathfrak{D}}$ yield $ab \in \bar{a} \circ \bar{b}$ and, moreover, $\boldsymbol{d}ab = \bar{a} \circ \bar{b} = \boldsymbol{d}a \circ \boldsymbol{d}b$ so that the mapping $\boldsymbol{d}$ actually preserves the multiplications in $\mathfrak{G}$ and $\bar{\mathfrak{D}}$. Consequently, $\mathfrak{G}$ may be deformed onto any factoroid $\bar{\mathfrak{D}}$ lying on $\mathfrak{G}$ in the way that every element of $\mathfrak{G}$ is mapped onto that element of $\bar{\mathfrak{D}}$ in which it is contained. Hence, $\mathfrak{G}$ may be deformed onto any groupoid $\mathfrak{G}^*$ isomorphic with some factoroid on $\mathfrak{G}$.

The above results are briefly summed up in *the first isomorphism theorem for groupoids*:

If a groupoid $\mathfrak{G}^*$ is homomorphic with a groupoid $\mathfrak{G}$, then it is isomorphic with a certain factoroid on $\mathfrak{G}$; if $\mathfrak{G}^*$ is isomorphic with some factoroid on $\mathfrak{G}$, then it is homomorphic with $\mathfrak{G}$. The mapping of the factoroid $\bar{\mathfrak{D}}$, belonging to the deformation $\boldsymbol{d}$ of $\mathfrak{G}$ onto $\mathfrak{G}^*$, under which every element $\bar{a} \in \bar{\mathfrak{D}}$ is mapped onto the $\boldsymbol{d}$-image of the points of $\bar{a}$ is an isomorphism.

2. *The second theorem.* Let $\bar{\mathfrak{A}}, \bar{\mathfrak{B}}$ stand for coupled factoroids in $\mathfrak{G}$.

Each element of $\bar{\mathfrak{A}}$ is incident with exactly one element of $\bar{\mathfrak{B}}$ and, simultaneously, each element of $\bar{\mathfrak{B}}$ is incident with exactly one element of $\mathfrak{A}$ (15.3.3). Associating, with every element $\bar{a} \in \bar{\mathfrak{A}}$, the element $\bar{b} \in \bar{\mathfrak{B}}$ incident with it, we obtain a simple mapping $\boldsymbol{i}$ of $\bar{\mathfrak{A}}$ onto $\bar{\mathfrak{B}}$. We shall show that $\boldsymbol{i}$ is an isomorphism of $\bar{\mathfrak{A}}$ onto $\bar{\mathfrak{B}}$. To that purpose, let us consider arbitrary elements $\bar{a}_1, \bar{a}_2 \in \bar{\mathfrak{A}}$ and the elements $\bar{b}_1, \bar{b}_2 \in \bar{\mathfrak{B}}$ incident with the former so that $\bar{b}_1 = \boldsymbol{i}\bar{a}_1$, $\bar{b}_2 = \boldsymbol{i}\bar{a}_2$. Set $\bar{x}_1 = \bar{a}_1 \cap \bar{b}_1$ $(\neq \emptyset)$, $\bar{x}_2 = \bar{a}_2 \cap \bar{b}_2$ $(\neq \emptyset)$. There obviously holds:

$$\bar{x}_1\bar{x}_2 \subset \bar{a}_1\bar{a}_2 \subset \bar{a}_1 \circ \bar{a}_2, \quad \bar{x}_1\bar{x}_2 \subset \bar{b}_1\bar{b}_2 \subset \bar{b}_1 \circ \bar{b}_2$$

where, of course, $\bar{a}_1 \circ \bar{a}_2 \in \bar{\mathfrak{A}}$ $(\bar{b}_1 \circ \bar{b}_2 \in \bar{\mathfrak{B}})$ is the product of $\bar{a}_1$ $(\bar{b}_1)$ and $\bar{a}_2$ $(\bar{b}_2)$. So we have: $\bar{x}_1\bar{x}_2 \subset \bar{a}_1 \circ \bar{a}_2 \cap \bar{b}_1 \circ \bar{b}_2$ and observe that $\bar{b}_1 \circ \bar{b}_2$ is incident with $\bar{a}_1 \circ \bar{a}_2$. Hence $\bar{b}_1 \circ \bar{b}_2 = \boldsymbol{i}(\bar{a}_1 \circ \bar{a}_2)$ and so $\boldsymbol{i}(\bar{a}_1 \circ \bar{a}_2) = \boldsymbol{i}\bar{a}_1 \circ \boldsymbol{i}\bar{a}_2$, which completes the proof.

The result we have arrived at is summed up in *the second isomorphism theorem for groupoids*:

Every two coupled factoroids $\overline{\mathfrak{A}}, \overline{\mathfrak{B}}$ *in* $\mathfrak{G}$ *are isomorphic,* hence $\overline{\mathfrak{A}} \simeq \overline{\mathfrak{B}}$. *The mapping of* $\overline{\mathfrak{A}}$ *onto* $\overline{\mathfrak{B}}$ *obtained by associating with every element of* $\overline{\mathfrak{A}}$ *the element of* $\overline{\mathfrak{B}}$ *incident with it is an isomorphism.*

A remarkable case (15.3.3) of the second isomorphism theorem concerns the isomorphism of the closure and the intersection of a subgroupoid $\mathfrak{X} \subset \mathfrak{G}$ and a factoroid $\overline{\mathfrak{Y}}$ in $\mathfrak{G}$: *if* $X \cap \boldsymbol{s}\overline{Y} \neq \emptyset$, *there holds*

$$\mathfrak{X} \sqsubset \overline{\mathfrak{Y}} \simeq \overline{\mathfrak{Y}} \sqcap \mathfrak{X}$$

while the isomorphism is realized by the incidence of elements. X and $\overline{Y}$ denote the fields of $\mathfrak{X}$ and $\overline{\mathfrak{Y}}$, respectively.

3. *The third theorem.* Let $\overline{\mathfrak{B}}$ and $\overline{\overline{\mathfrak{B}}}$ denote arbitrary factoroids on $\mathfrak{G}$ and $\overline{\mathfrak{B}}$, respectively. As we know (15.4.1), $\overline{\overline{\mathfrak{B}}}$ enforces a certain covering $\overline{\mathfrak{A}}$ of $\overline{\mathfrak{B}}$. Note that $\overline{\mathfrak{A}}$ is a factoroid on $\mathfrak{G}$ and each of its elements is the sum of all elements of $\overline{\mathfrak{B}}$ comprised in the same element of $\overline{\overline{\mathfrak{B}}}$. Associating, with every element $\bar{\bar{b}} \in \overline{\overline{\mathfrak{B}}}$, that element $\bar{a} \in \overline{\mathfrak{A}}$ which is the sum of all the elements of $\overline{\mathfrak{B}}$ lying in $\bar{\bar{b}}$, we obtain a mapping of $\overline{\overline{\mathfrak{B}}}$ onto $\overline{\mathfrak{A}}$; let us denote it by $\boldsymbol{i}$. We shall show that $\boldsymbol{i}$ is an isomorphism.

First, it is obvious that $\boldsymbol{i}$ is simple. To prove that it is a deformation, consider arbitrary elements $\bar{\bar{b}}_1, \bar{\bar{b}}_2 \in \overline{\overline{\mathfrak{B}}}$ and the product $\bar{\bar{b}}_3 \in \overline{\overline{\mathfrak{B}}}$ of $\bar{\bar{b}}_1$ and $\bar{\bar{b}}_2$. In accordance with the definition of the multiplication in $\overline{\overline{\mathfrak{B}}}$ there holds, for any $\bar{b}_1 \in \overline{\mathfrak{B}}$ of $\bar{\bar{b}}_1$ and any $\bar{b}_2 \in \overline{\mathfrak{B}}$ of $\bar{\bar{b}}_2$, the relation $\bar{b}_1 \circ \bar{b}_2 \in \bar{\bar{b}}_3$. Now let $\bar{a}_1$ be that element of $\overline{\mathfrak{A}}$ which is the sum of all $\bar{b}_1 \in \overline{\mathfrak{B}}$ contained in $\bar{\bar{b}}_1$, hence $\bar{a}_1 = \cup \bar{b}_1$ $(\bar{b}_1 \in \bar{\bar{b}}_1)$ and, analogously, let $\bar{a}_2 = \cup \bar{b}_2$ $(\bar{b}_2 \in \bar{\bar{b}}_2)$, $\bar{a}_3 = \cup \bar{b}_3$ $(\bar{b}_3 \in \bar{\bar{b}}_3)$ so that $\bar{a}_1 = \boldsymbol{i}\bar{\bar{b}}_1$, $\bar{a}_2 = \boldsymbol{i}\bar{\bar{b}}_2$, $\bar{a}_3 = \boldsymbol{i}\bar{\bar{b}}_3 \in \overline{\mathfrak{A}}$. Then the relation $\bar{b}_1 \circ \bar{b}_2 \in \bar{\bar{b}}_3$ $(\bar{b}_1 \in \bar{\bar{b}}_1, \bar{b}_2 \in \bar{\bar{b}}_2)$ yields $\bar{b}_1 \circ \bar{b}_2 \subset \bar{a}_3$ and, furthermore, $\bar{a}_1\bar{a}_2 = \cup \bar{b}_1\bar{b}_2 \subset \cup \bar{b}_1 \circ \bar{b}_2 \subset \bar{a}_3$; hence $\bar{a}_3$ is the element of $\overline{\mathfrak{A}}$ comprising $\bar{a}_1\bar{a}_2$ and we have $\bar{a}_3 = \bar{a}_1 \circ \bar{a}_2$. This equality may be written in the form $\boldsymbol{i}\bar{\bar{b}}_3 = \boldsymbol{i}\bar{\bar{b}}_1 \circ \boldsymbol{i}\bar{\bar{b}}_2$ and expresses that $\boldsymbol{i}$ is a deformation and therefore (since it is simple) an isomorphism. Thus we have arrived at the result summed up in *the third isomorphism theorem for groupoids*:

Any factoroid $\overline{\overline{\mathfrak{B}}}$ *on a factoroid* $\overline{\mathfrak{B}}$ *of* $\mathfrak{G}$ *and the covering* $\overline{\mathfrak{A}}$ *of* $\overline{\mathfrak{B}}$ *enforced by* $\overline{\overline{\mathfrak{B}}}$ *are isomorphic,* i.e., $\overline{\overline{\mathfrak{B}}} \simeq \overline{\mathfrak{A}}$. *The mapping of* $\overline{\overline{\mathfrak{B}}}$ *onto* $\overline{\mathfrak{A}}$ *under which every element* $\bar{\bar{b}} \in \overline{\overline{\mathfrak{B}}}$ *is mapped onto the sum of the elements of* $\overline{\mathfrak{B}}$ *contained in* $\bar{\bar{b}}$ *is an isomorphism.*

16.2. Extended deformations

Let $\boldsymbol{d}$ be a deformation of $\mathfrak{G}$ onto $\mathfrak{G}^*$.

From 16.1.1 we know that $\mathfrak{G}^*$ is isomorphic with the factoroid $\overline{\mathfrak{D}}$ corresponding to $\boldsymbol{d}$, i.e., with the factoroid on $\mathfrak{G}$ whose field is the decomposition $\overline{D}$ corresponding to $\boldsymbol{d}$.

In accordance with 7.1, $\boldsymbol{d}$ determines the extended mapping $\bar{\boldsymbol{d}}$ of the system of all subsets of $\mathfrak{G}$ into the system of all subsets of $\mathfrak{G}^*$; the $\bar{\boldsymbol{d}}$-image of every subset $A \subset \mathfrak{G}$ is the subset $\boldsymbol{d}A \subset \mathfrak{G}^*$ consisting of the $\boldsymbol{d}$-images of the individual elements $a \in A$; moreover, we put $\bar{\boldsymbol{d}}\emptyset = \emptyset$. Sometimes we write $\boldsymbol{d}$ instead of $\bar{\boldsymbol{d}}$, e.g., $\boldsymbol{d}A$ instead of $\bar{\boldsymbol{d}}A$.

Let us now consider an arbitrary factoroid $\bar{\mathfrak{A}}$ on $\mathfrak{G}$. Its field is a certain generating decomposition $\bar{A}$ of $\mathfrak{G}$.

With respect to the theorem in 7.2, $\boldsymbol{d}\bar{A}$ is a decomposition of $\mathfrak{G}^*$ if and only if $\bar{A}, \bar{D}$ are complementary, that is to say, if the factoroids $\bar{\mathfrak{A}}, \bar{\mathfrak{D}}$ are complementary.

Suppose this condition is satisfied.

1. It is easy to show that the *decomposition* $\boldsymbol{d}\bar{A}$ *is generating*. Indeed, let $\bar{a}^*, \bar{b}^* \in \boldsymbol{d}\bar{A}$ be arbitrary elements. Then there exist elements $\bar{a}, \bar{b}, \bar{c} \in \bar{A}$ such that $\boldsymbol{d}\bar{a} = \bar{a}^*$, $\boldsymbol{d}\bar{b} = \bar{b}^*$, $\bar{a}\bar{b} \subset \bar{c}$. By the theorem 13. 3. 2, we have $\boldsymbol{d}\bar{a} \,.\, \boldsymbol{d}\bar{b} \subset \boldsymbol{d}\bar{c}$ and observe that there exists an element $(\boldsymbol{d}\bar{c} =)\ \bar{c}^* \in \boldsymbol{d}\bar{A}$ such that $\bar{a}^*\bar{b}^* \subset \bar{c}^*$. Hence, $\boldsymbol{d}\bar{A}$ is generating.

The factoroid on $\mathfrak{G}^*$ whose field is the decomposition $\boldsymbol{d}\bar{A}$ is called the *image of* $\bar{\mathfrak{A}}$ *under the extended mapping* $\boldsymbol{d}$ and denoted by the symbol $\boldsymbol{d}\bar{\mathfrak{A}}$; $\bar{\mathfrak{A}}$ is called the *inverse image of* $\boldsymbol{d}\bar{\mathfrak{A}}$ *under the extended mapping* $\boldsymbol{d}$.

2. The extended mapping $\boldsymbol{d}$ determines a partial mapping of $\bar{\mathfrak{A}}$ onto $\boldsymbol{d}\bar{\mathfrak{A}}$ under which there corresponds, to every element $\bar{a} \in \bar{\mathfrak{A}}$, its image $\boldsymbol{d}\bar{a} \in \boldsymbol{d}\bar{\mathfrak{A}}$. By the mapping $\boldsymbol{d}$ of $\bar{\mathfrak{A}}$ onto $\boldsymbol{d}\bar{\mathfrak{A}}$ we shall, in what follows, mean this partial mapping.

We shall show that *the mapping* $\boldsymbol{d}$ *of* $\bar{\mathfrak{A}}$ *onto* $\boldsymbol{d}\bar{\mathfrak{A}}$ *is a deformation*. Indeed, from $\bar{a}, \bar{b}, \bar{c} \in \bar{\mathfrak{A}}$, $\bar{a} \circ \bar{b} = \bar{c}$ we have $\bar{a}\bar{b} \subset \bar{c}$ and, moreover, $\boldsymbol{d}\bar{a} \,.\, \boldsymbol{d}\bar{b} \subset \boldsymbol{d}\bar{c}$, hence $\boldsymbol{d}\bar{a} \circ \boldsymbol{d}\bar{b} = \boldsymbol{d}\bar{c} = \boldsymbol{d}(\bar{a} \circ \bar{b})$ and the proof is complete.

With regard to this result, the mapping $\boldsymbol{d}$ of $\bar{\mathfrak{A}}$ onto $\boldsymbol{d}\bar{\mathfrak{A}}$ is called the *extended deformation* $\boldsymbol{d}$.

3. To the extended deformation $\boldsymbol{d}$ of $\bar{\mathfrak{A}}$ onto $\boldsymbol{d}\bar{\mathfrak{A}}$ there corresponds a certain factoroid $\bar{\bar{\mathfrak{A}}}$ on $\bar{\mathfrak{A}}$. Its individual elements consist of all the elements of $\bar{\mathfrak{A}}$ that have the same image in the extended deformation $\boldsymbol{d}$.

In accordance with the theorem in 7.2, we conclude that the covering of $\bar{\mathfrak{A}}$ enforced by $\bar{\bar{\mathfrak{A}}}$ is the least common covering $[\bar{\mathfrak{A}}, \bar{\mathfrak{D}}]$ of $\bar{\mathfrak{A}}, \bar{\mathfrak{D}}$.

Associating with every element $\bar{u} \in [\bar{\mathfrak{A}}, \bar{\mathfrak{D}}]$ the element $\bar{\bar{a}} \in \bar{\bar{\mathfrak{A}}}$ that contains the elements of $\bar{\mathfrak{A}}$ lying in $\bar{u}$, we get an isomorphic mapping of $[\bar{\mathfrak{A}}, \bar{\mathfrak{D}}]$ onto $\bar{\bar{\mathfrak{A}}}$ (16.1.3); associating, on the other hand, with every $\bar{\bar{a}} \in \bar{\bar{\mathfrak{A}}}$ the element $\bar{a}^* \in \boldsymbol{d}\bar{\mathfrak{A}}$ that is the image of every $\bar{a} \in \bar{\mathfrak{A}}$ lying in $\bar{\bar{a}}$, we obtain an isomorphic mapping of $\bar{\bar{\mathfrak{A}}}$ onto $\boldsymbol{d}\bar{\mathfrak{A}}$ (16.1.1). Composing these two mappings, we get an isomorphic mapping of $[\bar{\mathfrak{A}}, \bar{\mathfrak{D}}$ onto $\boldsymbol{d}\bar{\mathfrak{A}}$ (13.4). Under this mapping there corresponds, to every element $\bar{u} \in [\bar{\mathfrak{A}}, \bar{\mathfrak{D}}]$, a certain element $\bar{a}^* \in \boldsymbol{d}\bar{\mathfrak{A}}$ which is the image of $\bar{u}$ under the extended mapping $\boldsymbol{d}$ (7.2).

The result:

If a factoroid $\bar{\mathfrak{A}}$ on $\mathfrak{G}$ is mapped, under the extended deformation $\boldsymbol{d}$, *onto some factoroid $\bar{\mathfrak{A}}^*$ on $\mathfrak{G}^*$, then the factoroids $[\bar{\mathfrak{A}}, \bar{\mathfrak{D}}]$, $\bar{\mathfrak{A}}^*$ are isomorphic. An isomorphic mapping of $[\bar{\mathfrak{A}}, \bar{\mathfrak{D}}]$ onto $\bar{\mathfrak{A}}^*$ is obtained by associating, with every element of $[\bar{\mathfrak{A}}, \bar{\mathfrak{D}}]$, its image under the extended mapping* $\boldsymbol{d}$.

In particular, *every factoroid which is a covering of the factoroid $\bar{\mathfrak{D}}$ is isomorphic with its own image under the extended deformation* $\boldsymbol{d}$; *an isomorphic mapping is obtained by associating, with every element of the covering, its own image under the extended mapping* $\boldsymbol{d}$.

16.3. Deformations of sequences of groupoids and α-grade groupoidal structures

In this chapter we shall be concerned with some more complicated situations in connection with deformations of sequences of groupoids and α-grade groupoidal structures. Our considerations naturally follow from situations treated in 6.9; we only add the algebraic part based on the multiplication.

1. *Mappings of sequences of groupoids.* Let $\alpha (\geqq 1)$ be an arbitrary positive integer.

Consider two α-membered sequences:

$$(\mathfrak{a}) = (\mathfrak{a}_1, \ldots, \mathfrak{a}_\alpha), \quad (\mathfrak{b}) = (\mathfrak{b}_1, \ldots, \mathfrak{b}_\alpha)$$

whose members $\mathfrak{a}_1, \ldots, \mathfrak{a}_\alpha$ and $\mathfrak{b}_1, \ldots, \mathfrak{b}_\alpha$ are groupoids.

a) The sequence $(\mathfrak{b})$ is said to be *isomorphic with* $(\mathfrak{a})$ if the following situation arises: There exists a mapping $\boldsymbol{a}$ of the sequence $(\mathfrak{a})$ onto the sequence $(\mathfrak{b})$ such that to every member $\mathfrak{a}_\gamma$ of $(\mathfrak{a})$ there corresponds an isomorphism $\boldsymbol{i}_\gamma$ of $\mathfrak{a}_\gamma$ onto $\mathfrak{b}_\delta = \boldsymbol{a}\mathfrak{a}_\gamma$ of $(\mathfrak{b})$.

If $(\mathfrak{b})$ is isomorphic with $(\mathfrak{a})$, then obviously $(\mathfrak{a})$ has the same property with respect to $(\mathfrak{b})$. Consequently, we speak about isomorphic sequences $(\mathfrak{a})$, $(\mathfrak{b})$.

b) Let us now assume that the members $\mathfrak{a}_1, \ldots, \mathfrak{a}_\alpha$ of $(\mathfrak{a})$ as well as the members $\mathfrak{b}_1, \ldots, \mathfrak{b}_\alpha$ of $(\mathfrak{b})$ are factoroids in $\mathfrak{G}$.

The sequence $(\mathfrak{b})$ is called *semi-coupled* or *loosely coupled* with the sequence $(\mathfrak{a})$ if the sequence $(b) = (b_1, \ldots, b_\alpha)$ consisting of the fields of the members of the sequence $(\mathfrak{b})$ is semi-coupled with the sequence $(a) = (a_1, \ldots, a_\alpha)$ consisting of the fields of the memberes of the sequence $(\mathfrak{a})$ (6.9.1c); the sequence $(\mathfrak{b})$ is called *coupled* with the sequence $(\mathfrak{a})$ if the sequence $(b) = (b_1, \ldots, b_\alpha)$ is coupled with $(a) = (a_1, \ldots, a_\alpha)$.

If $(\mathfrak{b})$ is loosely coupled (coupled) with $(\mathfrak{a})$, then $(\mathfrak{a})$ has the same property with regard to $(\mathfrak{b})$ and we speak about semi-coupled or loosely coupled (coupled) sequences $(\mathfrak{a})$, $(\mathfrak{b})$.

From the second isomorphism theorem for groupoids (16.1.2) we realize that *v ery two coupled sequences of factoroids in $\mathfrak{G}$ are isomorphic.*

2. *Deformations of α-grade groupoidal structures.* Let $\alpha(\geqq 1)$ be a positive integer and

$$\big((\mathfrak{A}) =\big)\, (\mathfrak{A}_1, \ldots, \mathfrak{A}_\alpha),$$
$$\big((\mathfrak{A}^*) =\big)\, (\mathfrak{A}_1^*, \ldots, \mathfrak{A}_\alpha^*)$$

arbitrary sequences of groupoids. Moreover, let $\bar{\mathfrak{A}}$ and $\bar{\mathfrak{A}}^*$ be arbitrary α-grade groupoidal structures with regard to $(\mathfrak{A})$ and $(\mathfrak{A}^*)$, respectively (15.5).

Note that every element $\bar{\bar{a}} \in \bar{\mathfrak{A}}$ $(\bar{\bar{a}}^* \in \bar{\mathfrak{A}}^*)$ is an α-membered sequence of sets, $\bar{\bar{a}} = (\bar{a}_1, \ldots, \bar{a}_\alpha)$ $\big(\bar{\bar{a}}^* = (\bar{a}_1^*, \ldots, \bar{a}_\alpha^*)\big)$ each member of which $\bar{a}_\gamma$ $(\bar{a}_\gamma^*)$ is a complex in the groupoid $\mathfrak{A}_\gamma$ $(\mathfrak{A}_\gamma^*)$; $\gamma = 1, \ldots, \alpha$.

Suppose there exists an isomorphism $\boldsymbol{i}$ of $\bar{\mathfrak{A}}$ onto $\bar{\mathfrak{A}}^*$. Then, for every two $\bar{\bar{a}} = (\bar{a}_1, \ldots, \bar{a}_\alpha)$, $\bar{\bar{b}} = (b_1, \ldots, b_\alpha) \in \bar{\boldsymbol{A}}$, we have: $\boldsymbol{i}\bar{\bar{a}} \,.\, \boldsymbol{i}\bar{\bar{b}} = \boldsymbol{i}(\bar{\bar{a}} \,.\, \bar{\bar{b}})$.

a) $\boldsymbol{i}$ is said to be a *strong isomorphism of $\bar{\mathfrak{A}}$ onto $\bar{\mathfrak{A}}^*$* if the following situation occurs: There exists a permutation $\boldsymbol{p}$ of the set $\{1, \ldots, \alpha\}$ with the following effect: To every member $\bar{a}_\gamma$ $(\gamma = 1, \ldots, \alpha)$ of an element $\bar{\bar{a}} = (\bar{a}_1, \ldots, \bar{a}_\alpha) \in \bar{\mathfrak{A}}$ there corresponds a simple function $\boldsymbol{a}_\gamma$ under which the set $\bar{a}_\gamma$ is mapped onto the set $\bar{a}_\delta^*$ which is the δ-th member of the element $\boldsymbol{i}\bar{\bar{a}} = \bar{\bar{a}}^* = (\bar{a}_1^*, \ldots, \bar{a}_\alpha^*) \in \bar{\mathfrak{A}}^*$; $\delta = \boldsymbol{p}\gamma$. Moreover, the following "deformation phenomenon" arises: Let

$$\bar{\bar{a}} = (\bar{a}_1, \ldots, \bar{a}_\alpha),\ \bar{\bar{b}} = (\bar{b}_1, \ldots, \bar{b}_\alpha) \in \bar{\mathfrak{A}}$$

be arbitrary elements and

$$\bar{\bar{a}}\bar{\bar{b}} = \bar{\bar{c}} = (\bar{c}_1, \ldots, \bar{c}_\alpha) \in \bar{\mathfrak{A}}$$

the corresponding product; by the definition of $\bar{\mathfrak{A}}$, we have:

$$\bar{a}_\gamma \bar{b}_\gamma \subset \bar{c}_\gamma.$$

Now let

$$\boldsymbol{i}\bar{\bar{a}} = \bar{\bar{a}}^* = (\bar{a}_1^*, \ldots, \bar{a}_\alpha^*), \quad \boldsymbol{i}\bar{\bar{b}} = \bar{\bar{b}}^* = (\bar{b}_1^*, \ldots, \bar{b}_\alpha^*)$$

be the $\boldsymbol{i}$-images of the elements $\bar{\bar{a}}$, $\bar{\bar{b}}$ and

$$\bar{\bar{a}}^* \bar{\bar{b}}^* = \bar{\bar{c}}^* = (\bar{c}_1^*, \ldots, \bar{c}_\alpha^*)$$

the corresponding product so that $\bar{a}_\gamma^* \bar{b}_\gamma^* \subset \bar{c}_\gamma^*$. Finally, let $\boldsymbol{a}_\gamma$, $\boldsymbol{b}_\gamma$, $\boldsymbol{c}_\gamma$ be the mentioned simple functions belonging to the members $\bar{a}_\gamma$, $\bar{b}_\gamma$, $\bar{c}_\gamma$; under these functions the sets $\bar{a}_\gamma$, $\bar{b}_\gamma$ are simply mapped onto $\bar{a}_\delta^*$, $\bar{b}_\delta^*$ and (since $\boldsymbol{i}\bar{\bar{c}} = \bar{\bar{c}}^*$) the set $\bar{c}_\gamma$ onto

$\bar{c}_\delta^*$ ($\delta = \boldsymbol{p}\gamma$). Then the deformation phenomenon can be described as follows: For every two points $a \in \bar{a}_\gamma$, $b \in \bar{b}_\gamma$ there holds: $\boldsymbol{a}_\gamma a \,.\, \boldsymbol{b}_\gamma b = \boldsymbol{c}_\gamma(ab)$.

We easily realize that the inverse mapping $\boldsymbol{i}^{-1}$ is a strong isomorphism of $\tilde{\mathfrak{A}}^*$ onto $\tilde{\mathfrak{A}}$.

If there exists a strong isomorphism of $\tilde{\mathfrak{A}}$ onto $\tilde{\mathfrak{A}}^*$, then $\tilde{\mathfrak{A}}^*$ is said to be *strongly isomorphic with* $\tilde{\mathfrak{A}}$. This notion applies, of course, equally to either $\tilde{\mathfrak{A}}$ and $\tilde{\mathfrak{A}}^*$; therefore we sometimes speak about strongly isomorphic groupoidal structures $\tilde{\mathfrak{A}}$, $\tilde{\mathfrak{A}}^*$.

b) Let us now assume that the sequences of the groupoids $(\mathfrak{A})$ and $(\mathfrak{A}^*)$ consist of factoroids $\bar{\mathfrak{A}}_1, \ldots, \bar{\mathfrak{A}}_\alpha$ and $\bar{\mathfrak{A}}_1^*, \ldots, \bar{\mathfrak{A}}_\alpha^*$ in $\mathfrak{G}$. In that case every element

$$\bar{\bar{a}} = (\bar{a}_1, \ldots, \bar{a}_\alpha) \in \tilde{\mathfrak{A}}$$

$$(\bar{\bar{a}}^* = (\bar{a}_1^*, \ldots, \bar{a}_\alpha^*) \in \tilde{\mathfrak{A}}^*)$$

is an α-membered sequence and each member $\bar{a}_\gamma$ ($\bar{a}_\gamma^*$) is a decomposition in $\mathfrak{G}$ which is a complex in the factoroid $\bar{\mathfrak{A}}_\gamma$ ($\bar{\mathfrak{A}}_\gamma^*$).

The mapping $\boldsymbol{i}$ is said to be an *isomorphism with semi-coupling* or *isomorphism with loose coupling* (*isomorphism with coupling*) of $\tilde{\mathfrak{A}}$ on $\tilde{\mathfrak{A}}^*$ under the following circumstances:

There exists a permutation $\boldsymbol{p}$ of the set $\{1, \ldots, \alpha\}$ with the following effect: Every member $\bar{a}_\gamma$ ($\gamma = 1, \ldots, \alpha$) of an arbitrary element $\bar{\bar{a}} = (\bar{a}_1, \ldots, \bar{a}_\alpha) \in \tilde{\mathfrak{A}}$ and the member $\bar{a}_\delta^*$, $\delta = \boldsymbol{p}\gamma$ of the corresponding element $\boldsymbol{i}\bar{\bar{a}} = \bar{\bar{a}}^* = (\bar{a}_1^*, \ldots, \bar{a}_\alpha^*)$ $\in \tilde{\mathfrak{A}}^*$ are semi-coupled (coupled) decompositions in $\mathfrak{G}$.

It is easy to see that the inverse mapping $\boldsymbol{i}^{-1}$ is an isomorphism of the same kind but in the opposite direction, i.e., of $\tilde{\mathfrak{A}}^*$ onto $\tilde{\mathfrak{A}}$.

Let $\boldsymbol{i}$ be an isomorphism with loose coupling of $\tilde{\mathfrak{A}}$ onto $\tilde{\mathfrak{A}}^*$. Consider arbitrary members $\bar{a}_\gamma$, $\bar{a}_\delta^*$ to which the above relation applies so that $\bar{a}_\gamma$ and $\bar{a}_\delta^*$ are members of $\bar{\bar{a}}$ and $\boldsymbol{i}\bar{\bar{a}} = \bar{\bar{a}}^*$, respectively, $\delta = \boldsymbol{p}\gamma$. In this situation the closures $\mathrm{H}\bar{a}_\gamma = \bar{a}_\delta^* \sqsubset$ $\sqsubset \bar{a}_\gamma$, $\mathrm{H}\bar{a}_\delta^* = \bar{a}_\gamma \sqsubset \bar{a}_\delta^*$ are nonempty and coupled (4.1).

Let $\boldsymbol{a}_\gamma$ denote the mapping of $\mathrm{H}\bar{a}_\gamma$ onto $\mathrm{H}\bar{a}_\delta^*$ defined by incidence of elements. In accordance with the second equivalence theorem (6.8), $\boldsymbol{a}_\gamma$ is simple. We observe that for every element $a \in \mathrm{H}\bar{a}_\gamma$ there holds $\boldsymbol{a}_\gamma a = a^*$ ($\in \mathrm{H}\bar{a}_\delta^*$) if and only if $a \cap a^*$ $\neq \emptyset$.

Let us show that, *for the mappings* $\boldsymbol{a}_\gamma$ *of the closures* $\mathrm{H}a_\gamma$ *corresponding to the individual members* $\bar{a}_\gamma$ ($\gamma = 1, \ldots, \alpha$) *of the elements* $\bar{\bar{a}} \in \tilde{\mathfrak{A}}$, *the above deformation phenomenon arises.*

Indeed, let

$$\bar{\bar{a}} = (\bar{a}_1, \ldots, \bar{a}_\alpha), \quad \bar{\bar{b}} = (\bar{b}_1, \ldots, \bar{b}_\alpha) \in \tilde{\mathfrak{A}}$$

be arbitrary elements and let

$$\bar{\bar{a}}\bar{\bar{b}} = \bar{\bar{c}} = (\bar{c}_1, \ldots, \bar{c}_\alpha) \in \tilde{\mathfrak{A}}$$

be their product; next, let

$$\boldsymbol{i}\bar{\bar{a}} = \bar{\bar{a}}^* = (\bar{a}_1^*, \ldots, \bar{a}_\alpha^*), \quad \boldsymbol{i}\bar{\bar{b}} = \bar{\bar{b}}^* = (\bar{b}_1^*, \ldots, \bar{b}_\alpha^*) \in \tilde{\mathfrak{A}}^*$$

stand for the images of $\bar{\bar{a}}$, $\bar{\bar{b}}$ under the isomorphism $\boldsymbol{i}$ and

$$\bar{\bar{a}}^*\bar{\bar{b}}^* = \bar{\bar{c}}^* = (\bar{c}_1^*, \ldots, \bar{c}_\alpha^*) \in \tilde{\mathfrak{A}}$$

for their product; finally, let $\boldsymbol{a}_\gamma$, $\boldsymbol{b}_\gamma$, $\boldsymbol{c}_\gamma$ denote the corresponding simple mappings of $\mathrm{H}\bar{a}_\gamma$, $\mathrm{H}\bar{b}_\gamma$, $\mathrm{H}\bar{c}_\gamma$, respectively.

Consider any two elements $a \in \mathrm{H}\bar{a}_\gamma$, $b \in \mathrm{H}\bar{b}_\gamma$, their images $\boldsymbol{a}_\gamma a = a^* \in \mathrm{H}\bar{a}_\delta{}^*$, $\boldsymbol{b}_\gamma b = b^* \in \mathrm{H}\bar{b}_\delta{}^*$ and the corresponding products $c = a \circ b \in \bar{c}_\gamma$, $c^* = a^* \circ b^* \in \bar{c}_\delta{}^*$. Then we have $a \cap a^* \neq \emptyset \neq b \cap b^*$ and, furthermore,

$$c = a \circ b \supset ab \supset (a \cap a^*)\,(b \cap b^*),$$

$$c^* = a^* \circ b^* \supset a^*b^* \supset (a^* \cap a)\,(b^* \cap b).$$

We see that c and c^* are incident. So we have: $c \in \mathrm{H}\bar{c}_\gamma$, $c^* \in \mathrm{H}\bar{c}_\delta{}^*$ and, moreover, $c^* = \boldsymbol{c}_\gamma c$. Consequently, $\boldsymbol{a}_\gamma a \circ \boldsymbol{b}_\gamma b = \boldsymbol{c}_\gamma(a \circ b)$, which completes the proof.

If $\boldsymbol{i}$ is, in particular, an isomorphism with coupling, then the considered closures coincide with the corresponding elements. We observe that *every isomorphism with coupling of* $\tilde{\mathfrak{A}}$ *onto* $\tilde{\mathfrak{A}}^*$ *is a strong isomorphism.*

If there exists an isomorphism with semi-coupling (isomorphism with coupling) of $\tilde{\mathfrak{A}}$ onto $\tilde{\mathfrak{A}}^*$, then $\tilde{\mathfrak{A}}^*$ is said to be *isomorphic and semi-coupled* or *isomorphic and loosely coupled* (*isomorphic and coupled*) *with* $\tilde{\mathfrak{A}}$. These notions are symmetric for both $\tilde{\mathfrak{A}}$ and $\tilde{\mathfrak{A}}^*$ and so we sometimes speak about isomorphic and semi-coupled or isomorphic and loosely coupled (isomorphic and coupled) groupoidal structures $\tilde{\mathfrak{A}}$, $\tilde{\mathfrak{A}}^*$. In particular, *every two isomorphic and coupled α-grade groupoidal structures are strongly isomorphic.*

16.4. Exercises

1. Consider the isomorphism theorems in connection with the groupoids $\mathfrak{Z}$, $\mathfrak{A}_m$, $\bar{\mathfrak{Z}}_n$, $\bar{\bar{\mathfrak{Z}}}_d$ dealt with in 15.2, 15.3.2, 15.4.1.
2. Let $\boldsymbol{i}$ be an isomorphism of $\mathfrak{G}$ onto $\mathfrak{G}^*$. The image of any factoroid $\overline{\mathfrak{A}}$ on $\mathfrak{G}$ under the extended mapping $\boldsymbol{i}$ is a factoroid $\boldsymbol{i}\overline{\mathfrak{A}}$ on $\mathfrak{G}^*$ and the partial extended mapping $\boldsymbol{i}$ of $\overline{\mathfrak{A}}$ onto $\boldsymbol{i}\overline{\mathfrak{A}}$ is an isomorphism.
3. Let $\boldsymbol{d}$ be a deformation of $\mathfrak{G}$ onto $\mathfrak{G}^*$. Every factoroid $\overline{\mathfrak{A}}^*$ on $\mathfrak{G}^*$ is the $\boldsymbol{d}$-image of a certain factoroid $\overline{\mathfrak{A}}$ which lies on $\mathfrak{G}$ and is a covering of the factoroid corresponding to $\boldsymbol{d}$.
4. Any two adjoint chains of factoroids in $\mathfrak{G}$ have coupled refinements. (Cf. 15.3.5; 6.10.9.)

17. Series of factoroids

In this chapter we shall develop a theory of the so-called series of factoroids. This theory is based on the properties we have verified for series of decompositions on sets, in chapter 10. But now our deliberations will be extended by including algebraic situations resulting from the multiplication. We shall often come across concepts connected with the properties of α-grade groupoidal structures.

17.1. Basic concepts

Let $\overline{\mathfrak{A}} \geqq \overline{\mathfrak{B}}$ denote arbitrary factoroids on $\mathfrak{G}$.

By a *series of factoroids on* $\mathfrak{G}$ *from* $\overline{\mathfrak{A}}$ *to* $\overline{\mathfrak{B}}$, briefly, a *series from* $\overline{\mathfrak{A}}$ *to* $\overline{\mathfrak{B}}$ we mean a finite α-membered ($\alpha \geqq 1$) sequence of factoroids $\overline{\mathfrak{A}}_1, \ldots, \overline{\mathfrak{A}}_\alpha$ on $\mathfrak{G}$ with the following properties: a) The first factoroid is $\overline{\mathfrak{A}}$, the last $\overline{\mathfrak{B}}$; hence $\overline{\mathfrak{A}}_1 = \overline{\mathfrak{A}}, \overline{\mathfrak{A}}_\alpha = \overline{\mathfrak{B}}$; b) every factoroid is a refinement of the preceding one and so:

$$(\overline{\mathfrak{A}} =)\ \overline{\mathfrak{A}}_1 \geqq \cdots \geqq \overline{\mathfrak{A}}_\alpha\ (= \overline{\mathfrak{B}}).$$

Such a series is briefly denoted by $(\overline{\mathfrak{A}})$. The factoroids $\overline{\mathfrak{A}}_1, \ldots, \overline{\mathfrak{A}}_\alpha$ are called *members of the series* $(\overline{\mathfrak{A}})$. $\overline{\mathfrak{A}}_1$ is the *initial*, $\overline{\mathfrak{A}}_\alpha$ the *final member of* $(\overline{\mathfrak{A}})$. By the *length of* $(\overline{\mathfrak{A}})$ we understand the number α of its members.

For example, the factoroid $\overline{\mathfrak{A}}$ is a series of length 1; the initial and the final member of this series coincide with the factoroid $\overline{\mathfrak{A}}$ itself.

The fields of the individual members of an arbitrary series $(\overline{\mathfrak{A}})$ on $\mathfrak{G}$ form a series of generating decompositions on $\mathfrak{G}$, $(\bar{A})$. The concepts and results valid for the series $(\bar{A})$ may directly be applied to the series $(\overline{\mathfrak{A}})$. In this way we may, for example, define the length of $(\overline{\mathfrak{A}})$ as the length of $(\bar{A})$. Naturally, as regards the theory of series of factoroids, we are particularly interested in situations connected with the multiplication.

The concepts, adopted by the theory of the series of factoroids in this way, will not be explicitely studied here, their meaning is obvious. For example: essential members, reduced length, shortening and lengthening, refinement of $(\overline{\mathfrak{A}})$, as well as the concepts of modular and complementary series of factoroids, etc.

17.2. Local chains

The following considerations are based on the concept of a local chain; it has also been adopted from the theory of series of decompositions (10.2) but will, however, be introduced here owing to its importance.

Let $\big((\overline{\mathfrak{A}}) =\big)\ \overline{\mathfrak{A}}_1 \geqq \cdots \geqq \overline{\mathfrak{A}}_\alpha$ be a series of factoroids on $\mathfrak{G}$, of an arbitrary length $\alpha \geqq 1$.

Let $\bar{a} \in \overline{\mathfrak{A}}_\alpha$ be an arbitrary element and $\bar{a}_\gamma$ denote that element of $\overline{\mathfrak{A}}_\gamma$ for which $\bar{a} \subset \bar{a}_\gamma$ $(\gamma = 1, \ldots, \alpha)$. Then we have:

$$\bar{a}_1 \supset \cdots \supset \bar{a}_\alpha \ (\bar{a}_\alpha = \bar{a}).$$

The intersection

$$\bar{K}_\gamma = \bar{a}_\gamma \sqcap \overline{\mathfrak{A}}_{\gamma+1}$$

coincides with the closure $\bar{a}_\gamma \sqsubset \overline{\mathfrak{A}}_{\gamma+1}$ and is a decomposition of the element $\bar{a}_\gamma$. It is a complex in $\overline{\mathfrak{A}}_{\gamma+1}$ such that $\bar{a}_{\gamma+1} \in \bar{K}_\gamma$ $(\bar{a}_{\alpha+1} = \bar{a}_\alpha)$.

The chain of decompositions in $\mathfrak{G}$ from $\bar{a}_1$ to $\bar{a}_{\alpha+1}$:

$$([\bar{K}] =)\ \bar{K}_1 \to \cdots \to \bar{K}_\alpha$$

is called the *local chain of the series* $(\overline{\mathfrak{A}})$ *corresponding to the base* $\bar{a}$, briefly, the *local chain with the base* $\bar{a}$. Notation as above or more accurately:

$$([\bar{K}\bar{a}] =)\ \bar{K}_1\bar{a} \to \cdots \to \bar{K}_\alpha\bar{a}.$$

In connection with the multiplication in $\mathfrak{G}$ it may happen that the base $\bar{a}$ and therefore even the elements $\bar{a}_\gamma \in \overline{\mathfrak{A}}_\gamma$ $(\gamma = 1, \ldots, \alpha)$ are groupoidal subsets (14.5.1). In that case the decompositions $\bar{K}_\gamma$ are generating (14.4.1). Such a local chain is called *groupoidal*. The factoroids $\bar{\mathfrak{K}}_\gamma$ in $\mathfrak{G}$, belonging to the individual generating decompositions $\bar{K}_\gamma$, form the *local chain of factoroids of the series* $(\overline{\mathfrak{A}})$, *corresponding to the base* $\bar{a}$, briefly, the *local chain of factoroids with the base* $\bar{a}$. Notation: $[\bar{\mathfrak{K}}]$ or $[\bar{\mathfrak{K}}\bar{a}]$.

17.3. The groupoid of local chains

Suppose that $\big((\overline{\mathfrak{A}}) =\big)\ \overline{\mathfrak{A}}_1 \geqq \cdots \geqq \overline{\mathfrak{A}}_\alpha$ $(\alpha \geqq 1)$ is an arbitrary series of factoroids on $\mathfrak{G}$.

To every element $\bar{a} \in \overline{\mathfrak{A}}_\alpha$ there corresponds a local chain $[\bar{K}\bar{a}]$ of the series $(\overline{\mathfrak{A}})$, with the base $\bar{a}$.

The set consisting of the local chains corresponding to the individual elements of the factoroid $\overline{\mathfrak{A}}_\alpha$ forms the *manifold of the local chains*, $\tilde{A}$, *corresponding to the series* $(\overline{\mathfrak{A}})$. It is obviously an α-grade structure with regard to the sequence of factoroids $\overline{\mathfrak{A}}_2, \ldots, \overline{\mathfrak{A}}_{\alpha+1}$ $(\overline{\mathfrak{A}}_{\alpha+1} = \overline{\mathfrak{A}}_\alpha)$.

The multiplication in $\tilde{A}$ may be defined as follows: The product $[\bar{K}\bar{a}]\,[\bar{K}\bar{b}]$ of every two elements $[\bar{K}\bar{a}], [\bar{K}\bar{b}] \in \tilde{A}$ is given by the formula:

$$[\bar{K}\bar{a}]\,[\bar{K}\bar{b}] = [\bar{K}\bar{a} \circ \bar{b}].$$

The manifold $\tilde{A}$ together with this multiplication forms a groupoid $\tilde{\mathfrak{A}}$, called the *groupoid of local chains, corresponding to the series* $(\overline{\mathfrak{A}})$.

Let us, first, show that the groupoid $\tilde{\mathfrak{A}}$ is an α-grade groupoidal structure with regard to the sequence of factoroids $\overline{\mathfrak{A}}_2, \ldots, \overline{\mathfrak{A}}_{\alpha+1}$ $(\overline{\mathfrak{A}}_{\alpha+1} = \overline{\mathfrak{A}}_\alpha)$.

In fact, every element of $\tilde{\mathfrak{A}}$ is an α-membered sequence each element of which, with an arbitrary index γ ($= 1, \ldots, \alpha$), is a decomposition in $\mathfrak{G}$ and is a complex in the factoroid $\overline{\mathfrak{A}}_{\gamma+1}$.

The multiplication in $\tilde{\mathfrak{A}}$ is such that for any two elements:

$$[\bar{K}\bar{a}] = \bar{K}_1\bar{a} \to \cdots \to \bar{K}_\alpha\bar{a}, \quad [\bar{K}\bar{b}] = \bar{K}_1\bar{b} \to \cdots \to \bar{K}_\alpha\bar{b} \in \tilde{\mathfrak{A}}$$

and their product

$$[\bar{K}\bar{a}]\,[\bar{K}\bar{b}] = [\bar{K}\bar{a} \circ \bar{b}] = \bar{K}_1\bar{a} \circ \bar{b} \to \cdots \to \bar{K}_\alpha\bar{a} \circ \bar{b} \in \tilde{\mathfrak{A}},$$

there holds (15.4.2):

$$\bar{K}_1\bar{a} \circ \bar{K}_1\bar{b} \subset \bar{K}_1\bar{a} \circ \bar{b}, \ldots, \bar{K}_\alpha\bar{a} \circ \bar{K}_\alpha\bar{b} \subset \bar{K}_\alpha\bar{a} \circ \bar{b}.$$

Associating, with every point $a \in \mathfrak{G}$, the local chain $[\bar{K}\bar{a}] \in \tilde{\mathfrak{A}}$ with the base $\bar{a} = \bar{a}_\alpha \in \overline{\mathfrak{A}}_\alpha$ containing the point a ($a \in \bar{a}$), we obtain a mapping $\boldsymbol{d}$ of $\mathfrak{G}$ onto the groupoid of local chains $\tilde{\mathfrak{A}}$; $\boldsymbol{d}$ is obviously a deformation. It is called the *natural deformation of* $\mathfrak{G}$ *onto* $\tilde{\mathfrak{A}}$. The factoroid corresponding to the deformation $\boldsymbol{d}$ coincides with the factoroid $\overline{\mathfrak{A}}_\alpha$. By the *local chain of* $(\overline{\mathfrak{A}})$, *corresponding to the point* a, we mean the local chain $[\bar{K}\bar{a}]$.

Now let:

$$(\overline{\mathfrak{A}}) =)\ \overline{\mathfrak{A}}_1 \geqq \cdots \geqq \overline{\mathfrak{A}}_\alpha,$$
$$(\overline{\mathfrak{B}}) =)\ \overline{\mathfrak{B}}_1 \geqq \cdots \geqq \overline{\mathfrak{B}}_\beta$$

be arbitrary series of factoroids on $\mathfrak{G}$ such that their end-members $\overline{\mathfrak{A}}_\alpha$, $\overline{\mathfrak{B}}_\beta$ coincide, hence $\overline{\mathfrak{A}}_\alpha = \overline{\mathfrak{B}}_\beta$.

Consider the groupoids of local chains, $\tilde{\mathfrak{A}}$, $\tilde{\mathfrak{B}}$, corresponding to $(\overline{\mathfrak{A}})$, $(\overline{\mathfrak{B}})$, respectively.

Associating, with every element $[\bar{K}\bar{a}] \in \tilde{\mathfrak{A}}$, the element $[\bar{L}\bar{a}] \in \tilde{\mathfrak{B}}$ with the same base $\bar{a}$, we get a simple mapping of $\tilde{\mathfrak{A}}$ onto $\tilde{\mathfrak{B}}$. This mapping is obviously isomorphic and is called the *co-basal isomorphism.*

We observe that *the groupoids of local chains corresponding to two series of factoroids with coinciding end-members are isomorphic, the deformation being the co-basal isomorphism.*

17.4. Chain-isomorphic series of factoroids

Assume

$$(\overline{\mathfrak{A}}) =)\ \overline{\mathfrak{A}}_1 \geqq \cdots \geqq \overline{\mathfrak{A}}_\alpha,$$
$$(\overline{\mathfrak{B}}) =)\ \overline{\mathfrak{B}}_1 \geqq \cdots \geqq \overline{\mathfrak{B}}_\alpha$$

to be arbitrary series of factoroids on $\mathfrak{G}$ of the same length α ($\geqq 1$).

Let $\bar{\mathfrak{A}}$, $\bar{\mathfrak{B}}$ stand for the groupoids of local chains, corresponding to the above series.

The series $(\overline{\mathfrak{B}})$ is said to be *chain-isomorphic with* $(\overline{\mathfrak{A}})$ if the groupoid $\bar{\mathfrak{B}}$ is strongly isomorphic with $\bar{\mathfrak{A}}$.

If $(\overline{\mathfrak{B}})$ is chain-isomorphic with $(\overline{\mathfrak{A}})$, then $(\overline{\mathfrak{A}})$ has the same property with respect to $(\overline{\mathfrak{B}})$ (16.3.1). Taking account of this symmetry, we sometimes use the term *chain-isomorphic series* $(\overline{\mathfrak{A}})$, $(\overline{\mathfrak{B}})$.

By the above definition, $(\overline{\mathfrak{B}})$ is chain-isomorphic with $(\overline{\mathfrak{A}})$ if there exists a strong isomorphism of $\bar{\mathfrak{A}}$ onto $\bar{\mathfrak{B}}$ (16.3.2). If, in particular, the end-members $\overline{\mathfrak{A}}_\alpha$, $\overline{\mathfrak{B}}_\alpha$ of the series $(\overline{\mathfrak{A}})$, $(\overline{\mathfrak{B}})$, respectively, coincide and the co-basal mapping of $\bar{\mathfrak{A}}$ onto $\bar{\mathfrak{B}}$ is a strong isomorphism, then $(\overline{\mathfrak{B}})$ is said to be *co-basally chain-isomorphic with* $(\overline{\mathfrak{A}})$ and we speak about *co-basally chain-isomorphic series* $(\overline{\mathfrak{A}})$, $(\overline{\mathfrak{B}})$.

Suppose that the series $(\overline{\mathfrak{A}})$, $(\overline{\mathfrak{B}})$ are chain-isomorphic.

This situation can briefly be described as follows:

There exists an isomorphic mapping $\boldsymbol{i}$ of $\bar{\mathfrak{A}}$ onto $\bar{\mathfrak{B}}$ and, moreover, a permutation $\boldsymbol{p}$ of the set $\{1, \ldots, \alpha\}$ with the following effect:

The permutation $\boldsymbol{p}$ determines, for every element $[\bar{K}]$ and its image $\boldsymbol{i}[\bar{K}]$ under the isomorphism $\boldsymbol{i}$, a simple function associating, with every member $\bar{K}_\gamma$ of the local chain $[\bar{K}]$ $(\gamma = 1, \ldots, \alpha)$, a member $\bar{L}_\delta$ of $\boldsymbol{i}[\bar{K}]$ with the index $\delta = \boldsymbol{p}\gamma$. Furthermore, to $\bar{K}_\gamma$ there corresponds a simple mapping $\boldsymbol{a}_\gamma$ of the set $\bar{K}_\gamma$ onto $\bar{L}_\delta$. The simple mappings $\boldsymbol{a}_\gamma$, $\boldsymbol{b}_\gamma$, $\boldsymbol{c}_\gamma$ corresponding to the members $\bar{K}_\gamma\bar{a}$, $\bar{K}_\gamma\bar{b}$ of arbitrary local chains $[\bar{K}\bar{a}]$, $[\bar{K}\bar{b}]$ and to the member $\bar{K}_\gamma\bar{a} \circ \bar{b}$ of the product $[\bar{K}\bar{a}]\,[\bar{K}\bar{b}] = [\bar{K}\bar{a} \circ \bar{b}]$ are of homomorphic character, i.e., for any elements $a \in \bar{K}_\gamma\bar{a}$, $b \in \bar{K}_\gamma\bar{b}$ there holds:

$$\boldsymbol{c}_\gamma(a \circ b) = (\boldsymbol{a}_\gamma a) \circ (\boldsymbol{b}_\gamma b).$$

It is obvious that $(\overline{\mathfrak{A}})$, $(\overline{\mathfrak{B}})$ are chain-equivalent so that our considerations concerning chain-equivalent series of decompositions of sets (10.5) may be applied to them. We observe, moreover, that $(\overline{\mathfrak{A}})$ *and* $(\overline{\mathfrak{B}})$ *are of the same reduced length.*

17.5. Semi-joint and joint series of factoroids

Considerations similar to those by which we have arrived at the notion of chain-isomorphic series of factoroids lead to semi-joint and joint series of factoroids.

Let us employ the same notation as above.

The series $(\overline{\mathfrak{B}})$ is said to be *semi-joint* or *loosely joint* (*joint*) with the series $(\overline{\mathfrak{A}})$ if the groupoid $\bar{\mathfrak{B}}$ is isomorphic and semi-coupled (isomorphic and coupled) with the groupoid $\bar{\mathfrak{A}}$.

If $(\overline{\mathfrak{B}})$ is loosely joint (joint) with $(\overline{\mathfrak{A}})$, then $(\overline{\mathfrak{A}})$ has the same property with regard to $(\overline{\mathfrak{B}})$. Accordingly, we also use the expression *semi-joint* or *loosely joint* (*joint*) *series* $(\overline{\mathfrak{A}})$, $(\overline{\mathfrak{B}})$.

By the above definition, $(\overline{\mathfrak{B}})$ is semi-joint (joint) with $(\overline{\mathfrak{A}})$ if there exists an isomorphism with loose coupling (an isomorphism with coupling) of $\tilde{\mathfrak{A}}$ onto $\tilde{B}$ (16.3.2). If, in particular, the end-members $\overline{\mathfrak{A}}_\alpha$, $\overline{\mathfrak{B}}_\alpha$ of $(\overline{\mathfrak{A}})$ and $(\overline{\mathfrak{B}})$, respectively, coincide and the co-basal mapping of $\tilde{\mathfrak{A}}$ onto $\tilde{\mathfrak{B}}$ is an isomorphism with loose coupling (isomorphism with coupling), then $(\overline{\mathfrak{B}})$ is said to be *co-basally semi-joint* or *co-basally loosely joint* (*co-basally joint*) *with* $(\overline{\mathfrak{A}})$; in that case we also speak about *co-basally semi-joint* or *co-basally loosely joint* (*co-basally joint*) *series* $(\overline{\mathfrak{A}})$, $(\overline{\mathfrak{B}})$.

This situation can briefly be described as follows:

There exists an isomorphic mapping $\boldsymbol{i}$ of $\tilde{\mathfrak{A}}$ onto $\tilde{\mathfrak{B}}$ and, moreover, a permutation $\boldsymbol{p}$ of the set $\{1, \ldots, \alpha\}$ with the following effect:

The permutation $\boldsymbol{p}$ determines, for every element $[\overline{K}] \in \tilde{\mathfrak{A}}$ and its image $\boldsymbol{i}[\overline{K}] \in \tilde{\mathfrak{B}}$ under the isomorphism $\boldsymbol{i}$, a simple function associating, with every member $\overline{K}_\gamma$ of the local chain $[K]$ ($\gamma = 1, \ldots, \alpha$), a member $\overline{L}_\delta$ of the local chain $\boldsymbol{i}[\overline{K}]$, while $\delta = \boldsymbol{p}\gamma$. Furthermore, to the closure $\mathrm{H}\overline{K}_\gamma = \overline{L}_\delta \sqsubset \overline{K}_\gamma$ there belongs a simple mapping $\boldsymbol{a}_\gamma$, given by the incidence of elements, which maps the closure $\mathrm{H}\overline{K}_\gamma$ onto $\mathrm{H}\overline{L}_\delta = \overline{K}_\gamma \sqsubset \overline{L}_\delta$. The mappings $\boldsymbol{a}_\gamma$, $\boldsymbol{b}_\gamma$ which belong to the closures $\mathrm{H}\overline{K}_\gamma\bar{a}$, $\mathrm{H}\overline{K}_\gamma\bar{b}$ corresponding to arbitrary local chains $[\overline{K}\bar{a}]$, $[\overline{K}\bar{b}] \in \tilde{\mathfrak{A}}$ and the mapping $\boldsymbol{c}_\gamma$ which belongs to the closure $\mathrm{H}\overline{K}_\gamma\bar{a} \circ \bar{b}$ corresponding to the product $[\overline{K}\bar{a}]\,[\overline{K}\bar{b}] = [\overline{K}\bar{a} \circ \bar{b}] \in \tilde{\mathfrak{A}}$ are of homomorphic character, i.e., for arbitrary elements $a \in \mathrm{H}\overline{K}_\gamma\bar{a}$, $b \in \mathrm{H}\overline{K}_\gamma\bar{b}$ there holds $\boldsymbol{c}_\gamma(a \circ b) = (\boldsymbol{a}_\gamma a) \circ (\boldsymbol{b}_\gamma b)$.

If, in particular, $(\overline{\mathfrak{A}})$ *and* $(\overline{\mathfrak{B}})$ *are joint, then they are chain-isomorphic and therefore of the same reduced length* (17.4).

17.6. Modular and complementary series of factoroids

Let

$$((\overline{\mathfrak{A}}) =)\ \overline{\mathfrak{A}}_1 \geqq \cdots \geqq \overline{\mathfrak{A}}_\alpha,$$

$$((\overline{\mathfrak{B}}) =)\ \overline{\mathfrak{B}}_1 \geqq \cdots \geqq \overline{\mathfrak{B}}_\beta$$

be modular series of factoroids on $\mathfrak{G}$ of lengths α, β ($\geqq 1$), respectively.

There holds the following theorem:

The series $(\overline{\mathfrak{A}})$, $(\overline{\mathfrak{B}})$ *have co-basally loosely joint refinements* $(\mathring{\mathfrak{A}})$, $(\mathring{\mathfrak{B}})$, *respectively, with the same initial and final members.*

Denote

$$[\overline{\mathfrak{A}}_1, \overline{\mathfrak{B}}_1] = \overline{\mathfrak{U}}, \quad (\overline{\mathfrak{A}}_\alpha, \overline{\mathfrak{B}}_\beta) = \overline{\mathfrak{V}},$$

$$\overline{\mathfrak{A}}_0 = \overline{\mathfrak{B}}_0 = \overline{\mathfrak{G}}_{\max}; \quad \overline{\mathfrak{A}}_{\alpha+1} = \overline{\mathfrak{B}}_{\beta+1} = \overline{\mathfrak{V}}$$

and, moreover, for $\gamma, \mu = 1, \ldots, \alpha + 1$; $\delta, \nu = 1, \ldots, \beta + 1$,

$$\mathring{\mathfrak{A}}_{\gamma,\nu} = [\overline{\mathfrak{A}}_\gamma, (\overline{\mathfrak{A}}_{\gamma-1}, \overline{\mathfrak{B}}_\nu)] = (\overline{\mathfrak{A}}_{\gamma-1}, [\overline{\mathfrak{A}}_\gamma, \overline{\mathfrak{B}}_\nu]),$$
$$\mathring{\mathfrak{B}}_{\delta,\mu} = [\overline{\mathfrak{B}}_\delta, (\overline{\mathfrak{B}}_{\delta-1}, \overline{\mathfrak{A}}_\mu)] = (\overline{\mathfrak{B}}_{\delta-1}, [\overline{\mathfrak{B}}_\delta, \overline{\mathfrak{A}}_\mu]).$$

Then the above co-basally loosely joint refinements of $(\overline{\mathfrak{A}})$, $(\overline{\mathfrak{B}})$ *are expressed by the following formulae*:

$$((\mathring{\mathfrak{A}}) =)\ \overline{\mathfrak{U}} = \mathring{\mathfrak{A}}_{1,1} \geqq \cdots \geqq \mathring{\mathfrak{A}}_{1,\beta+1} \geqq \mathring{\mathfrak{A}}_{2,1} \geqq \cdots \geqq \mathring{\mathfrak{A}}_{2,\beta+1} \geqq \cdots$$
$$\geqq \mathring{\mathfrak{A}}_{\alpha+1,1} \geqq \cdots \geqq \mathring{\mathfrak{A}}_{\alpha+1,\beta+1} = \overline{\mathfrak{B}},$$
$$((\mathring{\mathfrak{B}}) =)\ \overline{\mathfrak{U}} = \mathring{\mathfrak{B}}_{1,1} \geqq \cdots \geqq \mathring{\mathfrak{B}}_{1,\alpha+1} \geqq \mathring{\mathfrak{B}}_{2,1} \geqq \cdots \geqq \mathring{\mathfrak{B}}_{2,\alpha+1} \geqq \cdots$$
$$\geqq \mathring{\mathfrak{B}}_{\beta+1,1} \geqq \cdots \geqq \mathring{\mathfrak{B}}_{\beta+1,\alpha+1} = \overline{\mathfrak{B}}$$

If $(\overline{\mathfrak{A}})$, $(\overline{\mathfrak{B}})$ *are complementary, then the refinements* $(\mathring{\mathfrak{A}})$, $(\mathring{\mathfrak{B}})$ *are co-basally joint.*

The correctness of this theorem follows from 10.7, 10.8.

17.7. Exercises

1. If any two factoroids lying on $\mathfrak{G}$ are complementary, then any two series of factoroids on $\mathfrak{G}$ have co-basally joint refinements.

18. Remarkable kinds of groupoids

The study of some remarkable kinds of groupoids closely ties up with our considerations in chapter 11.2. We have not dealt with them before because we wish to emphasize that the preceding deliberations apply to all groupoids regardless of any particular properties. Now we shall be concerned with the groupoids that are of most importance to our theory, namely, the associative groupoids, the groupoids with uniquely defined division and the groupoids with a unit element.

Moreover, we shall pay a brief attention to the Brandt groupoids though they do not belong exactly within the range of our study.

18.1. Associative groupoids (semigroups)

1. *Definition.* The concept of an associative groupoid $\mathfrak{G}$ has already been determined in 12. 7. 2 by the property that *any three-membered sequence of elements of* $\mathfrak{G}$ *has only one product element*; that is to say, for any three elements $a_1, a_2, a_3 \in \mathfrak{G}$ there holds $a_1(a_2a_3) = (a_1a_2)a_3$. Associative groupoids are also called *semigroups.*

2. *The fundamental theorem of semigroups.* Now we shall show that *any associative groupoid* $\mathfrak{G}$ *has the property that every n-membered* ($n \geqq 2$) *sequence of elements of* $\mathfrak{G}$ *has only one product element*, i.e., the symbol $a_1 \dots a_n$ denotes, for $a_1, \dots, a_n \in \mathfrak{G}$ ($n \geqq 2$), exactly one element of $\mathfrak{G}$.

Let us consider an arbitrary associative groupoid and proceed by the method of complete induction. Our statement is correct if $n = 2$, for, in that case, it immediately follows from the definition of the multiplication in $\mathfrak{G}$. It remains to be shown that: if our statement applies to every, at most, $(n - 1)$-membered sequence of elements of $\mathfrak{G}$, n being a positive integer > 2, then it also holds for every n-membered sequence of elements of $\mathfrak{G}$.

Let $n > 2$. Suppose our statement holds for every, at most, $(n - 1)$-membered sequence of elements of $\mathfrak{G}$. Consider n arbitrary elements $a_1, \dots, a_n \in \mathfrak{G}$.

Then every symbol

$$a_1, a_2 \dots a_n, a_1a_2, a_3 \dots a_n, \dots, a_1 \dots a_{n-1}, a_n$$

denotes exactly one element of $\mathfrak{G}$ because, by our assumption, there exists, for example, only one product element $a_2 \dots a_n$ of the $(n - 1)$-membered sequence $a_2, \dots, a_n \in \mathfrak{G}$. Our object now is to show that all the elements

$$a_1(a_2 \dots a_n), (a_1a_2)(a_3 \dots a_n), \dots, (a_1 \dots a_{n-1})a_n \tag{1}$$

are equal. To that end let us, first, note that each of these elements is the product $(a_1 \dots a_k)(a_{k+1} \dots a_n)$ of the elements $a_1 \dots a_k, a_{k+1} \dots a_n \in \mathfrak{G}$, k being one of the numbers $1, \dots, n - 1$. In order to prove our statement we must verify that each of the elements (1) is equal to, e.g., $a_1(a_2 \dots a_n)$; that is to say, for every $k = 1, \dots, n - 1$ there holds

$$(a_1 \dots a_k)(a_{k+1} \dots a_n) = a_1(a_2 \dots a_n). \tag{2}$$

If $k = 1$, then this equality is obvious, hence we may restrict our attention to the case $k > 1$. In that case, $a_1 \dots a_k$ is the product element of an, at least, 2-membered and, at most, $(n - 1)$-membered sequence of elements $a_1, \dots, a_k$ and is therefore, by our assumption, equal to the element $a_1(a_2 \dots a_k)$; consequently, we have

$$(a_1 \dots a_k)(a_{k+1} \dots a_n) = \big(a_1(a_2 \dots a_k)\big)(a_{k+1} \dots a_n).$$

Since $\mathfrak{G}$ is associative, the element on the right-hand side of this equality is equal to the element $a_1((a_2 \dots a_k)(a_{k+1} \dots a_n))$, i.e., the element $a_1(a_2 \dots a_n)$ and we have (2), which completes the proof.

A similar result applies, of course, to finite sequences of the subsets of $\mathfrak{G}$.

3. *Effects of the fundamental theorem.* a) The uniqueness of a composite permutation. The result we have just arrived at is useful when we are to compose permutations of a (finite or infinite) set of elements. Let $\boldsymbol{p}_1, \dots, \boldsymbol{p}_n$ $(n \geqq 2)$ denote arbitrary permutations of a set H. What do we understand by a permutation composed of the permutations $\boldsymbol{p}_1, \dots, \boldsymbol{p}_n$ (in this order)? If $n = 2$, then it is, as we know, the composite mapping $\boldsymbol{p}_2\boldsymbol{p}_1$. If $n = 3$, then the concept of a composite permutation of $\boldsymbol{p}_1, \boldsymbol{p}_2, \boldsymbol{p}_3$ is defined as follows: By a permutation composed of $\boldsymbol{p}_1, \boldsymbol{p}_2, \boldsymbol{p}_3$ we mean either of the permutations $\boldsymbol{p}_3(\boldsymbol{p}_2\boldsymbol{p}_1)$, $(\boldsymbol{p}_3\boldsymbol{p}_2)\boldsymbol{p}_1$; notation $\boldsymbol{p}_3\boldsymbol{p}_2\boldsymbol{p}_1$. The symbol $\boldsymbol{p}_3\boldsymbol{p}_2\boldsymbol{p}_1$ therefore stands for the permutation composed of $\boldsymbol{p}_2\boldsymbol{p}_1$, $\boldsymbol{p}_3$ as well as for the permutation composed of $\boldsymbol{p}_1, \boldsymbol{p}_3\boldsymbol{p}_2$. If $n = 4$, then a permutation composed of $\boldsymbol{p}_1, \boldsymbol{p}_2, \boldsymbol{p}_3, \boldsymbol{p}_4$ is any of the permutations $\boldsymbol{p}_4(\boldsymbol{p}_3\boldsymbol{p}_2\boldsymbol{p}_1)$, $(\boldsymbol{p}_4\boldsymbol{p}_3)(\boldsymbol{p}_2\boldsymbol{p}_1)$, $(\boldsymbol{p}_4\boldsymbol{p}_3\boldsymbol{p}_2)\boldsymbol{p}_1$; it is denoted by the symbol $\boldsymbol{p}_4\boldsymbol{p}_3\boldsymbol{p}_2\boldsymbol{p}_1$ which stands for any of the following permutations of H: $\boldsymbol{p}_4(\boldsymbol{p}_3(\boldsymbol{p}_2\boldsymbol{p}_1))$, $\boldsymbol{p}_4((\boldsymbol{p}_3\boldsymbol{p}_2)\boldsymbol{p}_1)$, $(\boldsymbol{p}_4\boldsymbol{p}_3)(\boldsymbol{p}_2\boldsymbol{p}_1)$, $(\boldsymbol{p}_4(\boldsymbol{p}_3\boldsymbol{p}_2))\boldsymbol{p}_1$, $((\boldsymbol{p}_4\boldsymbol{p}_3)\boldsymbol{p}_2)\boldsymbol{p}_1$.

Generally, for $n \geqq 2$, a permutation composed of $\boldsymbol{p}_1, \dots, \boldsymbol{p}_n$ is defined as follows: It is any of the permutations

$$\boldsymbol{p}_n(\boldsymbol{p}_{n-1} \cdots \boldsymbol{p}_1),\ (\boldsymbol{p}_n\boldsymbol{p}_{n-1})(\boldsymbol{p}_{n-2} \cdots \boldsymbol{p}_1),\ \dots,\ (\boldsymbol{p}_n \cdots \boldsymbol{p}_2)\boldsymbol{p}_1,$$

where each symbol in parentheses stands for an arbitrary permutation composed of the involved permutations and ordered from right to left. A permutation composed of $\boldsymbol{p}_1, \dots, \boldsymbol{p}_n$ is denoted by $\boldsymbol{p}_n \dots \boldsymbol{p}_1$. The symbol $\boldsymbol{p}_n \dots \boldsymbol{p}_1$ therefore denotes, in accordance with the definition, a product element of the n-membered sequence of permutations $\boldsymbol{p}_1, \dots, \boldsymbol{p}_n$; the latter are elements of the groupoid consisting of all permutations of H, the multiplication being defined by the composition of permutations. By the associative law of composing permutations (8.7.3), the groupoid in question is associative and, according to the above result, *there exists only one permutation $\boldsymbol{p}_n \cdots \boldsymbol{p}_1$ composed of $\boldsymbol{p}_1, \dots, \boldsymbol{p}_n$*. This theorem may also be expressed in terms that, if the order of composing the permutations is the same, then the composite permutation does not depend on the way of composition. Consequently, we obtain the image $\boldsymbol{p}_n \dots \boldsymbol{p}_1 x$ of any element $x \in H$ by, e.g., the formula

$$\boldsymbol{p}_n \cdots \boldsymbol{p}_1 x = \boldsymbol{p}_n(\boldsymbol{p}_{n-1}(\cdots(\boldsymbol{p}_2(\boldsymbol{p}_1 x)) \dots)),$$

namely, by determining first the $\boldsymbol{p}_1$-image $\boldsymbol{p}_1 x$ of the element x, then the $\boldsymbol{p}_2$-image $\boldsymbol{p}_2(\boldsymbol{p}_1 x)$ of the element $\boldsymbol{p}_1 x$, etc. and, finally, the $\boldsymbol{p}_n$-image $\boldsymbol{p}_n(\boldsymbol{p}_{n-1}(\cdots(\boldsymbol{p}_2(\boldsymbol{p}_1 x))\dots))$ of the element $\boldsymbol{p}_{n-1}(\dots(\boldsymbol{p}_2(\boldsymbol{p}_1 x))\dots)$. From this it is immediately clear that if the permutations $\boldsymbol{p}_1, \dots, \boldsymbol{p}_n$ leave some element $x \in H$ invariant, then the same holds for the composite permutation $\boldsymbol{p}_n \cdots \boldsymbol{p}_1$.

b) The composition of a permutation of cyclic permutations. Let us make use of the above results to make a few remarks about permutations of a finite set. Suppose the set H is finite.

First we shall show that *any permutation of H is composed of a finite number of cyclic permutations whose cycles have no common elements.*

Consider a permutation $\boldsymbol{p}$ of the set H. As we know from 8.5, $\boldsymbol{p}$ is determined by a finite number of pure cyclic permutations $\boldsymbol{p}_{\bar{a}}, \ldots, \boldsymbol{p}_{\bar{m}}$, i.e., there exists a decomposition $\bar{H} = \{\bar{a}, \ldots, \bar{m}\}$ of the set H such that each of its elements $\bar{a}, \ldots, \bar{m}$ is invariant under $\boldsymbol{p}$ and the partial permutations $\boldsymbol{p}_{\bar{a}}, \ldots, \boldsymbol{p}_{\bar{m}}$ are pure cyclic permutations of the elements $\bar{a}, \ldots, \bar{m}$. Let $\bar{x}$ be an arbitrary element of $\bar{H}$ and $\boldsymbol{q}_{\bar{x}}$ the cyclic permutation of H that maps every element $x \in \bar{x}$ onto $\boldsymbol{p}_{\bar{x}}x$ and leaves all the other elements of H, if there are any, invariant. The cyclic permutation $\boldsymbol{q}_{\bar{x}}$ has therefore the same cycle as the pure cyclic permutation $\boldsymbol{p}_{\bar{x}}$ and so both $\boldsymbol{q}_{\bar{x}}$ and $\boldsymbol{p}_{\bar{x}}$ may be expressed by the same simplified symbol. To prove our statement we shall verify that the permutation $\boldsymbol{p}$ is composed of the cyclic permutations $\boldsymbol{q}_{\bar{a}}, \ldots, \boldsymbol{q}_{\bar{m}}$, i.e., $\boldsymbol{p} = \boldsymbol{q}_{\bar{m}} \ldots \boldsymbol{q}_{\bar{a}}$.

Let x denote an arbitrary element of H and $\bar{x}$ the element of $\bar{H}$ containing x so that the permutation $\boldsymbol{q}_{\bar{x}}$ maps x onto the element $\boldsymbol{q}_{\bar{x}}x$ but all the other permutations $\boldsymbol{q}_{\bar{a}}, \ldots, \boldsymbol{q}_{\bar{m}}$, if there are any, leave the element x invariant. Since the composite permutation does not depend, for the same ordering, on the way of composition, we have $\boldsymbol{q}_{\bar{m}} \ldots \boldsymbol{q}_{\bar{a}}x = (\boldsymbol{q}_{\bar{m}} \ldots)\boldsymbol{q}_{\bar{x}}(\ldots \boldsymbol{q}_{\bar{a}})x$; then of course, for $\bar{x} = \bar{m}$ and $\bar{x} = \bar{a}$, the symbols of the composite permutation, written in the first and the second parentheses, respectively, are left out. For $\bar{x} \neq \bar{a}$ we have $(\ldots \boldsymbol{q}_{\bar{a}})x = x$, since all the permutations of which $(\ldots \boldsymbol{q}_{\bar{a}})$ is composed leave the element x invariant. So we have, first, $\boldsymbol{q}_{\bar{m}} \ldots \boldsymbol{q}_{\bar{a}}x = (\boldsymbol{q}_{\bar{m}} \ldots)\boldsymbol{q}_{\bar{x}}x$. In a similar way we realize that the element on the right-hand side of this equation is $\boldsymbol{q}_{\bar{x}}x$, and so $\boldsymbol{q}_{\bar{m}} \ldots \boldsymbol{q}_{\bar{a}}x = \boldsymbol{q}_{\bar{x}}x$. By the definition of $\boldsymbol{q}_{\bar{x}}$ there holds $\boldsymbol{q}_{\bar{x}}x = \boldsymbol{p}_{\bar{x}}x$ and, furthermore, according to the definition of $\boldsymbol{p}_{\bar{x}}$ there is $\boldsymbol{p}_{\bar{x}}x = \boldsymbol{p}x$. So we have $\boldsymbol{q}_{\bar{m}} \ldots \boldsymbol{q}_{\bar{a}}x = \boldsymbol{p}x$ and the proof is complete.

Note that in the formula $\boldsymbol{p} = \boldsymbol{q}_{\bar{m}} \ldots \boldsymbol{q}_{\bar{a}}$ the order of the permutations $\boldsymbol{q}_{\bar{a}}, \ldots, \boldsymbol{q}_{\bar{m}}$ may be arbitrarily changed because, for every arrangement of $\boldsymbol{q}_{\bar{a}}, \ldots, \boldsymbol{q}_{\bar{m}}$, we may choose such a notation of the elements of $\bar{H}$ that the formula remains the same.

If we have any permutations $\boldsymbol{p}_1, \ldots, \boldsymbol{p}_n$ $(n \geqq 2)$ of H expressed by two-rowed or simplified symbols, then the composite permutation $\boldsymbol{p}_n \ldots \boldsymbol{p}_1$ is expressed by writing the symbols of the permutations $\boldsymbol{p}_1, \ldots, \boldsymbol{p}_n$ next to each other and in the inverse order. With regard to this and the way of expressing any permutations by pure cyclic permutations (8.6), we may understand, for example, the formula

$$\begin{pmatrix} a & b & c & d \\ d & c & b & a \end{pmatrix} = (a, d)\,(b, c)$$

either in the sense that the permutation of the set $\{a, b, c, d\}$ expressed by the symbol on the left-hand side is composed of cyclic permutations (b, c), (a, d) or in the sense that it is determined by pure cyclic permutations (a, d), (b, c).

4. *Weakly associative groupoids.* V. Devidé has generalized the concept of an associative groupoid as follows: The groupoid $\mathfrak{G}$ is called *weakly associative* if there exist simple mappings $\boldsymbol{f}, \boldsymbol{g}, \boldsymbol{h}$ of $\mathfrak{G}$ onto itself such that, for arbitrary elements $a, b, c \in \mathfrak{G}$, there holds: $(ab)c = \boldsymbol{f}a(\boldsymbol{g}b \,.\, \boldsymbol{h}c)$. Weakly associative groupoids may also be denoted as *weak semigroups*. It is obvious that if every mapping $\boldsymbol{f}, \boldsymbol{g}, \boldsymbol{h}$ is the identical mapping, then the concept of a weakly associative groupoid coincides with the concept of an associative groupoid.

Example. Suppose the field of $\mathfrak{G}$ is the set of all rational, real or complex numbers different from zero and let the multiplication in $\mathfrak{G}$ be defined by the division. Employ the symbol $\circ$ for the multiplication of numbers. Then, for $a, b, c \in \mathfrak{G}$, we have

$$(ab)c = \frac{a/b}{c} = \frac{a}{b \circ c} = a\Big/\left(b\Big/\frac{1}{c}\right) = a\left(b\,\frac{1}{c}\right).$$

We observe that the simple mappings of $\mathfrak{G}$ onto itself $\boldsymbol{f} = \boldsymbol{g} = \boldsymbol{e}$ (identical mapping) and the mapping $\boldsymbol{h}$ defined by the formula $\boldsymbol{h}c = 1/c$ satisfy the above condition.

18.2. Groupoids with cancellation laws

$\mathfrak{G}$ is said to be a *groupoid with cancellation laws* if it has the following property; If for certain elements $a, x, y \in \mathfrak{G}$ there holds $ax = ay$ or $xa = ya$, then $x = y$.

In a groupoid with cancellation laws we can therefore "cancel" the equality $ax = ay$ or $xa = ya$ by the element a.

A multiplication table of every finite groupoid $\mathfrak{G}$ with cancellation laws has the following characteristic property: In every row and every column of the table there occur, on the rigth of the vertical and under the horizontal heading, the symbols of all elements of $\mathfrak{G}$. In fact: if, for example, in some row $[a]$ (i.e., to the right of the letter a written in the vertical heading) there do not occur the symbols of all the elements of $\mathfrak{G}$, then in the row $[a]$ and in two different columns $[x]$, $[y]$ (i.e., under the symbols x, y of the horizontal heading) there occurs the symbol of the same element b; that means that the equalities $ax = ay = b$ are true and that there simultaneously holds $x \neq y$ which, however, contradicts the cancellation laws. If, conversely, the multiplication table of a certain groupoid $\mathfrak{G}$ has the above property, then for any elements $a, x, y \in \mathfrak{G}$, $x \neq y$, there holds: $ax \neq ay$, $xa \neq ya$. Hence, in $\mathfrak{G}$ the cancellation laws apply.

18.3. Groupoids with division

1. *Definition.* If a groupoid $\mathfrak{G}$ is such that to any two elements $a, b \in \mathfrak{G}$ there exist elements $x, y \in \mathfrak{G}$ satisfying the equalities

$$ax = b, \quad ya = b,$$

it is called a *groupoid with division.*

If there exists only a single element $x \in \mathfrak{G}$ and a single element $y \in \mathfrak{G}$ with the above property, then $\mathfrak{G}$ is called a *groupoid with unique division.*

Groupoids with unique division are also called *quasigroups.*

We leave it to the reader to verify that the theorems set out below are correct:

For every groupoid with division, $\mathfrak{G}$, there holds $\mathfrak{G}\mathfrak{G} = \mathfrak{G}$.

Every quasigroup is a groupoid with cancellation laws.

Every finite groupoid with cancellation laws is a quasigroup.

2. *Examples.* The groupoids $\mathfrak{Z}$, $\mathfrak{Z}_n$, $\mathfrak{S}_n$ $(n \geqq 1)$ are quasigroups: To every two elements $a, b \in \mathfrak{Z}$ there exists a single element $x \in \mathfrak{Z}$ as well as a single element $y \in \mathfrak{Z}$ such that $a + x = b$, $y + a = b$, namely: $x = -a + b$, $y = b - a$. Similarly, to every two elements $a, b \in \mathfrak{Z}_n$ there exists a single element $x \in \mathfrak{Z}_n$ as well as a single element $y \in \mathfrak{Z}_n$ such that the division of $a + x$ by n as well as the division of $y + a$ by n leaves the remainder b, namely: $x = y = -a + b$ and $x = y = n - a + b$ if $-a + b \geqq 0$ and $-a + b < 0$, respectively. To every two permutations $\boldsymbol{p}, \boldsymbol{q} \in \mathfrak{S}_n$ there exists a single permutation $\boldsymbol{x} \in \mathfrak{S}_n$ and a single permutation $\boldsymbol{y} \in \mathfrak{S}_n$ such that $\boldsymbol{p} \cdot \boldsymbol{x} = \boldsymbol{q}$, $\boldsymbol{y} \cdot \boldsymbol{p} = \boldsymbol{q}$, i.e., $\boldsymbol{x} = \boldsymbol{q}\boldsymbol{p}^{-1}$, $\boldsymbol{y} = \boldsymbol{p}^{-1}\,\boldsymbol{q}$ where $\boldsymbol{q}\boldsymbol{p}^{-1}$ denotes the permutation composed of $\boldsymbol{p}^{-1}$ and $\boldsymbol{q}$; similarly, $\boldsymbol{p}^{-1}\,\boldsymbol{q}$.

18.4. Groupoids with a unit element

1. *Definition.* If an element, let us denote it $\underline{1}$, of a groupoid $\mathfrak{G}$ has the property that the product of $\underline{1}$ and any element $a \in \mathfrak{G}$, in either order, is again a, then $\underline{1}$ is called a *unit element* or a *unit of* $\mathfrak{G}$.

A unit $\underline{1} \in \mathfrak{G}$ is therefore characterized by the equalities $\underline{1}a = a\underline{1} = a$ which hold for any element $a \in \mathfrak{G}$.

We can easily show that *every groupoid has at most one unit.* If $\underline{1}$, x denote units of a groupoid $\mathfrak{G}$, then there holds $\underline{1}x = x$, on the one hand, (since $\underline{1}a = a$ for every element $a \in \mathfrak{G}$), and $\underline{1}x = \underline{1}$, on the other hand (since $ax = a$ for every element $a \in \mathfrak{G}$). Hence $\underline{1} = x$.

If a groupoid $\mathfrak{G}$ has a unit element, then it is called a *groupoid with a unit element* or *with a unit.*

Let us note that the multiplication table of a finite groupoid with a unit has the following characteristic property: The row beginning with the unit contains, in the subsequent places, the same symbols in the same order as the horizontal heading of the table. Similarly, the column beginning with the unit contains, in the subsequent places, the same symbols in the same order as the vertical heading.

2. *Examples.* $\mathfrak{Z}$, $\mathfrak{Z}_n$, $\mathfrak{S}_n$ $(n \geqq 1)$ are groupoids with a unit. The unit of $\mathfrak{Z}$ is 0, since for every element $a \in \mathfrak{Z}$ there holds $0 + a = a + 0 = a$. The unit of $\mathfrak{Z}_n$ is also 0, since for every element $a \in \mathfrak{Z}_n$ the numbers $0 + a$, $a + 0$ divided by n leave the remainder a. The unit of $\mathfrak{S}_n$ is the identical permutation $\boldsymbol{e}$ of the set H, since for every element $\boldsymbol{p} \in \mathfrak{S}_n$ we have $\boldsymbol{pe} = \boldsymbol{ep} = \boldsymbol{p}$. On the other hand, e.g., the groupoid described in 14.5.3 has no unit element.

18.5. Further remarkable groupoids. Groups

1. Special groupoids may have some of the above properties, or even all of them, simultaneously. So we speak, for example, of *semigroups with cancellation laws*, of *quasigroups with a unit*, of *semigroups with division*, etc. Some of these groupoids have special names. Quasigroups with a unit are called *loops*.

Of particular importance to our further deliberations are the semigroups with division. Let us first show that *every semigroup with division contains a unit and its division is unique.*

Suppose $\mathfrak{G}$ is a semigroup with division.

a) Choose, in $\mathfrak{G}$, an element a. As $\mathfrak{G}$ is a groupoid with division, there exists an element $e_r \in \mathfrak{G}$ such that $ae_r = a$. We shall show that e_r is a unit of $\mathfrak{G}$. Let b denote an arbitrary element of $\mathfrak{G}$. Since $\mathfrak{G}$ is a groupoid with division, there exists an element $y \in \mathfrak{G}$ such that $ya = b$ and, $\mathfrak{G}$ being associative, there holds: $be_r = (ya)e_r = y(ae_r) = ya = b$. So we have $be_r = b$. In a similar way we find that for the element $e_l \in \mathfrak{G}$ such that $e_l a = a$ there holds $e_l b = b$. Since $e_l e_r = e_l$ (because $be_r = b$ for every $b \in \mathfrak{G}$) as well as $e_l e_r = e_r$ (because $e_l b = b$ for every element $b \in \mathfrak{G}$), we have $e_l = e_r$ and, consequently, $e_r = \underline{1}$.

b) Suppose $a, b \in \mathfrak{G}$ are arbitrary elements. We shall show that the relations $ax_1 = b = ax_2$ $(x_1, x_2 \in \mathfrak{G})$ yield $x_1 = x_2$. First, $\mathfrak{G}$ being a groupoid with division, there exists an element $u \in \mathfrak{G}$ such that $ua = \underline{1}$. Next (since the multiplication is associative), there holds: $ub = u(ax_1) = (ua)x_1 = \underline{1}x_1 = x_1$ and, similarly, $ub = x_2$. So we have, in fact, $x_1 = x_2$. In an analogous way $y_1 a = b = y_2 a$ $(y_1, y_2 \in \mathfrak{G})$ yield $y_1 = y_2$.

Semigroups with division are called *groups*. The above theorem may therefore be expressed by saying that every group is a loop. For example, $\mathfrak{Z}$, $\mathfrak{Z}_n$ and $\mathfrak{S}_n$ $(n \geqq 1)$ are groups. In particular, $\mathfrak{S}_n$ is called the *symmetric permutation group of grade n*

All the mentioned kinds of groupoids may, of course, have further properties, they can, for instance, be Abelian; in that case we speak, e.g., about Abelian associative groupoids with a unit, and similarly. Abelian semigroups all the elements of which are idempotent are called *semilattices*. An example of a semilattice is given by the groupoid whose field consists of all the decompositions in G and the product of the elements $\bar{A}$ and $\bar{B}$ is defined as the least common covering $[\bar{A}, \bar{B}]$ (3.7.4).

2. *Brandt groupoids*. In this chapter we shall briefly deal with the so-called Brandt groupoids, introduced into algebra in 1927 by the German mathematician H. Brandt. He was the first to use the term "groupoid". About ten years later the term groupoid entered into literature in the sense in which it is employed today and introduced in this book.

Brandt groupoids differ from those we are concerned with by the fact that the multiplication is not necessarily defined for *every* two-membered sequence of elements.

Let G be a nonempty set and suppose we are given a rule, let us again call it a multiplication or binary operation, that can be applied to certain two-membered sequences of the elements $a, b \in G$ and uniquely associates with each of them a certain element $c \in G$; to other sequences, however, it may not be applied. In the first case we say that *a can be multiplied by b* and c is called the *product of a and b*. In the second case we say that *a cannot be multiplied by b* and that *the product of a and b does not exist*.

This situation could be adapted so as to appear as one of those we have already considered, namely, when the multiplication is defined for all two-membered sequences of the elements of G. To that purpose it would suffice to introduce a new element for the non-existing products. But we shall not do that because it would only affect the formal part of our study without any particular effect.

The set G together with a multiplication (in the above sense) is called a *Brandt groupoid* if the four postulates set out below are satisfied:

1. If, for some elements a, b, c, there holds $ab = c$, then each of them is uniquely determined by the remaining two.

2. If there exist ab and bc, then the same holds for the products $(ab)c$ and $a(bc)$; if there exist ab and $(ab)c$, then there also exist bc and $a(bc)$; if there exist bc and $a(bc)$, then there also exist ab and $(ab)c$. In each of these cases there holds $(ab)c = a(bc)$; notation abc.

3. To every element a there exist the following uniquely determined elements: the *right-hand side unit* e, the *left-hand side unit* e' and the *inverse element* a^*; for these elements there holds:

$$ae = e'a = a, \quad a^*a = e.$$

4. To any two units e, e' there exist elements for which e and e' are the right-hand side unit and the left-hand side unit (further, briefly, right unit, left unit), respectively.

If the above postulates are satisfied, then, in particular,

$$aa^* = e', \quad ea^* = a^*, \quad a^*e' = a^*, \quad ee = e, \quad e'e' = e'.$$

This can easily be verified; for example, the first equality:

$$e'a = a = ae = a(a^*a) = (aa^*)a$$

yield $e'a = (aa^*)a$, hence $e' = aa^*$.

We see that a is the inverse of a^* and so a and a^* may be referred to as mutually inverse elements.

On passing to the in verse element, the right unit and the left unit interchange.

Furthermore, we observe that the equality $ee = e$ is a characteristic property of the units: Every right or left unit complies with it and, moreover, every element e satisfying it is both the right and the left unit of e. As regards the postulates 2 and 1, it is obvious that each unit e is the right (left) unit of each element a (b) for which there exists the product ae (eb). By means of the units it can easily be expressed when a may be multiplied by b: that occurs if and only if the right unit of a equals the left unit of b.

The existence of the inverse element implies that if, for certain elements a, b, c, there holds $ab = c$, then there simultaneously holds $a^*c = b$, $cb^* = a$, $b^*a^* = c^*$, $c^*a = b^*$, $bc^* = a^*$. The inverse element a^* is also denoted a^{-1}. The products aa^{-1} or $a^{-1}a$ are only important when they stand alone, otherwise they may be omitted. If $n = ab \ldots m$ is the product of a finite sequence of an arbitrary number of elements, then the inverse element n^{-1} is given by the formula: $n^{-1} = m^{-1} \ldots b^{-1}a^{-1}$.

We shall content ourselves with these remarks without studying the theory of Brandt groupoids in detail. The latter is, after all, closely related to the theory of groups which we are concerned with in Part III of this book. To illustrate the concept of the Brandt groupoid we introduce the following simple example.

Let G be the Cartesian square of a nonempty set A (1.8). The elements of G are therefore two-membered sequences (a, b) where a, b run over the individual elements of A. The multiplication in G is defined as follows: The element (a, b) may be multiplied by $(c, d) \in G$ if and only if $b = c$ and in that case the product is given by the formula:

$$(a, b)\,(b, d) = (a, d).$$

The set G with this multiplication is a Brandt groupoid. In fact, first, it is obvious that the postulates 1 and 2 are satisfied. Next, the same holds for 3 and 4: To every element $(a, b) \in G$ there exists the right unit (b, b), the left unit (a, a) and the inverse element (b, a); to every two units (b, b), (a, a) there exists an element $(a, b) \in G$ for which (b, b) is the right and (a, a) the left unit.

If, for example, A is the set of all points in a plane, then we can associate, with every element (a, b) $(a \neq b)$ or (a, a) of the Cartesian square of A, the oriented line segment $\overrightarrow{ab}$ or the point a, respectively. In this way we obtain a Brandt groupoid whose field consists of points and oriented line segments and whose multiplication is given by the connection of these elements.

18.6. Lattices

Let us conclude this chapter with a short exposition of lattices the concept of which closely ties up with our previous deliberations. Lattices are, essentially, pairs of co-field, that is to say, in the same field defined groupoids with special properties and with multiplications connected by certain laws. The theory of lattices plays an important part in modern mathematics not only for its extent and formal elegance but chiefly because it describes, from a unifying view, the properties of the special lattices actually occuring in various branches of mathematics.

Assume two given multiplications on G; let us call them the *upper* and the *lower multiplication*, respectively. The product of an element $a \in G$ and an element $b \in G$ under the upper (lower) multiplication is called the *meet* (the *join*) *of a and b* and is denoted by $a \cup b$ ($a \cap b$). The groupoid whose field is the set G and whose multiplication is the upper (lower) multiplication is called the *upper* (*lower*) *groupoid*. We shall make use of the same symbols as for the sum and intersection of sets (1.5, 1.6), i.e. $\cup$, $\cap$; there is no danger of misunderstanding.

1. *The definition of a lattice.* A pair consisting of an upper and a lower groupoid is called a *lattice on the field* G, briefly, a *lattice* if for any elements $a, b, c \in G$ the following equalities are true:

a) $a \cup b = b \cup a$, a′) $a \cap b = b \cap a$,

b) $a \cup a = a$, b′) $a \cap a = a$,

c) $a \cup (b \cup c) = (a \cup b) \cup c$, c′) $a \cap (b \cap c) = (a \cap b) \cap c$,

d) $a \cup (a \cap b) = a$, d′) $a \cap (a \cup b) = a$.

Either of the two groupoids of the lattice is therefore Abelian [a), a′)], associative [c), c′)] and all its elements are idempotent [b), b′)]. The multiplications in both groupoids are connected according to the formulae d), d)′; the latter express the *absorptive laws of the lattice.*

A lattice may also be described as a pair of semilattices defined in the same field and connected by the absorptive laws.

2. *Examples.* [1] G is the set of all positive integers 1, 2, 3, ... For $a, b \in G$, $a \cup b$ is the least common multiple and $a \cap b$ the greatest common divisor of a and b.

[2] G is the set of all subsets of a certain set. For $A, B \in G$, $A \cup B$ is the sum and $A \cap B$ the intersection of A and B.

[3] G is the set of all decompositions of a certain nonempty set. For $\bar{A}, \bar{B} \in G$, $\bar{A} \cup \bar{B}$ is the least common covering $[\bar{A}, \bar{B}]$ and $\bar{A} \cap \bar{B}$ the greatest common refinement $(\bar{A}, \bar{B})$ of $\bar{A}$ and $\bar{B}$.

3. *Fundamental partial ordering of a lattice.* Let Γ be a lattice on G and $a, b, c \in G$ denote arbitrary elements.

Note that *both relations*

$$a \cup b = b, \quad b \cap a = a \tag{u}$$

are simultaneously valid.

In fact, if $a \cup b = b$, then by a′) and d′):

$$b \cap a = (a \cup b) \cap a = a \cap (a \cup b) = a;$$

analogously, if $b \cap a = a$, then by a) and d):

$$a \cup b = (b \cap a) \cup b = b \cup (b \cap a) = b.$$

Associating with every element $a \in G$ any element $b \in G$ satisfying (u), we obtain a generalized mapping of G onto itself; notation $\boldsymbol{u}$.

The mapping $\boldsymbol{u}$ *is an antisymmetric congruence on* G. Indeed, from b) and b′) we see that it is reflexive. On taking account of c), we conclude that $a \cup b = b$, $b \cup c = c$ yield: $a \cup c = a \cup (b \cup c) = (a \cup b) \cup c = b \cup c = c$, hence $\boldsymbol{u}$ is transitive. Therefore it is a congruence on G. Finally, from $a \cup b = b$, $b \cup a = a$ there follows, by a), the equality $a = b$.

Thus we have verified that the congruence $\boldsymbol{u}$ is antisymmetric, that is to say, is a partial ordering of G. We call it the *upper partial ordering of the lattice* Γ.

Note that the following symbols have the same meaning: $a \leqq b$ $(\boldsymbol{u})$, $a \cup b = b$, $b \cap a = a$.

Analogous considerations are correct if the roles of the upper and the lower groupoids are exchanged. Then we have the following results:

Both relations

$$b \cup a = a, \quad a \cap b = b \tag{l}$$

are simultaneously valid.

Associating with every element $a \in G$ *any element* $b \in G$ *satisfying* (l), *we obtain, on* G, *an antisymmetric congruence* $\boldsymbol{l}$. It is called the *lower partial ordering of* Γ.

It is easy to see that the following symbols have the same meaning: $a \leqq b$ $(\boldsymbol{l})$, $b \cup a = a$, $a \cap b = b$.

The upper and the lower ordering of Γ are called the *fundamental partial orderings of* Γ.

The fundamental partial orderings of Γ are inverse of each other and so $\boldsymbol{l} = \boldsymbol{u}^{-1}$, $\boldsymbol{u} = \boldsymbol{l}^{-1}$ because, under $\boldsymbol{u}$, every $b \in G$ is the image of all elements $a \in G$ satisfying the equations (u); precisely these elements are, as we see from (l), images of b under the congruence $\boldsymbol{l}$.

The symbols $a \leqq b$ ($\boldsymbol{u}$) and $b \leqq a$ ($\boldsymbol{l}$) have the same meaning.

On the above lattice [1], for example, the upper or the lower partial ordering is obtained by associating, with every positive integer, each of its positive multiples or divisors, respectively; on lattice [2], by associating, with every set $A \in G$, each of its supersets or subsets $B \in G$, respectively; on lattice [3], by associating, with every decomposition $\bar{A} \in G$, each of its coverings or refinements $\bar{B} \in G$, respectively.

The elements $a \cup b$ and $a \cap b$ are, with regard to the upper (lower) partial ordering of the lattice, the least upper (the greatest lower) and the greatest lower (the least upper) bounds of the elements a, b, respectively.

Since the upper and the lower partial orderings are mutually inverse, it is sufficient to restrict the proof to the upper partial ordering (9.4.2). Let us consider the element $a \cup b$.

Our object is to show that, with regard to $\boldsymbol{u}$, there holds $a \leqq a \cup b$, $b \leqq a \cup b$ and, furthermore, that $a \leqq c$, $b \leqq c$ yield $a \cup b \leqq c$.

The correctness of $a \leqq a \cup b$, $b \leqq a \cup b$ follows from the formulae 18.6.1c), b), a):

$$a \cup (a \cup b) = (a \cup a) \cup b = a \cup b,$$

$$b \cup (a \cup b) = b \cup (b \cup a) = (b \cup b) \cup a = b \cup a = a \cup b.$$

The relations $a \leqq c$, $b \leqq c$ are expressed by the equalities

$$a \cup c = c, \quad b \cup c = c$$

which yield, by 18.6.1c),

$$(a \cup b) \cup c = a \cup (b \cup c) = a \cup c = c,$$

i.e., $a \cup b \leqq c$ and the proof is complete.

We observe that, *with regard to the upper (lower) partial ordering of a lattice, each pair of its elements has both the least upper and the greatest lower bounds; the least upper bound is the meet (join) of the pair and the greatest lower bound is its join (meet).*

4. *Comment on the definition of a lattice.* A lattice has been described as a pair of groupoids defined on the same field, with special properties and multiplications bound by certain laws. We have shown that on every lattice there are certain mutually inverse partial orderings with regard to which each pair of elements has a least upper bound and a greatest lower bound; both the least upper bound and

the greatest lower bound are the products of the relative elements under the multiplications in the groupoids of which the lattice consists.

The definition of a lattice may, conversely, be based on the concept of antisymmetric congruence. If we have, on G, an arbitrary antisymmetric congruence with regard to which each pair of elements $a, b \in G$ has a least upper bound $a \cup b$ and a least lower bound $a \cap b$, then two multiplications on G can be defined in the way that the product of the ordered pair of elements a, b is $a \cup b$ or $a \cap b$, respectively. It is easy to show that the pair of groupoids on G with these multiplications is a lattice and that the initial antisymmetric congruence and its inverse are the corresponding upper partial ordering and the lower partial ordering, respectively.

5. *Remarkable kinds of lattices.* Let Γ be a lattice on the field G.

a) Lattices with extreme elements. If some element $O \in G$ is such that for every element $a \in G$ there holds $a \leqq O$ $(\boldsymbol{u})$ [$a \leqq O$ $(\boldsymbol{l})$], then it is said to be the *greatest element with regard to the upper (lower) partial ordering*; any element $o \in G$ such that there always applies $o \leqq a$ $(\boldsymbol{u})$ [$o \leqq a$ $(\boldsymbol{l})$] is called the *least element with regard to the upper (lower) partial ordering.* Since the relations (u) or (l) (see 18.6.3) are simultaneously valid, it is easy to see that *the greatest (least) element with regard to the upper (lower) partial ordering is the least (greatest) with regard to the lower (upper) partial ordering.* It is also obvious that in a lattice there may be, with regard to the same fundamental partial ordering, at most one greatest and one least element. The greatest and the least elements with regard to either fundamental partial ordering of a lattice are called the *extreme elements.*

If, in the lattice Γ, both extreme elements with regard to the fundamental partial orderings exist, then Γ is said to be a *lattice with extreme elements.*

For example, the above lattice [1] has, with regard to the upper partial ordering, the least element 1 but has no greatest element; with regard to the lower partial ordering it therefore has the greatest element 1 but has no least element. [2] is a lattice with extreme elements; the greatest (least) element with regard to the upper partial ordering is the sum of all the elements of G (the empty set). Even [3] is a lattice with extreme elements; the greatest (least) element with regard to the upper partial ordering is the greatest (least) decomposition of the corresponding set.

b) Modular (Dedekind) lattices. If for some elements $a, b, c \in G$ such that $a \leqq c$ $(\boldsymbol{u})$ there holds

$$a \cup (b \cap c) = (a \cup b) \cap c,$$

then we say that the sequence a, b, c satisfies the *upper modular* or *upper Dedekind relation*; similarly, if $a \leqq c$ $(\boldsymbol{l})$ and there holds

$$a \cap (b \cup c) = (a \cap b) \cup c,$$

then the sequence a, b, c satisfies the *lower modular* or *lower Dedekind relation.*

It is clear that *if the sequence a, b, c satisfies the upper (lower) Dedekind relation, then the inverse sequence c, b, a satisfies the lower (upper) Dedekind relation.*

The lattice Γ is called *modular* or *Dedekind* if every sequence of elements $a, b, c \in G$ in which there is $a \leqq c$ ($\boldsymbol{u}$) $\left(a \leqq c\ (\boldsymbol{l})\right)$ satisfies the upper (lower) Dedekind relation.

For example, the above lattice [2] is a Dedekind lattice because, for any parts A, B, C of an arbitrary set such that $A \subset C$, there holds $A \cup (B \cap C) = (A \cup B) \cap C$ (1.10.5; 1.10.3). Note that this lattice has, at the same time, extreme elements.

6. *Homomorphic mappings (deformations) of lattices.* The notion of homomorphic mapping defined for groupoids (13.1) may easily be applied to lattices. Let Γ, Γ^* be arbitrary lattices.

The mapping $\boldsymbol{d}$ of the lattice Γ into Γ^* is called a *homomorphic mapping* or a *deformation* if it preserves both lattice multiplications, that is to say, if for any elements $a, b \in \Gamma$ there holds:

$$\boldsymbol{d}(a \cup b) = \boldsymbol{d}a \cup \boldsymbol{d}b, \qquad \boldsymbol{d}(a \cap b) = \boldsymbol{d}a \cap \boldsymbol{d}b.$$

In the same way further notions connected with the concept of deformation may be applied to lattices. In particular, a simple deformation of Γ into Γ^* is called an *isomorphic mapping* and, in case of a mapping onto Γ^*, an *isomorphism*. If Γ is, under an isomorphism $\boldsymbol{i}$, mapped onto Γ^*, then Γ^* is said to be the *isomorphic image of Γ under the isomorphism $\boldsymbol{i}$* or the *$\boldsymbol{i}$-image of Γ*: $\Gamma^* = \boldsymbol{i}\Gamma$.

18.7. Exercises

1. If a permutation $\boldsymbol{p}$ of a set is composed of the permutations $\boldsymbol{p}_1, \ldots, \boldsymbol{p}_n$ ($n \geqq 2$), then the inverse permutation $\boldsymbol{p}^{-1}$ is composed of $\boldsymbol{p}_n^{-1}, \ldots, \boldsymbol{p}_1^{-1}$.
2. Every cyclic permutation of a finite set whose cycle consists of at least two members may be composed of transpositions as follows: $(a, b, c, \ldots, k, l, m) = (a, b)\,(b, c) \ldots (k, l)\,(l, m)$.
3. Denote, for convenience, the elements of a set H of order n ($\geqq 2$) by the numbers $1, \ldots, n$. Every transposition $(i, i + j)$ of H may be composed of some transpositions $(1, 2), (2, 3), \ldots, (n - 1, n)$ as follows:

$$(i, i + j) = (i + j - 1, i + j) \ldots (i + 1, i + 2)\,(i, i + 1)\,(i + 1, i + 2) \ldots (i + j - 1, i + j).$$

Every permutation of H may be composed of some transpositions

$$(1, 2), (2, 3), \ldots, (n - 1, n).$$

4. If the groupoid $\mathfrak{G}$ has a unit, then the image of the latter under any deformation $\boldsymbol{d}$ of $\mathfrak{G}$ into $\mathfrak{G}^*$ is a unit of $\boldsymbol{d}\mathfrak{G}$.
5. Every factoroid $\overline{\mathfrak{G}}$ on an arbitrary groupoid with a unit, $\mathfrak{G}$, has a unit; the element $\bar{a} \in \overline{\mathfrak{G}}$ containing the unit of $\mathfrak{G}$ is the unit of $\overline{\mathfrak{G}}$.

6. Give examples of groupoids that have only one or (with the exception of groups) exactly two properties described in 18.1, 18.3 and 18.4.

7. Every finite semigroup is a group.

8. In a semigroup the product of an arbitrary n-membered sequence of elements $a_1, \ldots, a_n$ ($n \geqq 2$) does not depend on their order if any two elements a_i, a_j are interchangeable. In an Abelian semigroup the product of a finite sequence of elements does not depend on their order.

9. In a Brandt groupoid there follow, from $ab = c$, for the right and the left units of the elements a, b, c; $a^{-1}, b^{\ 1}, c^{-1}$ the relations: b and c, c^{-1} and a^{-1}, a and b^{-1} have the same right units; b^{-1} and c^{-1}, c and a, a^{-1} and b have the same left units (equal to the corresponding right units). Each of these relations is sufficient for $ab = c$ to apply.

10. In any Brandt groupoid the elements having simultaneously the same unit on the right and on the left are called *doubly corresponding*. All elements doubly corresponding to some unit form a group. The sets of doubly corresponding elements form again a Brandt groupoid; its units are groups consisting of elements doubly corresponding to the units.

11. The properties of the upper and the lower groupoid required in the definition of the lattice (18.6.1) are not independent, since the properties b), b′) are a consequence of the others. Make sure of this by applying the equality d′) [d)] to the elements a, $b = a$ and the equality d) [d′)] to the elements a, $b = a \cup a$ [a, $b = a \cap a$].

12. If a lattice consists of decompositions on G every two of which are complementary and if the multiplications are defined as in 18.6.2. [3], then it is modular.

13. A lattice is modular if and only if any three of its elements a, b, c satisfy the equality:
$$(a \cup b) \cap [c \cup (a \cap b)] = (a \cap b) \cup [c \cap (a \cup b)].$$

14. An isomorphic image of a modular lattice is again modular.

15. Let Γ be an arbitrary lattice of decompositions on G with lattice operations [ ] and ( ). A series of decompositions on G all the members of which are elements of Γ is called a *main series of* Γ if each of its refinements containing only elements of Γ is its lengthening. The following theorems apply: a) *if Γ contains the greatest* (*least*) *element, then the latter is the initial* (*final*) *element of every main series of* Γ; b) *all mutually complementary main series of Γ have the same reduced length.*

III. GROUPS

19. Basic concepts relative to groups

19.1. Axioms of a group

The object of our further study are groups. By the definition in 18.5.1, a group is a semigroup with division.

More accurately:

A groupoid $\mathfrak{G}$ is called a *group* if the following *axioms of a group* are satisfied:

1. *For any elements $a, b, c \in \mathfrak{G}$ there holds $a(bc) = (ab)c$.*

2. *To any elements $a, b \in \mathfrak{G}$ there exists an element $x \in \mathfrak{G}$ satisfying $ax = b$ and an element $y \in \mathfrak{G}$ satisfying $ya = b$.*

These axioms are briefly called the *associative law* and the *axiom of division.* From these, as we have shown in 18.5.1, there follows the existence of a unit, i.e., an element $\underline{1}$ such that for $a \in \mathfrak{G}$ there holds $\underline{1}a = a\underline{1} = a$ and, moreover, the uniqueness of the division in $\mathfrak{G}$. Hence *every group is a quasigroup with a unit (loop).*

In what follows, $\mathfrak{G}$ denotes an arbitrary group.

19.2. Inverse elements. Inversion

Since $\mathfrak{G}$ is a quasigroup with a unit, there exists to every element $a \in \mathfrak{G}$ a unique element $x \in \mathfrak{G}$ such that $ax = \underline{1}$ and a unique element $y \in \mathfrak{G}$ such that $ya = \underline{1}$; the symbol $\underline{1}$ denotes (in our study) the unit of $\mathfrak{G}$.

It is easy to show that, in consequence of the associative law, both elements x and y are equal. In fact, first, the product of the element y and the element ax $(= \underline{1})$ is $y(ax) = y\underline{1} = y$. Next, by the associative law there holds $y(ax) = (ya)x = \underline{1}x = x$ and we actually have $x = y$.

Thus *there exists, to every element $a \in \mathfrak{G}$, a unique element a^{-1} such that $aa^{-1} = a^{-1}a = \underline{1}$.* It is called *the inverse element of a* or *the inverse of a.*

The inverse of a is therefore, by the definition, the only solution of the equation $ax = \underline{1}$ and, simultaneously, the only solution of the equation $ya = \underline{1}$. Since the element a satisfies the equation $a^{-1}x = \underline{1}$, it is the inverse of a^{-1}, i.e., $(a^{-1})^{-1} = a$. We also say that the elements a, a^{-1} are inverse of each other. Note that the inverse of a may be a itself because, e.g., $\underline{1} = \underline{1}$.

On the group $\mathfrak{G}$ we therefore have an important decomposition each element of which consists either of one element only, i.e., the inverse of itself, or of two elements inverse of each other.

For example, in the group $\mathfrak{Z}$ we have the unit 0 and the element inverse of an arbitrary element a is $-a$. The mentioned decomposition of $\mathfrak{Z}$ is: $\{0\}, \{1, -1\}, \{2, -2\}, \ldots$

Let a, b denote arbitrary elements of $\mathfrak{G}$. From $aa^{-1} = \underline{1}$ and in accordance with the associative law, we have

$$a(a^{-1}b) = (aa^{-1})b = \underline{1}b = b$$

so that the element $a^{-1}b$ is the (only) solution of the equation $ax = b$. In a similar way we ascertain that ba^{-1} is the (only) solution of $ya = b$. Furthermore, it is easy to verify that the element inverse of the product ab is $b^{-1}a^{-1}$; it is sufficient to realize that $b^{-1}a^{-1}$ is the solution of the equation $(ab)x = \underline{1}$. That follows from:

$$(ab)\,(b^{-1}a^{-1}) = a(bb^{-1}a^{-1}) = a(bb^{-1})a^{-1} = a(\underline{1}a^{-1}) = aa^{-1} = \underline{1}\,.$$

Analogously we arrive at a more general result, namely, that *the element inverse of the product* $a_1a_2 \ldots a_n$ *of an arbitrary* $n(\geqq 2)$*-membered sequence of elements* $a_1, a_2, \ldots, a_n \in \mathfrak{G}$ *is* $a_n^{-1} \ldots a_2^{-1}a_1^{-1}$.

Remark. The existence of the inverse of any element follows, as we have seen, from the characteristic properties of a group. Conversely, if any element $a \in \mathfrak{G}$ of an associative groupoid with a unit $\underline{1}$, $\mathfrak{G}$, has an inverse a^{-1}, i.e., the element satisfying $aa^{-1} = a^{-1}a = \underline{1}$, then $\mathfrak{G}$ is a quasigroup and therefore (since it is associative) a group. For in that case there exist, to any two elements $a, b \in \mathfrak{G}$, elements $x, y \in \mathfrak{G}$ such that $ax = b$, $ya = b$, i.e., $x = a^{-1}b$, $y = ba^{-1}$; it can easily be verified that x and y are the only elements with this property.

The property that to each element a of a group there exists an inverse a^{-1} is characteristic of groups and distinguishes them among the associative groupoids with a unit.

Making use of the inverse elements, we may define a certain simple mapping of $\mathfrak{G}$ onto itself, important for the following considerations. It is done by way of associating with every $a \in \mathfrak{G}$ the inverse element $a^{-1} \in \mathfrak{G}$. Thus we obtain a simple mapping of $\mathfrak{G}$ onto itself, hence a permutation on $\mathfrak{G}$, uniquely determined by $\mathfrak{G}$. It is called the *inversion of* $\mathfrak{G}$ and denoted by $\boldsymbol{n}$. We observe that $\boldsymbol{n}$ is an involutary mapping (6.7).

19.3. Powers of elements

Let a be an element of $\mathfrak{G}$ and n an arbitrary positive integer. Since $\mathfrak{G}$ is associative, there exists only one element $\underbrace{aa \ldots a}_{n}$ called the nth *power of* a and denoted by a^n. For $n = 1$ we have $a^1 = a$. Similarly, the element $\underbrace{a^{-1} \ldots a^{-1}}_{n}$ is called the $-n^{th}$ *power of* a and denoted by a^{-n}. By these definitions there holds $a^{-n} = (a^{-1})^n$, $a^{-n} = (a^n)^{-1}$. Thus we have defined the positive and the negative powers of a. It is useful to define even the *zero*th *power* a^0 *of* a as the unit of $\mathfrak{G}$ so that $a^0 = \underline{1}$.

With each element $a \in \mathfrak{G}$ we have thus associated infinitely many powers of a: $\ldots, a^{-2}, a^{-1}, a^0, a^1, a^2, \ldots,$ with the exponents $\ldots, -2, -1, 0, 1, 2, \ldots$; some of these elements may, of course, be equal.

For the powers of an element $a \in \mathfrak{G}$ there holds:

$$a^m a^n = a^{m+n}, \qquad (a^m)^n = a^{mn} \tag{1}$$

for all the integers m, n.

For brevity, we shall only prove the first formula and leave the proof of the second to the reader. If one or both numbers m, n are 0, then the above formula is obviously correct. If both m and n are positive, then we have

$$a^m a^n = \underbrace{(a \ldots a)}_{m} \underbrace{(a \ldots a)}_{n} = \underbrace{a \ldots a}_{m+n} = a^{m+n},$$

and the formula again applies. If both m and n are negative, then we denote $m' = -m$, $n' = -n$, hence m', n' are positive integers and we have

$$a^m a^n = a^{-m'} a^{-n'} = \underbrace{(a^{-1} \ldots a^{-1})}_{m'} \underbrace{(a^{-1} \ldots a^{-1})}_{n'} = \underbrace{a^{-1} \ldots a^{-1}}_{m'+n'}$$
$$= a^{-(m'+n')} = a^{-m'-n'} = a^{m+n}.$$

It remains to consider the case when one of the numbers m, n is positive and the other negative. If m is positive and n negative, we denote $n' = -n$ so that m, n' are positive integers and we have

$$a^m a^n = a^m a^{-n'} = \underbrace{(a \ldots a)}_{m} \underbrace{(a^{-1} \ldots a^{-1})}_{n'}$$
$$= \begin{cases} \underbrace{a \ldots a}_{m-n'} = a^{m-n'} = a^{m+n} & \text{if } m > n'; \\ \underline{1} = a^0 = a^{m-n'} = a^{m+n} & \text{if } m = n'; \\ \underbrace{a^{-1} \ldots a^{-1}}_{n'-m} = a^{-(n'-m)} = a^{m+n} & \text{if } m < n'. \end{cases}$$

Finally, if m is negative and n positive, then we write $m' = -m$, so that m', n are positive integers and we have:

$$a^m a^n = a^{-m'} a^n = (a^{-1})^{m'} [(a^{-1})^{-1}]^n = (a^{-1})^{m'} (a^{-1})^{-n} = (a^{-1})^{m'-n}$$
$$= a^{-(m'-n)} = a^{-m'+n} = a^{m+n},$$

which completes the proof.

If, for example, a stands for an arbitrary element of $\mathfrak{Z}$, then the individual powers of a are: ..., $-2a$, $-a$, 0, a, $2a$, ...; in particular, for $a = 1$ we have: ..., -2, -1, 0, 1, 2, ... and we observe that the set of all the powers of $1 \in \mathfrak{Z}$ coincides with the field of $\mathfrak{Z}$.

19.4. Subgroups and supergroups

1. *Definition.* Let $\mathfrak{A}$ be a subgroupoid of $\mathfrak{G}$. By 12.9.8, $\mathfrak{A}$ is an associative groupoid. If $\mathfrak{A}$ is a group, then we say that $\mathfrak{A}$ is a *subgroup of* $\mathfrak{G}$ or that $\mathfrak{G}$ is a *supergroup of* $\mathfrak{A}$ and write, as usual, $\mathfrak{A} \subset \mathfrak{G}$ or $\mathfrak{G} \supset \mathfrak{A}$.

$\mathfrak{A} \subset \mathfrak{G}$ is called a *proper subgroup of* $\mathfrak{G}$ if the field A of $\mathfrak{A}$ is a proper subset of $\mathfrak{G}$. Then we say that $\mathfrak{G}$ is a *proper supergroup of* $\mathfrak{A}$. There exist at least two subgroups of $\mathfrak{G}$, namely, the *greatest subgroup* which is identical with $\mathfrak{G}$ and the *least subgroup* $\mathfrak{G}$ whose field consists of the single element $\underline{1}$. These are the *extreme subgroups of* $\mathfrak{G}$.

To any groups $\mathfrak{A}$, $\mathfrak{B}$, $\mathfrak{G}$ there evidently apply the following statements:

a) *If $\mathfrak{B}$ is a subgroup of $\mathfrak{A}$ and $\mathfrak{A}$ a subgroup of $\mathfrak{G}$, then $\mathfrak{B}$ is a subgroup of $\mathfrak{G}$.*

b) *If $\mathfrak{A}$, $\mathfrak{B}$ are subgroups of $\mathfrak{G}$ and for their fields A, B there holds $B \subset A$, then $\mathfrak{B}$ is a subgroup of $\mathfrak{A}$.*

2. *Characteristic properties of subgroups.* Consider a subgroup $\mathfrak{A}$ of $\mathfrak{G}$. Denote by j the unit of $\mathfrak{A}$. Is there any relation between the unit $\underline{1}$ of $\mathfrak{G}$ and the unit j of $\mathfrak{A}$? By the definition of j there applies to every element $a \in \mathfrak{A}$ the equality $a = ja$ and, simultaneously, there of course holds $a = \underline{1}a$. Hence, in accordance with 19.1.2, we have $j = \underline{1}$. We see that *the unit of $\mathfrak{G}$ is, at the same time, the unit of $\mathfrak{A}$.* Consequently, *the inverse of an arbitrary element $a \in \mathfrak{A}$ is the element a^{-1}*, namely, the inverse of a in $\mathfrak{G}$.

Thus, *if a subgroupoid of $\mathfrak{G}$ is a subgroup of $\mathfrak{G}$, it contains the unit of $\mathfrak{G}$ and with each of its elements a, the element a^{-1}; conversely, if a subgroupoid of $\mathfrak{G}$ has these properties, then it is a subgroup of $\mathfrak{G}$.*

Owing to this result we can easily deduce a certain property of subgroups by which they are distinguishable among the subgroupoids. A subgroup $\mathfrak{A} \subset \mathfrak{G}$

contains, as we know, with each of its elements even the inverse of the latter, and so, if it contains any elements a, b, then it also contains the element ab^{-1}. If, conversely, a certain subgroupoid of $\mathfrak{G}$ contains with every two elements a, b even the element ab^{-1}, then it contains (for $b = a$) the unit $\underline{1}$ of $\mathfrak{G}$ and (for $a = \underline{1}$) the element b^{-1}; hence it is a subgroup of $\mathfrak{G}$.

The subgroups of $\mathfrak{G}$ are distinguishable among the subgroupoids of $\mathfrak{G}$ by that they contain, with every two elements a, b, even the element ab^{-1}.

Note, moreover, that any nonempty subset $A \subset \mathfrak{G}$ containing, with every two elements a, b, even the element ab^{-1} is groupoidal and therefore is the field of a subgroup of $\mathfrak{G}$. A similar observation applies to the element $a^{-1}b$.

3. *Examples.* Let us again consider the group $\mathfrak{Z}$ and let $\mathfrak{A}$ be a subgroup of $\mathfrak{Z}$. Since $\mathfrak{A}$ contains, with each of its elements b even the inverse, $-b$, $\mathfrak{A}$ consists either only of the element 0 or comprises both negative and positive numbers. In the first case, $\mathfrak{A}$ is the least subgroup of $\mathfrak{Z}$. In the second case, denote by a the least positive integer contained in $\mathfrak{A}$. The subgroup $\mathfrak{A}$, naturally, comprises all the powers of a, i.e., the multiples of a:

$$\ldots, -3a, -2a, -a, 0, a, 2a, 3a, \ldots$$

Let b denote an arbitrary element of $\mathfrak{A}$. As we know, there exist integers q, r such that $b = qa + r, 0 \leqq r \leqq a - 1$. Since $\mathfrak{A}$ contains b, qa, it also includes $b - qa = r$ and, as it contains no positive integers $<a$, we have $r = 0$. Hence $b = qa$ so that $\mathfrak{A}$ contains only multiples of a certain non-negative integer. Conversely, it is obvious that the set of all multiples of an arbitrary non-negative integer together with the adequate multiplication is a subgroup of $\mathfrak{Z}$.

The result: *all subgroups of $\mathfrak{Z}$ consist of all the multiples of the single non-negative integers.* Note that all *positive* multiples of a *positive* integer form a subgroupoid but not a subgroup of $\mathfrak{Z}$. Thus groups may comprise subgroupoids that are not subgroups.

4. *Remark.* Though we have succeeded in determining all the subgroups of $\mathfrak{Z}$, we must not expect a similar success in case of other groups where the multiplication is more complicated. No law by which it would be possible to determine all the subgroups of any group has, so far, been found.

19.5. The intersection and the product of subgroups

1. *The intersection of subgroups.* Consider two arbitrary subgroups $\mathfrak{A}, \mathfrak{B} \subset \mathfrak{G}$. Since both $\mathfrak{A}$ and $\mathfrak{B}$ contain the element $\underline{1} \in \mathfrak{G}$, there exists, as we know from our study of the groupoids, their intersection $\mathfrak{A} \cap \mathfrak{B}$. It is easy to show that $\mathfrak{A} \cap \mathfrak{B}$ is again a subgroup of $\mathfrak{G}$. $\mathfrak{A} \cap \mathfrak{B}$ is evidently an associative subgroupoid

of $\mathfrak{G}$ with the unit $\underline{1}$; it will therefore be sufficient to make sure that it contains, with every element a, even the inverse element a^{-1}. If $a \in \mathfrak{A} \cap \mathfrak{B}$, then simultaneously $a \in \mathfrak{A}$, $a \in \mathfrak{B}$ and, as $\mathfrak{A}$, $\mathfrak{B}$ are subgroups, there follows $a^{-1} \in \mathfrak{A}$, $a^{-1} \in \mathfrak{B}$ whence $a^{-1} \in \mathfrak{A} \cap \mathfrak{B}$ and the proof is complete. Consequently, *every two subgroups of $\mathfrak{G}$ have an intersection which is a subgroup of* $\mathfrak{G}$. It is also, as we see, a subgroup of each of the mentioned subgroups. This result may easily be applied to any number of subgroups of $\mathfrak{G}$.

2. *The product of subgroups.* Suppose the subgroups $\mathfrak{A}$, $\mathfrak{B}$ are interchangeable, i.e., $AB = BA$ where A and B stand for the fields of $\mathfrak{A}$ and $\mathfrak{B}$, respectively. Under this assumption there exists the product $\mathfrak{A}\mathfrak{B}$ of the subgroups $\mathfrak{A}$, $\mathfrak{B}$ (12.9.9) which is a subgroup of $\mathfrak{G}$. In fact, it is associative and, in accordance with the relations: $\underline{1} \in \mathfrak{A}$, $\underline{1} \in \mathfrak{B}$, $\underline{1} = \underline{1}\underline{1} \in \mathfrak{A}\mathfrak{B}$, comprises the unit $\underline{1}$ of $\mathfrak{G}$. Moreover, each element of $\mathfrak{A}\mathfrak{B}$ is the product ab of an element $a \in \mathfrak{A}$ and an element $b \in \mathfrak{B}$. The inverse of ab is $b^{-1}a^{-1}$ which lies, by the relation $BA = AB$, in the subgroupoid $\mathfrak{A}\mathfrak{B}$. Hence $\mathfrak{A}\mathfrak{B}$ is a subgroup of $\mathfrak{G}$. Note that there also holds 19.7.6. Furthermore, $\mathfrak{A}\mathfrak{B} \supset \mathfrak{A}$, $\mathfrak{A}\mathfrak{B} \supset \mathfrak{B}$; in particular $\mathfrak{A}^2$, i.e., $\mathfrak{A}\mathfrak{A}$ is the subgroup $\mathfrak{A}$ of $\mathfrak{G}$. It is also important to realize that in every Abelian group (any two subgroups are interchangeable and therefore) there exists a product of any two subgroups which is again a subgroup.

3. *Example.* Any two subgroups of $\mathfrak{Z}$ have an intersection and a product. Determine, for example, the intersection and the product of the subgroups $\mathfrak{A}$, $\mathfrak{B}$ whose fields are

$$\{\ldots, -8, -4, 0, 4, 8, \ldots\},$$
$$\{\ldots, -14, -7, 0, 7, 14, \ldots\}.$$

Every element of the intersection $\mathfrak{A} \cap \mathfrak{B}$ is, simultaneously, a multiple of the numbers 4 and 7; hence it is a multiple of the least common multiple of 4 and 7, i.e., of 28. The intersection $\mathfrak{A} \cap \mathfrak{B}$ therefore consists of the members

$$\ldots, -56, -28, 0, 28, 56, \ldots$$

As for the product $\mathfrak{A}\mathfrak{B}$, it obviously contains $4 + 7 = 11$. Moreover, as a subgroup of $\mathfrak{Z}$, $\mathfrak{A}\mathfrak{B}$ consists of all multiples of a certain non-negative integer a (19.4.3). Consequently, 11 is a multiple of a and therefore $a = 1$ or $a = 11$. Since $\mathfrak{A}\mathfrak{B}$ obviously comprises even, e.g., the number 4, there holds $a = 1$ because 4 is not a multiple of 11. It follows that the subgroup $\mathfrak{A}\mathfrak{B}$ consists of all multiples of 1, hence it is equal to $\mathfrak{Z}$.

19.6. Comments on the multiplication tables of finite groups

1. *Characteristic properties of the tables.* Let $\mathfrak{G}$ denote an arbitrary finite group and consider the corresponding multiplication table. Since there hold, in $\mathfrak{G}$, the

cancellation laws (18.3.1), we find in every row and every column of the multiplication table, to the right of the vertical and under the horizontal heading, the symbols of all the elements of $\mathfrak{G}$. There occurs, in particular, $\underline{1}$ and simultaneously with each element even the symbol of its inverse. These properties are characteristic of the multiplication table of a finite group only if there simultaneously applies the associative law. For example, the multiplication tables for groups of the order 1, 2, 3 whose members have been denoted by $\underline{1}$, a, b are:

	$\underline{1}$
$\underline{1}$	$\underline{1}$

	$\underline{1}$	a
$\underline{1}$	$\underline{1}$	a
a	a	$\underline{1}$

	$\underline{1}$	a	b
$\underline{1}$	$\underline{1}$	a	b
a	a	b	$\underline{1}$
b	b	$\underline{1}$	a

For groups of order 4 whose elements have been denoted by $\underline{1}$, a, b, c we have two different multiplication tables:

	$\underline{1}$	a	b	c
$\underline{1}$	$\underline{1}$	a	b	c
a	a	$\underline{1}$	c	b
b	b	c	a	$\underline{1}$
c	c	b	$\underline{1}$	a

	$\underline{1}$	a	b	c
$\underline{1}$	$\underline{1}$	a	b	c
a	a	$\underline{1}$	c	b
b	b	c	$\underline{1}$	a
c	c	b	a	$\underline{1}$

The above multiplication tables are found in the following way: one considers all the products of two equal or different elements and (taking account of the fact that in the multiplication table there occur, in every row and every column, the symbols of all the elements of the group, each only once) one decides which of the elements the product could be; finally, one verifies that the associative law is satisfied. But this procedure is, without further knowledge of the groups, rather tedious. Though we know rules by means of which the multiplication tables of all groups of certain orders may be determined, the main and hitherto unsolved problem is the enumeration of all finite groups of an arbitrary order.

2. *Normal tables.* Every multiplication table of a group of an arbitrary order may, first, be simplified by omitting both headings. Then we write, as the first, that row which contains the symbol $\underline{1}$ in the first place; as the second, that row which contains the symbol $\underline{1}$ in the second place, and so on until, as the last, that row which contains the symbol $\underline{1}$ in the last place. Such a multiplication table is called *normal*. Examples of normal multiplication tables of groups of the orders 1, 2, 3, 4 with the symbols $\underline{1}$, a, b, c of their elements are set out below:

$\underline{1}$

$\underline{1}$	a
a	$\underline{1}$

$\underline{1}$	a	b
b	$\underline{1}$	a
a	b	$\underline{1}$

$\underline{1}$	a	b	c		$\underline{1}$	a	b	c
a	$\underline{1}$	c	b		a	$\underline{1}$	c	b
c	b	$\underline{1}$	a		b	c	$\underline{1}$	a
b	c	a	$\underline{1}$		c	b	a	$\underline{1}$

3. *The rectangle rule.* In every normal multiplication table there is, at each place of the main diagonal, the symbol of the unit. Consider the normal multiplication table of a finite group. The symbol of the product of two elements a and b is, naturally, at the intersection of the row beginning with a and the column beginning with b. If a and b are symmetrically placed with regard to the main diagonal, then we have $ab = \underline{1}$ and, of course, a, b are inverse of each other. We observe that in the first row of the table there are the symbols of the elements inverse of those written in the first column.

Now consider any three elements x, y, z whose symbols together with $\underline{1}$ form the vertices of a rectangle so that, for example, x is in the same column and y in the same row as $\underline{1}$ and so z is in the same row as x and in the same column as y. Let a and b stand for the first letters of the rows containing $\underline{1}$ and x, respectively, and let, similarly, c and d be the first letters of the columns containing $\underline{1}$ and y, respectively. Then, for example, x lies at the intersection of the row beginning with b and the column beginning with c, so that $x = bc$ and, similarly, we have $y = ad$, $z = bd$, $\underline{1} = ac$. From $\underline{1} = ac$ we see that a and c are inverse of each other, hence there simultaneously holds $ca = \underline{1}$. So we have: $xy = (bc)\,(ad) = b(ca)d = b\underline{1}d = bd = z$, hence $xy = z$, which expresses the rectangle rule:

If, on a normal multiplication table, the symbols of four elements one of which is $\underline{1}$ form the vertices of a rectangle, then the element lying on the vertex opposite to $\underline{1}$ is the product of the element lying on the vertex in the same column as $\underline{1}$ and the element lying on the remaining vertex.

For example, in the last normal multiplication table introduced in 19.6.2, the elements $\underline{1}$, c in the second row together with the elements b, a in the fourth row form the vertices of a rectangle. In accordance with the above rule, we have $bc = a$ and, in fact, at the intersection of the row beginning with b and the column beginning with c there is a.

19.7. Exercises

1. A groupoid whose field is the set of all Euclidean motions on a straight line $\mathbf{f}[a]$, $\mathbf{g}[a]$ or in a plane $\mathbf{f}[\alpha; a, b]$, $\mathbf{g}[\alpha; a, b]$, the multiplication being defined by the composition of the motions (11.5.1), is a group called the *complete group of Euclidean motions on a straight line* or *in a plane*, respectively. In the latter all the elements $\mathbf{f}[a]$ or $\mathbf{f}[\alpha; a, b]$ form a subgroup. Find some further subgroups of the mentioned groups.
   Remark. The Euclidian geometry in a plane describes the properties of figures consisting of points and straight lines such as configurations of points and straight lines,

triangles, conics, etc. It is based on the complete group of Euclidean motions in a plane in the sense that two figures are considered equal if one can be mapped onto the other under an Euclidean motion.

2. The groupoid whose field is the set of $2n$ permutations of the vertices of a regular n-gon in a plane ($n \geqq 3$), described in Exercise 8.8.4, and whose multiplication is defined by the composition of permutations, is a group called the *diedric permutation group of order* $2n$. The latter contains, besides the least subgroup, further proper subgroups: the subgroup of order n which consists of all the elements corresponding to the rotations of the vertices about the center; n subgroups of order 2 each of which consists of the identical permutation and the permutation associating, with the vertices of the n-gon, the vertices symmetric with regard to an axis of symmetry.
3. In every finite group of an even or an odd order, the number of elements inverse of themselves is even or odd, respectively.
4. Every subgroupoid of a finite group is a subgroup (cf. 18.7.7).
5. The inversion of every Abelian group is an automorphism of itself.
6. In every Abelian group all the elements inverse of themselves form a subgroup.
7. In every Abelian group $\mathfrak{G}$ there applies, to any two elements $a, b \in \mathfrak{G}$ and an integer n, the equality $(ab)^n = a^n b^n$. Show that in a non-Abelian group this formula is not necessarily true.
8. If $\mathfrak{A}$, $\mathfrak{B}$ are subgroups of $\mathfrak{G}$ and the product of their fields, AB, is the field of a subgroup of $\mathfrak{G}$, then $\mathfrak{A}$, $\mathfrak{B}$ are interchangeable.
9. If $\mathfrak{A} \subset \mathfrak{B}$ are subgroups of $\mathfrak{G}$, then $\mathfrak{A}\mathfrak{B} = \mathfrak{B}\mathfrak{A} = \mathfrak{B}$, $\mathfrak{A} \cap \mathfrak{B} = \mathfrak{A}$. If even $\mathfrak{C}$ is a subgroup of $\mathfrak{G}$ and is interchangeable with $\mathfrak{A}$, then the subgroup $\mathfrak{C} \cap \mathfrak{B}$ is also interchangeable with $\mathfrak{A}$.
10. Every group has a center.
11. Assuming $p \in \mathfrak{G}$ to be an arbitrary element of $\mathfrak{G}$, let a new multiplication in $\mathfrak{G}$, marked with the sign $\circ$, be defined as follows: For arbitrary elements $x, y \in \mathfrak{G}$, the product $x \circ y$ is given by $x \circ y = xp^{-1}y$. Then there holds: a) the groupoid $\mathring{\mathfrak{G}}$ whose field is the field G of $\mathfrak{G}$, the multiplication being defined in the mentioned way, is a group; b) the unit of $\mathring{\mathfrak{G}}$ is the element p, the inverse of $x \in \mathring{\mathfrak{G}}$ is $px^{-1}p$.

    Remark. The group $\mathring{\mathfrak{G}}$ will henceforth be called the *(p)-group associated with the group* $\mathfrak{G}$.

20. Cosets of subgroups

20.1. Definition

Consider a group $\mathfrak{G}$ and let $\mathfrak{A}$ be a subgroup of $\mathfrak{G}$. Let, moreover, $p \in \mathfrak{G}$ be an arbitrary element of $\mathfrak{G}$.

The subset $p\mathfrak{A}$ of $\mathfrak{G}$, i.e., the set of the products of the elements p and each element of $\mathfrak{A}$, is called the *left coset* or *left class of p with regard to* $\mathfrak{A}$ or (if we know it is a question of $\mathfrak{A}$), briefly, the *left coset* or *left class of p*.

Similarly, the subset $\mathfrak{A}p$, i.e., the set of the products of each element of $\mathfrak{A}$ and the element p is called the *right coset* or *right class of p with regard to* $\mathfrak{A}$, briefly the *right coset* or *right class of p*.

Note that the field A of the subgroup $\mathfrak{A}$ is simultaneously both the left and the right coset of the element $\underline{1}$ with regard to $\mathfrak{A}$.

We shall first describe, in a few simple theorems, the properties of the left cosets; as the properties of the right cosets are analogous, we shall not deal with them here and leave it to the reader to consider them himself.

20.2. Properties of the left (right) cosets

Let $\mathfrak{A} \subset \mathfrak{G}$ be a subgroup and p, q arbitrary elements of $\mathfrak{G}$. Then the following theorems are true:

1. *The left coset $p\mathfrak{A}$ contains the element p.*

Indeed, since $\mathfrak{A}$ is a subgroup, there holds $\underline{1} \in \mathfrak{A}$ and we have $p = p\underline{1} \in p\mathfrak{A}$.

2. *If and only if $p \in \mathfrak{A}$, then $p\mathfrak{A} = A$.*

To prove that the above statement applies, let us first assume $p \in \mathfrak{A}$. Since $\mathfrak{A}$ is a subgroup, the product of p and any element of $\mathfrak{A}$ is again included in $\mathfrak{A}$, hence $p\mathfrak{A} \subset A$. Moreover, $p^{-1} \in \mathfrak{A}$ and, for any element $a \in \mathfrak{A}$, there is $p^{-1}a \in \mathfrak{A}$ so that $p(p^{-1}a) \in p\mathfrak{A}$; but $p(p^{-1}a) = (pp^{-1})a = \underline{1}a = a$; hence $a \in p\mathfrak{A}$ and we have $A \subset p\mathfrak{A}$. Consequently, $p\mathfrak{A} = A$. Let us now suppose that for some element $p \in \mathfrak{G}$ there holds $p\mathfrak{A} = A$. Then the product pa, for every $a \in \mathfrak{A}$, is contained in $\mathfrak{A}$ and therefore, in particular (for $a = \underline{1}$), p is an element of $\mathfrak{A}$.

Theorem 2 may be generalized in terms of:

3. *If and only if $p^{-1}q \in \mathfrak{A}$, then $p\mathfrak{A} = q\mathfrak{A}$.*

In fact, if $p^{-1}q \in \mathfrak{A}$, then in accordance with theorem 2, $p^{-1}q\mathfrak{A} = A$ and, consequently, $q\mathfrak{A} = (pp^{-1})q\mathfrak{A} = p(p^{-1}q)\mathfrak{A} = p(p^{-1}q\mathfrak{A}) = p\mathfrak{A}$. Conversely, from $q\mathfrak{A} = p\mathfrak{A}$ there follows $(p^{-1}q)\mathfrak{A} = A$ and so $p^{-1}q \in \mathfrak{A}$, which we were to prove.

4. *The left cosets $p\mathfrak{A}$, $q\mathfrak{A}$ are either disjoint or identical.*

This remarkable property of the left cosets may be verified in the following way: If both left cosets $p\mathfrak{A}$ and $q\mathfrak{A}$ have a common element x and so $x \in p\mathfrak{A}$, $x \in q\mathfrak{A}$, then $p^{-1}x \in \mathfrak{A}$, $q^{-1}x \in \mathfrak{A}$. Hence, in accordance with theorem 3, we have $p\mathfrak{A} = x\mathfrak{A} = q\mathfrak{A}$ and both left cosets $p\mathfrak{A}$ and $q\mathfrak{A}$ are identical.

5. *The left cosets $p\mathfrak{A}$, $q\mathfrak{A}$ are equivalent sets.*

Our object now is to show that there exists a simple mapping of the set $p\mathfrak{A}$ onto $q\mathfrak{A}$. Each element of $p\mathfrak{A}$ ($q\mathfrak{A}$) is the product pa (qa) of the element p (q) and a convenient element $a \in \mathfrak{A}$. Since $pa = pb$ ($qa = qb$) yields $a = b$, the element a is uniquely determined. Conversely, if $a \in \mathfrak{A}$, then $pa \in p\mathfrak{A}$ ($qa \in q\mathfrak{A}$). We observe that $\begin{pmatrix} pa \\ a \end{pmatrix}$ is a simple mapping of the set $p\mathfrak{A}$ onto $\mathfrak{A}$ and, similarly, $\begin{pmatrix} a \\ qa \end{pmatrix}$ is a simple mapping of the subgroup $\mathfrak{A}$ onto $q\mathfrak{A}$. Hence $\begin{pmatrix} a \\ qa \end{pmatrix}\begin{pmatrix} pa \\ a \end{pmatrix}$ is a simple mapping of the set $p\mathfrak{A}$ onto $q\mathfrak{A}$, which was to be proved.

Let us now proceed to the case when $\mathfrak{G}$ contains, besides $\mathfrak{A}$, a further subgroup, $\mathfrak{B}$.

6. *If the left cosets $p\mathfrak{A}$, $q\mathfrak{B}$ are incident, then their intersection $p\mathfrak{A} \cap q\mathfrak{B}$ is the left coset of each of its own elements with regard to the subgroup $\mathfrak{A} \cap \mathfrak{B}$.*

In fact, if the cosets $p\mathfrak{A}$, $q\mathfrak{B}$ have a common element $c \in \mathfrak{G}$, then by theorem 1 and theorem 2, there holds: $p\mathfrak{A} = c\mathfrak{A}$, $q\mathfrak{B} = c\mathfrak{B}$ whence $p\mathfrak{A} \cap q\mathfrak{B} = c\mathfrak{A} \cap c\mathfrak{B}$. Every element $x \in c\mathfrak{A} \cap c\mathfrak{B}$ is the product of c and a convenient element $a \in \mathfrak{A}$ and, at the same time, the product of c and a convenient element $b \in \mathfrak{B}$ and so $x = ca = cb$. Consequently, $a = b \in \mathfrak{A} \cap \mathfrak{B}$ and, therefore, $x \in c(\mathfrak{A} \cap \mathfrak{B})$. Thus we have $p\mathfrak{A} \cap q\mathfrak{B} \subset c(\mathfrak{A} \cap \mathfrak{B})$. Moreover, every element $x \in c(\mathfrak{A} \cap \mathfrak{B})$ is the product of the element c and an element $a \in \mathfrak{A} \cap \mathfrak{B}$, so that $x = ca \in c\mathfrak{A} \cap c\mathfrak{B}$. Consequently, $c(\mathfrak{A} \cap \mathfrak{B}) \subset p\mathfrak{A} \cap q\mathfrak{B}$ and we have the required result.

7. *If $\mathfrak{A} \subset \mathfrak{B}$, then from the incidence of the left cosets $p\mathfrak{A}$, $q\mathfrak{B}$ there follows $p\mathfrak{A} \subset q\mathfrak{B}$.*

Indeed, by 1.10.3, $\mathfrak{A} \subset \mathfrak{B}$ yields $\mathfrak{A} \cap \mathfrak{B} = \mathfrak{A}$; in accordance with theorems 6 and 4, there applies $p\mathfrak{A} \cap q\mathfrak{B} = p\mathfrak{A}$ and, consequently, $p\mathfrak{A} \subset q\mathfrak{B}$ (1.10.3).

As we have already mentioned, the properties of the right cosets with regard to $\mathfrak{A}$ are analogous. Between the left and the right cosets with regard to $\mathfrak{A}$ there holds the following relation:

8. *The left coset $p\mathfrak{A}$ is mapped, under the inversion $\boldsymbol{n}$ of the group $\mathfrak{G}$, onto the right coset $\mathfrak{A}p^{-1}$: $\boldsymbol{n}(p\mathfrak{A}) = \mathfrak{A}p^{-1}$. Simultaneously there holds the analogous formula $\boldsymbol{n}(\mathfrak{A}p) = p^{-1}\mathfrak{A}$.*

From $x \in p\mathfrak{A}$ there follows $x = pa$ ($a \in \mathfrak{A}$) and $x^{-1} = a^{-1}p^{-1}$ ($a^{-1} \in \mathfrak{A}$) yields $x^{-1} \in \mathfrak{A}p^{-1}$. Hence $\boldsymbol{n}(p\mathfrak{A}) \subset \mathfrak{A}p^{-1}$. Moreover, every point $y = a'p^{-1} \subset \mathfrak{A}p^{-1}$ ($a' \in \mathfrak{A}$)

is the $\boldsymbol{n}$-image of the element $pa'^{-1} \in p\mathfrak{A}$ $(a'^{-1} \in \mathfrak{A})$. Thus we have $\mathfrak{A}p^{-1} \subset \boldsymbol{n}(p\mathfrak{A})$ and, consequently, $\boldsymbol{n}(p\mathfrak{A}) = \mathfrak{A}p^{-1}$, which completes the proof.

Remark. Both $p\mathfrak{A}$ and $\mathfrak{A}p^{-1}$ are referred to as *mutually inverse cosets*. If one of them is denoted e.g. by $\bar{a}$, then the other is $\bar{a}^{-1}$.

9. *The left coset* $p\mathfrak{A}$ *and the right coset* $\mathfrak{A}q$ *are equivalent sets.*

We are to prove that there exists a simple mapping of the set $p\mathfrak{A}$ onto $\mathfrak{A}q$. In accordance with theorem 8 and 7.3.4, the sets $p\mathfrak{A}$ and $\mathfrak{A}p^{-1}$ are equivalent; by the theorem analogous to theorem 5 and valid for the right cosets, $\mathfrak{A}p^{-1}$ and $\mathfrak{A}q$ have the same property. Consequently, by 6.10.7, the assertion is correct.

20.3. Exercises

1. If $\mathfrak{G}$ is Abelian, then the left coset of an element $p \in \mathfrak{G}$ with regard to a subgroup $\mathfrak{A} \subset \mathfrak{G}$ is, at the same time, the right coset and so $p\mathfrak{A} = \mathfrak{A}p$.
2. Let $\mathfrak{A}, \mathfrak{B}$ denote arbitrary subgroups and C a complex in $\mathfrak{G}$. Prove that there holds: a) the sum of all left (right) cosets with regard to $\mathfrak{A}$ which are incident with C coincides with the complex $C\mathfrak{A}$ $(\mathfrak{A}C)$; b) the sum $\mathfrak{B}p\mathfrak{A}$ of all left cosets with regard to $\mathfrak{A}$ which are incident with some right coset $\mathfrak{B}p$ $(p \in \mathfrak{G})$ coincides with the sum of all right cosets with regard to $\mathfrak{B}$ which are incident with the left coset $p\mathfrak{A}$.
3. Let $p \in \mathfrak{G}$ be an arbitrary element and $\mathring{\mathfrak{G}}$ the (p)-group associated with $\mathfrak{G}$ (19.7.11). Next, let $\mathfrak{A}$ be an arbitrary subgroup of $\mathfrak{G}$. Prove that: a) the left (right) coset $p\mathfrak{A}$ $(\mathfrak{A}p)$ of p with regard to $\mathfrak{A}$ is the field of a subgroup $\mathring{\mathfrak{A}}_l \subset \mathring{\mathfrak{G}}$ $(\mathring{\mathfrak{A}}_r \subset \mathring{\mathfrak{G}})$ of $\mathring{\mathfrak{G}}$; b) the left (right) coset $x \circ \mathring{\mathfrak{A}}_l$ $(\mathring{\mathfrak{A}}_r \circ x)$ coincides, for each element x of $\mathring{\mathfrak{G}}$, with the left (right) coset $x\mathfrak{A}$ $(\mathfrak{A}x)$.

21. Decompositions generated by subgroups

A most remarkable property of groups is that every subgroup of an arbitrary group determines certain decompositions on the latter.

21.1. Left and right decompositions

Consider the system of all the subsets of the group $\mathfrak{G}$ given by the left cosets with regard to $\mathfrak{A}$. By 20.2.1, every element $p \in \mathfrak{G}$ is included in the left coset $p\mathfrak{A}$ which is, of course, an element of the considered system. By 20.2.4, every two elements of the system are disjoint. The system in question is therefore a decomposition of $\mathfrak{G}$, called the *decomposition of $\mathfrak{G}$ into left cosets, generated by $\mathfrak{A}$*, briefly, the *left decomposition of $\mathfrak{G}$ generated by $\mathfrak{A}$*. Notation: $\mathfrak{G}/_l\mathfrak{A}$.

Analogously, the system of all subsets of $\mathfrak{G}$ given by the right cosets with regard to $\mathfrak{A}$ is the *decomposition of $\mathfrak{G}$ into right cosets, generated by $\mathfrak{A}$*, briefly, the *right decomposition generated by $\mathfrak{A}$*. Notation: $\mathfrak{G}/_r\mathfrak{A}$:

We have, for instance, the formulas: $\mathfrak{G}/_l\mathfrak{G} = \mathfrak{G}/_r\mathfrak{G} = \bar{G}_{\max}$, $\mathfrak{G}/_l\{\underline{1}\} = \mathfrak{G}/_r\{\underline{1}\} = \bar{G}_{\min}$; $\bar{G}_{\max}$, $\bar{G}_{\min}$ are, of course, the greatest and the least decomposition of $\mathfrak{G}$, respectively.

In the following theorems we shall describe the properties of the left decompositions of a group. The properties of the right decompositions are analogous and will therefore be omitted. Finally, we shall deal with the relations between the left and the right decompositions of the group $\mathfrak{G}$ with regard to the same subgroup $\mathfrak{A}$.

21.2. Intersections and closures in connection with left decompositions

1. Let $\mathfrak{A} \supset \mathfrak{B}, \mathfrak{C}$ be arbitrary subgroups of $\mathfrak{G}$. Consider the intersection $\mathfrak{A}/_l\mathfrak{B} \sqcap \mathfrak{C}$ and the closure $\mathfrak{C} \sqsubset \mathfrak{A}/_l\mathfrak{B}$. Since $A \cap C \neq \emptyset$, neither of these figures is empty; $A \supset B, C$ denote, of course, the fields of the corresponding subgroups.

We shall prove: *There holds*

$$\mathfrak{A}/_l\mathfrak{B} \sqcap \mathfrak{C} = (\mathfrak{A} \cap \mathfrak{C})/_l(\mathfrak{B} \cap \mathfrak{C}). \tag{1}$$

If the subgroups $\mathfrak{A} \cap \mathfrak{C}$, $\mathfrak{B}$ are interchangeable, then there also holds:

$$\mathfrak{C} \sqsubset \mathfrak{A}/_l\mathfrak{B} = (\mathfrak{C} \cap \mathfrak{A})\mathfrak{B}/_l\mathfrak{B}. \tag{2}$$

Proof. a) We shall show that each element of the decomposition on the right- or the left-hand side of the formula (1) is an element of the decomposition on the left- or the right-hand side, respectively. Every element $\bar{a} \in (\mathfrak{A} \cap \mathfrak{C})/_l(\mathfrak{B} \cap \mathfrak{C})$ has the form

$$\bar{a} = a(\mathfrak{B} \cap \mathfrak{C}) = a\mathfrak{B} \cap a\mathfrak{C},$$

where $a \in \mathfrak{A} \cap \mathfrak{C}$. From $a \in \mathfrak{A}$ and $\mathfrak{A} \supset \mathfrak{B}$ there follows $a\mathfrak{B} \in \mathfrak{A}/_l\mathfrak{B}$ and from $a \in \mathfrak{C}$ we have $a\mathfrak{C} = C$. So there holds:

$$\bar{a} = a\mathfrak{B} \cap \mathfrak{C} \in \mathfrak{A}/_l\mathfrak{B} \sqcap \mathfrak{C}.$$

Now let $\bar{a} \in \mathfrak{A}/_l\mathfrak{B} \sqcap \mathfrak{C}$ be an arbitrary element and so $\bar{a} = a\mathfrak{B} \cap \mathfrak{C}$ ($\neq \emptyset$), $a \in \mathfrak{A}$. Moreover, let $x \in \bar{a}$ be an arbitrary element. From $x \in a\mathfrak{B}$ there follows $a\mathfrak{B} = x\mathfrak{B}$ and, since $x \in \mathfrak{C}$, there holds $C = x\mathfrak{C}$ and therefore $\bar{a} = x\mathfrak{B} \cap x\mathfrak{C} = x(\mathfrak{B} \cap \mathfrak{C})$. Since $a \in \mathfrak{A}$, $\mathfrak{A} \supset \mathfrak{B}$ yields $a\mathfrak{B} \subset \mathfrak{A}$, we have $x \in \mathfrak{A} \cap \mathfrak{C}$ so that $\bar{a} \in (\mathfrak{A} \cap \mathfrak{C})/_l(\mathfrak{B} \cap \mathfrak{C})$ and the proof of the formula (1) is complete.

b) Let us now assume that the subgroups $\mathfrak{A} \cap \mathfrak{C}$, $\mathfrak{B}$ are interchangeable. That occurs if, for example, the subgroups $\mathfrak{B}$, $\mathfrak{C}$ are interchangeable (22.2.1).

To prove the formula (2) we shall proceed analogously as in the case a). Every element $\bar{a} \in (\mathfrak{C} \cap \mathfrak{A})\mathfrak{B}/_l\mathfrak{B}$ has the form $x\mathfrak{B}$ where $x \in (\mathfrak{C} \cap \mathfrak{A})\mathfrak{B}$; we observe that the element x is the product ab of a point $a \in \mathfrak{C} \cap \mathfrak{A}$ and a point $b \in \mathfrak{B}$. Hence $\bar{a} = (ab)\,\mathfrak{B} = a(b\mathfrak{B}) = a\mathfrak{B}$ (the last equality is true with regard to the relation $b\mathfrak{B} = B$, correct by 20.2.2). From $a \in \mathfrak{A}$, $\mathfrak{A} \supset \mathfrak{B}$ we have $a\mathfrak{B} \in \mathfrak{A}/_l\mathfrak{B}$ and, since $a \in \mathfrak{C}$, the left coset $a\mathfrak{B}$ is incident with $\mathfrak{C}$. Thus we have $\bar{a} \in \mathfrak{C} \sqsubset \mathfrak{A}/_l\mathfrak{B}$. Let now $\bar{a}$ be an arbitrary element of $\mathfrak{C} \sqsubset \mathfrak{A}/_l\mathfrak{B}$ and so $\bar{a} = a\mathfrak{B}$ where a is a point of $\mathfrak{A}$ and $a\mathfrak{B}$ is incident with $\mathfrak{C}$; furthermore. let $c \in \mathfrak{C} \cap a\mathfrak{B}$ be an arbitrary point. From $c \in a\mathfrak{B}$ there follows, by the theorems 20.2.1 and 20.2.4, $\bar{a} = c\mathfrak{B}$ which yields (since $\bar{a} \subset \mathfrak{A}$) $c \in \mathfrak{A}$. So we have $c \in \mathfrak{C} \cap \mathfrak{A}$ and, consequently, $c = c\,.\,\underline{1} \in (\mathfrak{C} \cap \mathfrak{A})\mathfrak{B}$. From this and $\mathfrak{B} \subset (\mathfrak{C} \cap \mathfrak{A})\mathfrak{B}$ we have $\bar{a} \in (\mathfrak{C} \cap \mathfrak{A})\mathfrak{B}/_l\mathfrak{B}$ and the proof is accomplished.

Let us note, in particular, the case *when the subgroup $\mathfrak{A}$ coincides with $\mathfrak{G}$. Then we have*:

$$\mathfrak{G}/_l\mathfrak{B} \sqcap \mathfrak{C} = \mathfrak{C}/_l(\mathfrak{B} \cap \mathfrak{C}) \tag{1'}$$

and, moreover, if the subgroups $\mathfrak{B}$, $\mathfrak{C}$ are interchangeable:

$$\mathfrak{C} \sqsubset \mathfrak{G}/_l\mathfrak{B} = \mathfrak{C}\mathfrak{B}/_l\mathfrak{B}. \tag{2'}$$

2. The above deliberations will now be extended in the sense that the subgroup $\mathfrak{C}$ will be replaced by the left decomposition of a subgroup of $\mathfrak{G}$.

Let $\mathfrak{A} \supset \mathfrak{B}$ and $\mathfrak{C} \supset \mathfrak{D}$ be arbitrary subgroups of $\mathfrak{G}$. Consider the intersection $\mathfrak{A}/_l\mathfrak{B} \sqcap \mathfrak{C}/_l\mathfrak{D}$ and the closure $\mathfrak{C}/_l\mathfrak{D} \sqsubset \mathfrak{A}/_l\mathfrak{B}$. Since $A \cap C \neq \emptyset$, neither of these figures is empty. $A \supset B$, $C \supset D$ are, of course, the fields of the corresponding subgroups.

We shall show that *there holds*

$$\mathfrak{A}/_l\mathfrak{B} \sqcap \mathfrak{C}/_l\mathfrak{D} = (\mathfrak{A} \cap \mathfrak{C})/_l(\mathfrak{B} \cap \mathfrak{D}) \tag{3}$$

and, moreover, if the subgroups $\mathfrak{A} \cap \mathfrak{C}$, $\mathfrak{B}$ are interchangeable, even

$$\mathfrak{C}/_l\mathfrak{D} \sqsubset \mathfrak{A}/_l\mathfrak{B} = (\mathfrak{C} \cap \mathfrak{A})\mathfrak{B}/_l\mathfrak{B}. \tag{4}$$

Proof. a) Every element $\bar{a} \in (\mathfrak{A} \cap \mathfrak{C})/_l(\mathfrak{B} \cap \mathfrak{D})$ has the form $\bar{a} = a(\mathfrak{B} \cap \mathfrak{D}) = a\mathfrak{B} \cap a\mathfrak{D}$ where $a \in \mathfrak{A} \cap \mathfrak{C}$. From $a \in \mathfrak{A}$, $\mathfrak{A} \supset \mathfrak{B}$ there follows $a\mathfrak{B} \in \mathfrak{A}/_l\mathfrak{B}$. Analogously, from $a \in \mathfrak{C}$, $\mathfrak{C} \supset \mathfrak{D}$ we have $a\,\mathfrak{D} \subset \mathfrak{C}/_l\mathfrak{D}$. It is easy to see that $\bar{a}$ is the (nonempty) intersection of the elements $a\mathfrak{B}$ and $a\mathfrak{D}$ of the decompositions $\mathfrak{A}/_l\mathfrak{B}$ and $\mathfrak{C}/_l\mathfrak{D}$, respectively, so we have $\bar{a} \in \mathfrak{A}/_l\mathfrak{B} \sqcap \mathfrak{C}/_l\mathfrak{D}$.

Now let $\bar{a} \in \mathfrak{A}/_l\mathfrak{B} \sqcap \mathfrak{C}/_l\mathfrak{D}$ be an arbitrary element, hence

$$\bar{a} = a\mathfrak{B} \cap c\mathfrak{D}\ (\neq \emptyset),\ a \in \mathfrak{A},\ c \in \mathfrak{C};$$

furthermore, let $x \in \bar{a}$ denote an arbitrary point. From $x \in a\mathfrak{B}$ we have $a\mathfrak{B} = x\mathfrak{B}$ and, analogously, $x \in c\mathfrak{D}$ yields $c\mathfrak{D} = x\mathfrak{D}$; hence

$$\bar{a} = a\mathfrak{B} \cap c\mathfrak{D} = x\mathfrak{B} \cap x\mathfrak{D} = x(\mathfrak{B} \cap \mathfrak{D}).$$

Since $a \in \mathfrak{A} \supset \mathfrak{B}$, $c \in \mathfrak{C} \supset \mathfrak{D}$, we have $a\mathfrak{B} \subset \mathfrak{A}$, $c\mathfrak{D} \subset \mathfrak{C}$ and, consequently, $x \in \mathfrak{A} \cap \mathfrak{C}$. Thus we arrive at the result:

$$\bar{a} \in (\mathfrak{A} \cap \mathfrak{C})/_l(\mathfrak{B} \cap \mathfrak{D})$$

and there follows (3).

b) The formula (4) directly follows from

$$\mathfrak{C}/_l\mathfrak{D} \sqsubset \mathfrak{A}/_l\mathfrak{B} = \mathbf{s}(\mathfrak{C}/_l\mathfrak{D}) \sqsubset \mathfrak{A}/_l\mathfrak{B}, \quad \mathbf{s}(\mathfrak{C}/_l\mathfrak{D}) = C$$

and from the formula (2).

In the particular case when the subgroups $\mathfrak{A}, \mathfrak{C}$ coincide with $\mathfrak{G}$ and, consequently, the decompositions $\mathfrak{A}/_l\mathfrak{B}$ $(= \mathfrak{G}/_l\mathfrak{B})$, $\mathfrak{C}/_l\mathfrak{D}$ $(= \mathfrak{G}/_l\mathfrak{D})$ lie on $\mathfrak{G}$, the intersection $\mathfrak{G}/_l\mathfrak{B} \sqcap \mathfrak{G}/_l\mathfrak{D}$ of the latter coincides with the greatest common refinement $(\mathfrak{G}/_l\mathfrak{B}, \mathfrak{G}/_l\mathfrak{D})$ (3.5). Hence

$$(\mathfrak{G}/_l\mathfrak{B}, \mathfrak{G}/_l\mathfrak{D}) = \mathfrak{G}/_l(\mathfrak{B} \cap \mathfrak{D}).$$

21.3. Coverings and refinements of the left decompositions

Given two subgroups $\mathfrak{A}$, $\mathfrak{B}$ in $\mathfrak{G}$, let us ascertain when the left decomposition of $\mathfrak{G}$ generated by $\mathfrak{A}$ ($\mathfrak{B}$) is a covering (refinement) of the left decomposition generated by $\mathfrak{B}$ ($\mathfrak{A}$), i.e., $\mathfrak{G}/_l\mathfrak{A} \geqq \mathfrak{G}/_l\mathfrak{B}$.

If the left decomposition of $\mathfrak{G}$ generated by $\mathfrak{A}$ is a covering of the left decomposition generated by $\mathfrak{B}$ then, in particular, the field A of $\mathfrak{A}$ is the sum of certain left cosets with regard to $\mathfrak{B}$. Among the latter there is the field B of $\mathfrak{B}$ because both A and B have a common element $\underline{1}$. Consequently, $\mathfrak{A}$ is a supergroup of $\mathfrak{B}$, i.e., $\mathfrak{A} \supset \mathfrak{B}$. Conversely, if $\mathfrak{A}$ is a supergroup of $\mathfrak{B}$, then (by 20.2.7) every left coset with regard to $\mathfrak{A}$ is the sum of all the left cosets with regard to $\mathfrak{B}$ that are incident with it. We observe that the left decomposition of $\mathfrak{G}$ generated by $\mathfrak{A}$ ($\mathfrak{B}$) is a covering (refinement) of the left decomposition generated by $\mathfrak{B}$ ($\mathfrak{A}$).

The result: *The left decomposition of $\mathfrak{G}$ generated by the subgroup $\mathfrak{A}$ ($\mathfrak{B}$) is a covering (refinement) of the left decomposition generated by $\mathfrak{B}$ ($\mathfrak{A}$) if and only if $\mathfrak{A}$ is a supergroup of $\mathfrak{B}$.* In other words: *$\mathfrak{G}/_l\mathfrak{A} \geqq \mathfrak{G}/_l\mathfrak{B}$ holds if and only if $\mathfrak{A} \supset \mathfrak{B}$.*

21.4. The greatest common refinement of two left decompositions

Let $\mathfrak{A}, \mathfrak{B} \supset \mathfrak{G}$ be subgroups of $\mathfrak{G}$.

The greatest common refinement of the left decompositions of $\mathfrak{G}$, *generated by* $\mathfrak{A}$, $\mathfrak{B}$, *is the left decomposition generated by the subgroup* $\mathfrak{A} \cap \mathfrak{B}$, i.e., $(\mathfrak{G}/_l\mathfrak{A}, \mathfrak{G}/_l\mathfrak{B}) = \mathfrak{G}/_l(\mathfrak{A} \cap \mathfrak{B})$.

Indeed, the greatest common refinement of the decompositions $\mathfrak{G}/_l\mathfrak{A}$, $\mathfrak{G}/_l\mathfrak{B}$ is the system of all nonempty intersections of the left cosets $p\mathfrak{A}$ and the left cosets $q\mathfrak{B}$ (3.5). Every nonempty intersection $p\mathfrak{A} \cap q\mathfrak{B}$ is the left coset of each of its elements with regard to the subgroup $\mathfrak{A} \cap \mathfrak{B}$. Every left coset $c(\mathfrak{A} \cap \mathfrak{B})$ is the intersection of the left cosets $c\mathfrak{A}$ and $c\mathfrak{B}$ (20.2.6), which accomplishes the proof. (Cf. the result in 21.2.)

21.5. The least common covering of two left decompositions

Suppose $\mathfrak{A}, \mathfrak{B}$ are two interchangeable subgroups of $\mathfrak{G}$.

Then there exists the product $\mathfrak{A}\mathfrak{B}$ of $\mathfrak{A}$ and $\mathfrak{B}$ which is a subgroup of $\mathfrak{G}$.

The least common covering of the left decompositions of $\mathfrak{G}$, *generated by* $\mathfrak{A}$, $\mathfrak{B}$, *is the left decomposition generated by the subgroup* $\mathfrak{A}\mathfrak{B}$, i.e., $[\mathfrak{G}/_l\mathfrak{A}, \mathfrak{G}/_l\mathfrak{B}] = \mathfrak{G}/_l\mathfrak{A}\mathfrak{B}$.

In fact, first, with regard to $\mathfrak{A} \subset \mathfrak{A}\mathfrak{B}$, $\mathfrak{B} \subset \mathfrak{A}\mathfrak{B}$ and to the theorem in 21.3, the decomposition $\mathfrak{G}/_l\mathfrak{A}\mathfrak{B}$ is a common covering of the decompositions $\mathfrak{G}/_l\mathfrak{A}$, $\mathfrak{G}/_l\mathfrak{B}$. We are to show that two cosets $c\mathfrak{A}, p\mathfrak{A} \in \mathfrak{G}/_l\mathfrak{A}$ can be connected in $\mathfrak{G}/_l\mathfrak{B}$ if and only if they lie in the same element of $\mathfrak{G}/_l\mathfrak{A}\mathfrak{B}$.

a) If the left cosets $c\mathfrak{A}$, $p\mathfrak{A}$ lie in the same element of $\mathfrak{G}/_l\mathfrak{A}\mathfrak{B}$, then $p = cba$, while $b \in \mathfrak{B}$, $a \in \mathfrak{A}$ denote convenient elements. Both $c\mathfrak{A}$ and $p\mathfrak{A}$ are incident with $c\mathfrak{B} \in \mathfrak{G}/_l\mathfrak{B}$ and so they can be connected in $\mathfrak{G}/_l\mathfrak{B}$.

b) If there exists a binding $\{\mathfrak{G}/_l\mathfrak{A}, \mathfrak{G}/_l\mathfrak{B}\}$ from $c\mathfrak{A}$ to $p\mathfrak{A}$,

$$c_1\mathfrak{A}, \ldots, c_\alpha\mathfrak{A} \quad (c_1 = c,\ c_\alpha = p),$$

then every two neighbouring cosets $c_\beta\mathfrak{A}$, $c_{\beta+1}\mathfrak{A}$ are incident with a certain coset $d_\beta\mathfrak{B}$; therefore there exist elements

$$x_\beta \in c_\beta\mathfrak{A} \cap d_\beta\mathfrak{B}, \quad y_\beta \in d_\beta\mathfrak{B} \cap c_{\beta+1}\mathfrak{A} \quad (\beta = 1, \ldots, \alpha - 1).$$

The elements $x_\gamma, y_{\gamma-1}$ ($\gamma = 1, \ldots, \alpha$; $y_0 = c_1$, $x_\alpha = c_\alpha$) lie in the same coset $c_\gamma\mathfrak{A}$ and, similarly, the elements x_β, y_β lie in the same coset $d_\beta\mathfrak{B}$. Consequently, there holds $x_\gamma = y_{\gamma-1}\ a_\gamma$, $y_\beta = x_\beta b_\beta$ where $a_\gamma \in \mathfrak{A}$, $b_\beta \in \mathfrak{B}$ denote convenient elements. Thus,

$$c_\alpha = c_1 a_1 b_1 \ldots b_{\alpha-1} a_\alpha \in c_1\mathfrak{A}\mathfrak{B}$$

from which it is clear that the left cosets $c\mathfrak{A}$, $p\mathfrak{A}$ lie in the same coset $c\mathfrak{A}\mathfrak{B} \in \mathfrak{G}/_l\mathfrak{A}\mathfrak{B}$.

21.6. Complementary left decompositions

Consider arbitrary subgroups $\mathfrak{A}, \mathfrak{B} \subset \mathfrak{G}$ of $\mathfrak{G}$.

The left decompositions $\mathfrak{G}/_l\mathfrak{A}, \mathfrak{G}/_l\mathfrak{B}$ of $\mathfrak{G}$ are complementary if and only if the subgroups $\mathfrak{A}, \mathfrak{B}$ are interchangeable.

Proof. a) Suppose $\mathfrak{G}/_l\mathfrak{A}, \mathfrak{G}/_l\mathfrak{B}$ are complementary. Let $\bar{u} \in [\mathfrak{G}/_l\mathfrak{A}, \mathfrak{G}/_l\mathfrak{B}]$ be the element containing the unit $\underline{1} \in \mathfrak{G}$. From $\underline{1} \in \mathfrak{A} \cap \mathfrak{B}$ it is obvious that the fields of $\mathfrak{A}$ and $\mathfrak{B}$ are parts of $\bar{u}$. Consider arbitrary points $a \in \mathfrak{A}$, $b \in \mathfrak{B}$ and the left cosets $b\mathfrak{A} \in \mathfrak{G}/_l\mathfrak{A}$, $a^{-1}\mathfrak{B} \in \mathfrak{G}/_l\mathfrak{B}$. The latter are incident with the subgroups $\mathfrak{B}$ or $\mathfrak{A}$, respectively, hence they are subsets of $\bar{u}$ and we have $b\mathfrak{A} \subset \bar{u}$, $a^{-1}\mathfrak{B} \subset \bar{u}$. But, since $\mathfrak{G}/_l\mathfrak{A}$ and $\mathfrak{G}/_l\mathfrak{B}$ are complementary, there holds $b\mathfrak{A} \cap a^{-1}\mathfrak{B} \neq \emptyset$. Consequently, there exist points $a' \in \mathfrak{A}$, $b' \in \mathfrak{B}$ such that $ba' = a^{-1}b'$. Hence $ab = b'a'^{-1} \in \mathfrak{B}\mathfrak{A}$ and we have $\mathfrak{A}\mathfrak{B} \subset \mathfrak{B}\mathfrak{A}$. Analogously, we may show that $\mathfrak{B}\mathfrak{A} \subset \mathfrak{A}\mathfrak{B}$. Thus $\mathfrak{A}\mathfrak{B} = \mathfrak{B}\mathfrak{A}$.

b) Suppose the subgroups $\mathfrak{A}, \mathfrak{B}$ are interchangeable.

By the above theorem (21.5), the least common covering of $\mathfrak{G}/_l\mathfrak{A}$ and $\mathfrak{G}/_l\mathfrak{B}$ is $\mathfrak{G}/_l\mathfrak{A}\mathfrak{B}$. Let $c\mathfrak{A}\mathfrak{B} \in \mathfrak{G}/_l\mathfrak{A}\mathfrak{B}$ be an arbitrary element. Every element of $\mathfrak{G}/_l\mathfrak{A}$ lying in $c\mathfrak{A}\mathfrak{B}$ is $cb\mathfrak{A}$ where $b \in \mathfrak{B}$ is a convenient element. Similarly, every element of $\mathfrak{G}/_l\mathfrak{B}$ lying in $c\mathfrak{A}\mathfrak{B}$ is $ca\mathfrak{B}$, where $a \in \mathfrak{A}$ is a convenient element. We are to show that every two left cosets $cb\mathfrak{A}$ and $ca\mathfrak{B}$ lying in $c\mathfrak{A}\mathfrak{B}$ are incident, that is to say, that there exist elements $a_1 \in \mathfrak{A}$, $b_1 \in \mathfrak{B}$ such that $ba_1 = ab_1$. That is easy: Since the subgroups $\mathfrak{A}$ and $\mathfrak{B}$ are interchangeable, there exist elements $a_1 \in \mathfrak{A}$, $b_1 \in \mathfrak{B}$ satisfying the equality $a^{-1}b = b_1a_1^{-1}$. Hence $ba_1 = ab_1$ and the proof is complete.

21.7. Relations between the left and the right decompositions

Let $\mathfrak{A}, \mathfrak{B}$ stand for arbitrary subgroups of $\mathfrak{G}$.

1. *The left or the right decomposition $\mathfrak{G}/_l\mathfrak{A}$ or $\mathfrak{G}/_r\mathfrak{A}$, respectively, is mapped, under the extended inversion $\boldsymbol{n}$ of $\mathfrak{G}$, onto the right or the left decomposition $\mathfrak{G}/_r\mathfrak{A}$ or $\mathfrak{G}/_l\mathfrak{A}$ and so*

$$\boldsymbol{n}(\mathfrak{G}/_l\mathfrak{A}) = \mathfrak{G}/_r\mathfrak{A}, \quad \boldsymbol{n}(\mathfrak{G}/_r\mathfrak{A}) = \mathfrak{G}/_l\mathfrak{A}.$$

The decompositions $\mathfrak{G}/_l\mathfrak{A}, \mathfrak{G}/_r\mathfrak{A}$ are therefore equivalent sets:

$$\mathfrak{G}/_l\mathfrak{A} \simeq \mathfrak{G}/_r\mathfrak{A}.$$

Proof. In accordance with 7.3.4, the set $\boldsymbol{n}(\mathfrak{G}/_l\mathfrak{A})$ is a decomposition of $\mathfrak{G}$ equivalent to $\mathfrak{G}/_l\mathfrak{A}$. By 20.2.8, each element of $\boldsymbol{n}(\mathfrak{G}/_l\mathfrak{A})$ is an element of $\mathfrak{G}/_r\mathfrak{A}$. Hence $\boldsymbol{n}(\mathfrak{G}/_l\mathfrak{A}) = \mathfrak{G}/_r\mathfrak{A}$. Analogously we arrive at $\boldsymbol{n}(\mathfrak{G}/_r\mathfrak{A}) = \mathfrak{G}/_l\mathfrak{A}$.

2. *The least common covering of the left decomposition $\mathfrak{G}/_l\mathfrak{A}$ and the right decomposition $\mathfrak{G}/_r\mathfrak{B}$ is the set consisting of all the complexes $\mathfrak{B}p\mathfrak{A} \subset \mathfrak{G}$ ($p \in \mathfrak{G}$). The decompositions $\mathfrak{G}/_l\mathfrak{A}$, $\mathfrak{G}/_r\mathfrak{B}$ are complementary.*

Proof. Let us associate, with each point $p \in \mathfrak{G}$, the complex $\mathfrak{B}p\mathfrak{A} \subset \mathfrak{G}$ and consider the set $\bar{C}$ consisting of all these complexes. We observe, first, that each point of $\mathfrak{G}$ lies at least in one element of $\bar{C}$. Next, we shall show that two different elements of $\bar{C}$ are disjoint. Indeed, if any elements $\mathfrak{B}p\mathfrak{A}$, $\mathfrak{B}q\mathfrak{A} \in \bar{C}$ are incident, then there exist points $a, a' \in \mathfrak{A}$, $b, b' \in \mathfrak{B}$ such that $bpa = b'qa'$. Hence we have

$$(\mathfrak{B}b)p(a\mathfrak{A}) = (\mathfrak{B}b')q(a'\mathfrak{A})$$

and, moreover (by 20.2.2 and by the analogous theorem on right cosets), $\mathfrak{B}p\mathfrak{A} = \mathfrak{B}q\mathfrak{A}$. Thus the set $\bar{C}$ is a decomposition of $\mathfrak{G}$. Furthermore, by 20.3.2, each element $\mathfrak{B}p\mathfrak{A} \in \bar{C}$ is the sum of all elements of the left decomposition $\mathfrak{G}/_l\mathfrak{A}$ that are incident with the right coset $\mathfrak{B}p$ and, at the same time, the sum of all elements of the right decomposition $\mathfrak{G}/_r\mathfrak{B}$ incident with $p\mathfrak{A}$. We observe that the decomposition $\bar{C}$ is a common covering of the decompositions $\mathfrak{G}/_l\mathfrak{A}$, $\mathfrak{G}/_r\mathfrak{B}$. Let $\bar{u} = \mathfrak{B}p\mathfrak{A} \in \bar{C}$ be an arbitrary element and $\bar{a} \in \mathfrak{G}/_l\mathfrak{A}$, $\bar{b} \in \mathfrak{G}/_r\mathfrak{B}$ arbitrary cosets lying in $\bar{u}$. Then we have $\bar{a} = bp\mathfrak{A}$, $\bar{b} = \mathfrak{B}pa$ where $a \in \mathfrak{A}$, $b \in \mathfrak{B}$. Since $bpa \in \bar{a} \cap \bar{b}$, the sets $\bar{a}, \bar{b}$ are incident. Consequently, by 5.2, we have:

$$\bar{C} = [\mathfrak{G}/_l\mathfrak{A}, \mathfrak{G}/_r\mathfrak{B}].$$

Hence $\mathfrak{G}/_l\mathfrak{A}$, $\mathfrak{G}/_r\mathfrak{B}$ are complementary and the proof is accomplished.

For $\mathfrak{B} = \mathfrak{A}$, in particular, there applies:

The system of sets $\mathfrak{A}p\mathfrak{A} \subset \mathfrak{G}$, where $p \in \mathfrak{G}$, is for each subgroup $\mathfrak{A} \subset \mathfrak{G}$ the least common covering of the left and the right decompositions $\mathfrak{G}/_l\mathfrak{A}$, $\mathfrak{G}/_r\mathfrak{A}$ of $\mathfrak{G}$. The decompositions $\mathfrak{G}/_l\mathfrak{A}$, $\mathfrak{G}/_r\mathfrak{A}$ are complementary.

21.8. Exercises

1. In every Abelian group $\mathfrak{G}$, the left and the right decompositions with regard to any subgroup $\mathfrak{A} \subset \mathfrak{G}$ coincide: $\mathfrak{G}/_l\mathfrak{A} = \mathfrak{G}/_r\mathfrak{A}$.
2. The left (and, simultaneously, the right) decomposition of the group $\mathfrak{Z}$ with regard to the subgroup $\mathfrak{A}$ consisting of all the multiples of some natural number n is the decomposition $\bar{Z}_n$ described in 15.2.
3. Give an example to show that the left decomposition of a group $\mathfrak{G}$ with regard to a given subgroup $\mathfrak{A} \subset \mathfrak{G}$ need not coincide with the right decomposition.

4. Suppose $\mathfrak{A} \supset \mathfrak{B}$ are subgroups of $\mathfrak{G}$. Consider arbitrary left and right cosets $\bar{a}_l$, $\bar{c}_l$ and $\bar{a}_r$, $\bar{c}_r$ with regard to $\mathfrak{A}$, respectively, and denote:

$$\bar{A}_l = \bar{a}_l \sqcap \mathfrak{G}/_l\mathfrak{B} \; (= \bar{a}_l \sqsubset \mathfrak{G}/_l\mathfrak{B}), \qquad \bar{C}_l = \bar{c}_l \sqcap \mathfrak{G}/_l\mathfrak{B} \; (= \bar{c}_l \sqsubset \mathfrak{G}/_l\mathfrak{B}),$$
$$\bar{A}_r = \bar{a}_r \sqcap \mathfrak{G}/_r\mathfrak{B} \; (= \bar{a}_r \sqsubset \mathfrak{G}/_r\mathfrak{B}), \qquad \bar{C}_r = \bar{c}_r \sqcap \mathfrak{G}/_r\mathfrak{B} \; (= \bar{c}_r \sqsubset \mathfrak{G}/_r\mathfrak{B}).$$

Each element of the decompositions $\bar{A}_l, \bar{C}_l$ or $\bar{A}_r, \bar{C}_r$ is a left or a right coset with regard to $\mathfrak{B}$, respectively. Moreover, there holds: $\bar{A}_l \simeq \bar{C}_l$, $\bar{A}_r \simeq \bar{C}_r$.

5. Let $\mathfrak{A} \supset \mathfrak{B}$ be subgroups of $\mathfrak{G}$. Consider arbitrary cosets $\bar{a} \in \mathfrak{G}/_l\mathfrak{A}$, $\bar{a}^{-1} \in \mathfrak{G}/_r\mathfrak{A}$ inverse of each other and, on the latter, the decompositions set out below:

$$\bar{A}_l = \bar{a} \sqcap \mathfrak{G}/_l\mathfrak{B} \; (= \bar{a} \sqsubset \mathfrak{G}/_l\mathfrak{B}), \quad \bar{A}_r = \bar{a}^{-1} \sqcap \mathfrak{G}/_r\mathfrak{B} \; (= \bar{a}^{-1} \sqsubset \mathfrak{G}/_r\mathfrak{B}).$$

Either of the decompositions $\bar{A}_l$, $\bar{A}_r$ is, under the extended inversion $\boldsymbol{n}$ of $\mathfrak{G}$, mapped onto the other. $\bar{A}_l$, $\bar{A}_r$ are equivalent sets, hence: $\bar{A}_l \simeq \bar{A}_r$.

6. If $\bar{A}_l$ and $\bar{C}_r$ are the same as in exercise 4, there holds $\bar{A}_l \simeq \bar{C}_r$.

7. Let $p \in \mathfrak{G}$ denote an arbitrary element and $\mathring{\mathfrak{G}}$ the p-group associated with $\mathfrak{G}$ (19.7.11). Moreover, let $\mathfrak{A} \subset \mathfrak{G}$ be a subgroup of $\mathfrak{G}$ and $\mathring{\mathfrak{A}}_l \subset \mathring{\mathfrak{G}}$ ($\mathring{\mathfrak{A}}_r \subset \mathring{\mathfrak{G}}$) the subgroup of $\mathring{\mathfrak{G}}$ on the field $p\mathfrak{A}$ ($\mathfrak{A}p$) (20.3.3). Show that the left (right) decomposition of the group $\mathring{\mathfrak{G}}$ with regard to the subgroup $\mathring{\mathfrak{A}}_l$ ($\mathring{\mathfrak{A}}_r$) coincides with the left (right) decomposition of $\mathfrak{G}$ with regard to $\mathfrak{A}$, that is to say:

$$\mathring{\mathfrak{G}}/_l\mathring{\mathfrak{A}}_l = \mathfrak{G}/_l\mathfrak{A}, \quad \mathring{\mathfrak{G}}/_r\mathring{\mathfrak{A}}_r = \mathfrak{G}/_r\mathfrak{A}.$$

22. Consequences of the properties of decompositions generated by subgroups

22.1. Lagrange's theorem

Assuming $\mathfrak{A} \subset \mathfrak{G}$ to be an arbitrary subgroup of $\mathfrak{G}$, we shall now consider the consequences of the properties of the decompositions $\mathfrak{G}/_l\mathfrak{A}$ and $\mathfrak{G}/_r\mathfrak{A}$.

Suppose $\mathfrak{G}$ is finite.

Let us denote by N and n the order of $\mathfrak{G}$ and $\mathfrak{A}$, respectively, so that N is the number of the elements of G and n the number of the elements of $\mathfrak{A}$. One of the elements of $\mathfrak{G}/_l\mathfrak{A}$ is the field A of $\mathfrak{A}$. This element therefore consists of n elements of $\mathfrak{G}$ and, consequently (by 20.2.5), each element of $\mathfrak{G}/_l\mathfrak{A}$ consists of n elements of $\mathfrak{G}$. Hence $N = qn$, q denoting the number of the elements of $\mathfrak{G}/_l\mathfrak{A}$. Thus we have arrived at the following result:

The order of each subgroup $\mathfrak{A}$ of an arbitrary finite group $\mathfrak{G}$ is a divisor of the order of $\mathfrak{G}$.

This is *Lagrange's theorem*, considered to be one of the most important in the theory of finite groups. The number q, i.e., the number of the elements of the decomposition $\mathfrak{G}/_l\mathfrak{A}$ and, at the same time, even the quotient of N and n is called the *index of* $\mathfrak{A}$ *in* $\mathfrak{G}$. Since $\mathfrak{G}/_l\mathfrak{A}$ and $\mathfrak{G}/_r\mathfrak{A}$ are equivalent sets, the index of $\mathfrak{A}$ in $\mathfrak{G}$ simultaneously indicates the number of the elements of $\mathfrak{G}/_r\mathfrak{A}$. According to Lagrange's theorem, e.g., an arbitrary finite group whose order is a prime number does not contain any proper subgroup different from the least subgroup.

Lagrange's theorem applies even if $\mathfrak{G}$ is infinite ($N = 0$).

Example. Consider the group $\mathfrak{S}_3$ whose elements are denoted by $\underline{1}, \boldsymbol{a}, \boldsymbol{b}, \boldsymbol{c}, \boldsymbol{d}, \boldsymbol{f}$, as in 11.4. From the multiplication table of the group $\mathfrak{S}_3$ (11.4) we see that the elements $\underline{1}, \boldsymbol{f}$ form a subgroup of $\mathfrak{S}_3$. Let us denote it by $\mathfrak{A}$.

The left cosets of the individual elements with respect to $\mathfrak{A}$ are:

$$\underline{1}\mathfrak{A} = \boldsymbol{f}\mathfrak{A} = \{\underline{1}, \boldsymbol{f}\}, \quad \boldsymbol{a}\mathfrak{A} = \boldsymbol{c}\mathfrak{A} = \{\boldsymbol{a}, \boldsymbol{c}\}, \quad \boldsymbol{b}\mathfrak{A} = \boldsymbol{d}\mathfrak{A} = \{\boldsymbol{b}, \boldsymbol{d}\}.$$

The right cosets are:

$$\mathfrak{A}\underline{1} = \mathfrak{A}\boldsymbol{f} = \{\underline{1}, \boldsymbol{f}\}, \quad \mathfrak{A}\boldsymbol{a} = \mathfrak{A}\boldsymbol{d} = \{\boldsymbol{a}, \boldsymbol{d}\}, \quad \mathfrak{A}\boldsymbol{b} = \mathfrak{A}\boldsymbol{c} = \{\boldsymbol{b}, \boldsymbol{c}\}.$$

The left decomposition of the group $\mathfrak{S}_3$, generated by $\mathfrak{A}$, therefore consists of the elements $\{\underline{1}, \boldsymbol{f}\}$, $\{\boldsymbol{a}, \boldsymbol{c}\}$, $\{\boldsymbol{b}, \boldsymbol{d}\}$, whereas the right decomposition comprises the elements $\{\underline{1}, \boldsymbol{f}\}$, $\{\boldsymbol{a}, \boldsymbol{d}\}$, $\{\boldsymbol{b}, \boldsymbol{c}\}$. Note that these two decompositions are different. The order of $\mathfrak{S}_3$ is 6, the order of $\mathfrak{A}$ is 2, the index of $\mathfrak{A}$ in $\mathfrak{S}_3$ is $6 : 2 = 3 =$ the number of the elements of the left and, simultaneously, even of the right decomposition of $\mathfrak{S}_3$ generated by $\mathfrak{A}$.

22.2. Relations between interchangeable subgroups

The result arrived at in 21.6 and the properties of complementary decompositions (5) lead to a number of consequences as regards interchangeable subgroups. We shall restrict our attention to a few of them and leave further initiative to the reader. The formulae we shall obtain can mostly be verified directly. Owing to our method we can not only find them but even get a better understanding of their structure.

1. *Let* $\mathfrak{A} \supset \mathfrak{B}, \mathfrak{D}$ *be arbitrary subgroups of* $\mathfrak{G}$ *and suppose* $\mathfrak{B}$ *and* $\mathfrak{D}$ *are interchangeable. Then even the subgroups* $\mathfrak{B}, \mathfrak{A} \cap \mathfrak{D}$ *are interchangeable and there holds*

$$\mathfrak{A} \cap \mathfrak{D}\mathfrak{B} = (\mathfrak{A} \cap \mathfrak{D})\mathfrak{B}. \tag{1}$$

In fact, by 21.6, the decompositions $\mathfrak{G}/_l\mathfrak{B}$, $\mathfrak{G}/_l\mathfrak{D}$ are complementary. Since $\mathfrak{A} \supset \mathfrak{B}$, we have $\mathfrak{G}/_l\mathfrak{A} \geqq \mathfrak{G}/_l\mathfrak{B}$ (21.3). In accordance with 5.3, $\mathfrak{G}/_l\mathfrak{B}$ is complementary to $(\mathfrak{G}/_l\mathfrak{A}, \mathfrak{G}/_l\mathfrak{D})$ and, by 21.4, there holds $(\mathfrak{G}/_l\mathfrak{A}, \mathfrak{G}/_l\mathfrak{D}) = \mathfrak{G}/_l(\mathfrak{A} \cap \mathfrak{D})$. We observe that the decompositions $\mathfrak{G}/_l\mathfrak{B}$, $\mathfrak{G}/_l(\mathfrak{A} \cap \mathfrak{D})$ are complementary; from 21.6 we conclude that the subgroups $\mathfrak{B}$, $\mathfrak{A} \cap \mathfrak{D}$ are interchangeable.

By 5.4, $\mathfrak{G}/_l\mathfrak{D}$ is modular with regard to $\mathfrak{G}/_l\mathfrak{A}$, $\mathfrak{G}/_l\mathfrak{B}$ and so

$$(\mathfrak{G}/_l\mathfrak{A}, [\mathfrak{G}/_l\mathfrak{B}, \mathfrak{G}/_l\mathfrak{D}]) = [\mathfrak{G}/_l\mathfrak{B}, (\mathfrak{G}/_l\mathfrak{A}, \mathfrak{G}/_l\mathfrak{D})].$$

Hence, on taking account of 21.4 and 21.5, there follows

$$(\mathfrak{G}/_l\mathfrak{A}, \mathfrak{G}/_l\mathfrak{D}\mathfrak{B}) = \left[\mathfrak{G}_l\mathfrak{B}, \mathfrak{G}/_l(\mathfrak{A} \cap \mathfrak{D})\right]$$

as well as

$$\mathfrak{G}/_l(\mathfrak{A} \cap \mathfrak{D}\mathfrak{B}) = \mathfrak{G}/_l(\mathfrak{A} \cap \mathfrak{D})\mathfrak{B}.$$

It is easy to see that the element of this decomposition, containing the unit $\underline{1}$ of $\mathfrak{G}$, is the field of the subgroup $\mathfrak{A} \cap \mathfrak{D}\mathfrak{B}$ and, at the same time, the field of the subgroup $(\mathfrak{A} \cap \mathfrak{D})\mathfrak{B}$. The above formula is therefore correct.

2. *Let $\mathfrak{A} \supset \mathfrak{B}$, $\mathfrak{C} \supset \mathfrak{D}$ be arbitrary subgroups of $\mathfrak{G}$ and suppose $\mathfrak{B}$, $\mathfrak{D}$ are interchangeable. Then $\mathfrak{B}$, $\mathfrak{A} \cap \mathfrak{D}$ and $\mathfrak{D}$, $\mathfrak{C} \cap \mathfrak{B}$ are interchangeable as well. Simultaneously, even $\mathfrak{A} \cap \mathfrak{D}$, $\mathfrak{C} \cap \mathfrak{B}$ have the same property and there holds*

$$\mathfrak{A} \cap \mathfrak{C} \cap \mathfrak{D}\mathfrak{B} = (\mathfrak{A} \cap \mathfrak{D})\,(\mathfrak{C} \cap \mathfrak{B}). \tag{2}$$

Indeed, the first part of this statement immediately follows from the above theorem. Moreover, there holds (by 5.6.1)

$$\big((\mathfrak{G}/_l\mathfrak{A}, \mathfrak{G}/_l\mathfrak{C}), [\mathfrak{G}/_l\mathfrak{B}, \mathfrak{G}/_l\mathfrak{D}]\big) = [(\mathfrak{G}/_l\mathfrak{A}, \mathfrak{G}/_l\mathfrak{D}), (\mathfrak{G}/_l\mathfrak{C}, \mathfrak{G}/_l\mathfrak{B})] \tag{3}$$

and the decompositions $(\mathfrak{G}/_l\mathfrak{A}, \mathfrak{G}/_l\mathfrak{D})$, $(\mathfrak{G}/_l\mathfrak{C}, \mathfrak{G}/_l\mathfrak{B})$, i.e., $\mathfrak{G}/_l(\mathfrak{A} \cap \mathfrak{D})$, $\mathfrak{G}/_l(\mathfrak{C} \cap \mathfrak{B})$ are complementary. Consequently (21.6), the subgroups $\mathfrak{A} \cap \mathfrak{D}$, $\mathfrak{C} \cap \mathfrak{B}$ are interchangeable and so (3) yields (2).

3. *In the situation described by Theorem 2 there also holds:*

$$(\mathfrak{A} \cap \mathfrak{D})\mathfrak{B} \cap \mathfrak{C} = (\mathfrak{C} \cap \mathfrak{B})\mathfrak{D} \cap \mathfrak{A} = (\mathfrak{A} \cap \mathfrak{D})\,(\mathfrak{C} \cap \mathfrak{B}). \tag{4}$$

We know that $\mathfrak{G}/_l\mathfrak{B}$ and $\mathfrak{G}/_l\mathfrak{D}$ are complementary; moreover, there holds $\mathfrak{G}/_l\mathfrak{A} \geqq \mathfrak{G}/_l\mathfrak{B}$, $\mathfrak{G}/_l\mathfrak{C} \geqq \mathfrak{G}/_l\mathfrak{D}$. Note that the fields of the subgroups $\mathfrak{A}$, $\mathfrak{B}$, $\mathfrak{C}$, $\mathfrak{D}$ are elements of the corresponding left decompositions and contain the unit $\underline{1}$ of $\mathfrak{G}$.

Let us now use the result of 5.5 by which the decompositions

$$\mathfrak{A}/_l\mathfrak{B} = \mathfrak{A} \sqsubset \mathfrak{G}/_l\mathfrak{B}\ (= \mathfrak{G}/_l\mathfrak{B} \sqcap \mathfrak{A}),$$

$$\mathfrak{C}/_l\mathfrak{D} = \mathfrak{C} \sqsubset \mathfrak{G}/_l\mathfrak{D}\ (= \mathfrak{G}/_l\mathfrak{D} \sqcap \mathfrak{C})$$

are adjoint with respect to $\mathfrak{B}$, $\mathfrak{D}$. Thus there holds

$$\mathbf{s}(\mathfrak{D} \sqsubset \mathfrak{A}/_l\mathfrak{B} \sqcap \mathfrak{C}) = \mathbf{s}(\mathfrak{B} \sqsubset \mathfrak{C}/_l\mathfrak{D} \sqcap \mathfrak{A}). \tag{5}$$

By 2.6.5 a) we have

$$\mathfrak{D} \sqsubset \mathfrak{A}/_l\mathfrak{B} \sqcap \mathfrak{C} = (\mathfrak{D} \sqsubset \mathfrak{A}/_l\mathfrak{B}) \sqcap \mathfrak{C} = \mathfrak{D} \sqsubset (\mathfrak{A}/_l\mathfrak{B} \sqcap \mathfrak{C}),$$
$$\mathfrak{B} \sqsubset \mathfrak{C}/_l\mathfrak{D} \sqcap \mathfrak{A} = (\mathfrak{B} \sqsubset \mathfrak{C}/_l\mathfrak{D}) \sqcap \mathfrak{A} = \mathfrak{B} \sqsubset (\mathfrak{C}/_l\mathfrak{D} \sqcap \mathfrak{A})$$

and the results from 21.2.1 yield the formulas

$$(\mathfrak{D} \sqsubset \mathfrak{A}/_l\mathfrak{B}) \sqcap \mathfrak{C} = \big((\mathfrak{A} \cap \mathfrak{D})\mathfrak{B} \cap \mathfrak{C}\big)/_l(\mathfrak{C} \cap \mathfrak{B}),$$
$$\mathfrak{D} \sqsubset (\mathfrak{A}/_l\mathfrak{B} \sqcap \mathfrak{C}) = (\mathfrak{A} \cap \mathfrak{D})(\mathfrak{C} \cap \mathfrak{B})/_l(\mathfrak{C} \cap \mathfrak{B}),$$
$$(\mathfrak{B} \sqsubset \mathfrak{C}/_l\mathfrak{D}) \sqcap \mathfrak{A} = \big((\mathfrak{C} \cap \mathfrak{B})\mathfrak{D} \cap \mathfrak{A}\big)/_l(\mathfrak{A} \cap \mathfrak{D}),$$
$$\mathfrak{B} \sqsubset (\mathfrak{C}/_l\mathfrak{D} \sqcap \mathfrak{A}) = (\mathfrak{C} \cap \mathfrak{B})(\mathfrak{A} \cap \mathfrak{D})/_l(\mathfrak{A} \cap \mathfrak{D}).$$

So we have

$$\mathbf{s}(\mathfrak{D} \sqsubset \mathfrak{A}/_l\mathfrak{B} \sqcap \mathfrak{C}) = (\mathfrak{A} \cap \mathfrak{D})\mathfrak{B} \cap \mathfrak{C} = (\mathfrak{A} \cap \mathfrak{D})(\mathfrak{C} \cap \mathfrak{B}),$$
$$\mathbf{s}(\mathfrak{B} \sqsubset \mathfrak{C}/_l\mathfrak{D} \sqcap \mathfrak{A}) = (\mathfrak{C} \cap \mathfrak{B})\mathfrak{D} \cap \mathfrak{A} = (\mathfrak{C} \cap \mathfrak{B})(\mathfrak{A} \cap \mathfrak{D})$$

which, together with (5), yield the formulas (4).

22.3. Modular lattices of subgroups and of decompositions generated by subgroups

Consider an arbitrary nonempty system O of subgroups of the group $\mathfrak{G}$. Assume every two subgroups of the system O to be interchangeable and O to be closed with regard to the intersections and the products of the pairs of subgroups: the intersection and the product of any pair of subgroups $\mathfrak{A}, \mathfrak{B} \in O$ also belong to O, hence $\mathfrak{A} \cap \mathfrak{B}, \mathfrak{A}\mathfrak{B} \in O$.

Let us associate, with every two-membered sequence of subgroups $\mathfrak{A}, \mathfrak{B} \in O$, first, the intersection $\mathfrak{A} \cap \mathfrak{B}$ and, next, the product $\mathfrak{A}\mathfrak{B}$ of $\mathfrak{A}$ and $\mathfrak{B}$. Thus we have defined two multiplications in the system O, hence a pair of groupoids on the field O. Each of the two groupoids is Abelian (1.6), associative (1.10.4; 18.1.1) and all its elements are idempotent (1.10.1; 15.6.4). Moreover, the multiplications in both groupoids are connected by the formulae:

$$\mathfrak{A}(\mathfrak{A} \cap \mathfrak{B}) = \mathfrak{A}, \quad \mathfrak{A} \cap \mathfrak{A}\mathfrak{B} = \mathfrak{A}.$$

It follows that the above pair of groupoids is a lattice, Ω.

Let us now choose the upper (lower) multiplication in the lattice Ω in the manner that, to every two-membered sequence of subgroups $\mathfrak{A}, \mathfrak{B} \in \Omega$, there corresponds their product $\mathfrak{AB}$ (intersection $\mathfrak{A} \cap \mathfrak{B}$):

$$\mathfrak{A} \cup \mathfrak{B} = \mathfrak{AB}, \quad \mathfrak{A} \cap \mathfrak{B} = \mathfrak{A} \cap \mathfrak{B}.$$

Then we obtain the upper (lower) partial ordering $\boldsymbol{u}$ ($\boldsymbol{l}$) of Ω by associating, with each $\mathfrak{A} \in \Omega$, all its supergroups (subgroups) $\mathfrak{B} \in \Omega$. From 22.2.1 it is evident that every three-membered sequence of subgroups $\mathfrak{A}, \mathfrak{B}, \mathfrak{C} \in \Omega$ for which $\mathfrak{A} \leq \mathfrak{C}$ ($\boldsymbol{u}$), satisfies the upper modular relation. Consequently, Ω is modular.

Let us, furthermore, associate with every subgroup $\mathfrak{A} \in \Omega$ the decomposition $\mathfrak{G}/_l\mathfrak{A}$ and denote the corresponding system of the left decompositions of $\mathfrak{G}$ by the symbol O^*. Considering 21.4, 21.5, we realize that the system O^* is closed with respect to the operations (), [] and therefore includes, with every pair of left decompositions $\mathfrak{G}/_l\mathfrak{A}, \mathfrak{G}/_l\mathfrak{B} \in O^*$, even their greatest common refinement and their least common covering:

$$(\mathfrak{G}/_l\mathfrak{A}, \mathfrak{G}/_l\mathfrak{B}), [\mathfrak{G}/_l\mathfrak{A}, \mathfrak{G}/_l\mathfrak{B}] \in O^*.$$

Two multiplications in O^* may be defined by associating, with each two-membered sequence of the left decompositions $\mathfrak{G}/_l\mathfrak{A}, \mathfrak{G}/_l\mathfrak{B} \in O^*$, first, the greatest common refinement and, next, the least common covering of these decompositions. Thus we obtain, on O^*, a pair of groupoids Ω^* which, as it can again be verified, is a lattice.

The function $\boldsymbol{i}$, associating with each subgroup $\mathfrak{A} \in \Omega$ the left decomposition $\mathfrak{G}/_l\mathfrak{A} \in \Omega^*$, is clearly a simple mapping of Ω onto Ω^* such that for every $\mathfrak{A}, \mathfrak{B} \in \Omega$ there holds

$$\boldsymbol{i}(\mathfrak{A} \cap \mathfrak{B}) = \mathfrak{G}/_l(\mathfrak{A} \cap \mathfrak{B}), \quad \boldsymbol{i}\mathfrak{AB} = \mathfrak{G}/_l\mathfrak{AB},$$

i.e.,

$$\boldsymbol{i}(\mathfrak{A} \cap \mathfrak{B}) = \boldsymbol{i}\mathfrak{A} \cap \boldsymbol{i}\mathfrak{B}, \quad \boldsymbol{i}(\mathfrak{A} \cup \mathfrak{B}) = \boldsymbol{i}\mathfrak{A} \cup \boldsymbol{i}\mathfrak{B}.$$

The mapping $\boldsymbol{i}$ is therefore an isomorphism of Ω onto Ω^*. Since Ω is modular, Ω^* is modular as well (18.7.14).

The result:

A nonempty system of subgroups of $\mathfrak{G}$ any two elements of which are interchangeable and which is closed with respect to the intersections and the products of any two subgroups forms — together with the multiplications defined by the forming of the intersections and the products — a modular lattice. The system of the left (right) decompositions of $\mathfrak{G}$, generated by the individual elements of this lattice is, with respect to the operations (), [], closed and forms — with the multiplications defined by these operations — also a modular lattice which is isomorphic with the former.

22.4. Exercises

1. The order of any group consisting of permutations of a finite set of order n is a divisor of $n!$.
2. In every finite Abelian group of order N the number of elements inverse of themselves is a divisor of N.

23. Special decompositions of groups, generated by subgroups

23.1. Semi-coupled and coupled left decompositions

Consider the subgroups $\mathfrak{A} \supset \mathfrak{B}, \mathfrak{C} \supset \mathfrak{D}$ of $\mathfrak{G}$. Their fields are denoted by A, B, C, D.

We first ask under what conditions the left decompositions $\mathfrak{A}/_l\mathfrak{B}, \mathfrak{C}/_l\mathfrak{D}$ are semi-coupled or coupled.

Since the intersection $A \cap B$ contains the unit of $\mathfrak{G}$ and therefore is not empty, it is obvious, with respect to 4.1, that the mentioned decompositions are semi-coupled if and only if

$$\mathfrak{A}/_l\mathfrak{B} \sqcap \mathfrak{C} = \mathfrak{C}/_l\mathfrak{D} \sqcap \mathfrak{A}.$$

In accordance with 21.2.1, this may be written

$$(\mathfrak{A} \cap \mathfrak{C})/_l (\mathfrak{C} \cap \mathfrak{B}) = (\mathfrak{A} \cap \mathfrak{C})/_l (\mathfrak{A} \cap \mathfrak{D}).$$

This equality is evidently true if and only if

$$\mathfrak{A} \cap \mathfrak{D} = \mathfrak{C} \cap \mathfrak{B}. \tag{1}$$

Thus we have verified that *the left decompositions $\mathfrak{A}/_l\mathfrak{B}, \mathfrak{C}/_l\mathfrak{D}$ are semi-coupled if and only if the subgroups $\mathfrak{A} \cap \mathfrak{D}, \mathfrak{C} \cap \mathfrak{B}$ coincide, i.e., if $\mathfrak{A} \cap \mathfrak{D} = \mathfrak{C} \cap \mathfrak{B}$.*

Now suppose the left decompositions $\mathfrak{A}/_l\mathfrak{B}, \mathfrak{C}/_l\mathfrak{D}$ are coupled. Then (by 4.1; 20.3.2) we have, besides (1), even:

$$A = (A \cap C)B, \quad C = (C \cap A)D,$$

from which it follows (19.7.8) that $\mathfrak{A} \cap \mathfrak{C}$ is interchangeable with both $\mathfrak{B}$ and $\mathfrak{D}$ and so:

$$\mathfrak{A} = (\mathfrak{A} \cap \mathfrak{C})\mathfrak{B}, \quad \mathfrak{C} = (\mathfrak{C} \cap \mathfrak{A})\,\mathfrak{D}. \tag{2}$$

Conversely, if (1) and (2) simultaneously apply, then with respect to 4.1 and 21.2.1, the left decompositions $\mathfrak{A}/_l\mathfrak{B}, \mathfrak{C}/_l\mathfrak{D}$ are coupled.

Consequently, *the left decompositions* $\mathfrak{A}/_l\mathfrak{B}$, $\mathfrak{C}/_l\mathfrak{D}$ *are coupled if and only if there simultaneously holds*:

$$\mathfrak{A} \cap \mathfrak{D} = \mathfrak{C} \cap \mathfrak{B},$$
$$\mathfrak{A} = (\mathfrak{A} \cap \mathfrak{C})\mathfrak{B}, \quad \mathfrak{C} = (\mathfrak{C} \cap \mathfrak{A})\mathfrak{D}.$$

23.2. The general five-group theorem

Consider arbitrary subgroups $\mathfrak{A} \supset \mathfrak{B}$, $\mathfrak{C} \supset \mathfrak{D}$ of $\mathfrak{G}$.

Suppose the subgroups $\mathfrak{A} \cap \mathfrak{D}$, $\mathfrak{C} \cap \mathfrak{B}$ are interchangeable. Moreover, let $\mathfrak{U}$ be a subgroup of $\mathfrak{G}$ such that

$$\mathfrak{A} \cap \mathfrak{C} \supset \mathfrak{U} \supset (\mathfrak{A} \cap \mathfrak{D})(\mathfrak{C} \cap \mathfrak{B})$$

and let $\mathfrak{A} \cap \mathfrak{C}$ and $\mathfrak{U}$ be interchangeable with both $\mathfrak{B}$ and $\mathfrak{D}$.

Then there holds *the general five-group theorem*:

The left decompositions $(\mathfrak{A} \cap \mathfrak{C})\mathfrak{B}/_l\mathfrak{U}\mathfrak{B}$, $(\mathfrak{C} \cap \mathfrak{A})\mathfrak{D}/_l\mathfrak{U}\mathfrak{D}$ *are coupled and therefore equivalent, whence*

$$(\mathfrak{A} \cap \mathfrak{C})\mathfrak{B}/_l\mathfrak{U}\mathfrak{B} \simeq (\mathfrak{C} \cap \mathfrak{A})\mathfrak{D}/_l\mathfrak{U}\mathfrak{D}.$$

Moreover, there holds:

$$(\mathfrak{A} \cap \mathfrak{C})\mathfrak{B} \cap \mathfrak{U}\mathfrak{D} = \mathfrak{U} = (\mathfrak{C} \cap \mathfrak{A})\mathfrak{D} \cap \mathfrak{U}\mathfrak{B}. \tag{1}$$

Proof. Denote $\mathfrak{A}' = (\mathfrak{A} \cap \mathfrak{C})\mathfrak{B}$, $\mathfrak{C}' = (\mathfrak{C} \cap \mathfrak{A})\mathfrak{D}$ and, furthermore, $\bar{A} = \mathfrak{A}'/_l\mathfrak{B}$, $\bar{C} = \mathfrak{C}'/_l\mathfrak{D}$. Then we have $\mathfrak{A}' \supset \mathfrak{B}$, $\mathfrak{C}' \supset \mathfrak{D}$ and, moreover (20.3.2): $\bar{A} = \bar{C} \sqsubset \bar{A}$, $\bar{C} = \bar{A} \sqsubset \bar{C}$.

Consider the decompositions:

$$\bar{A} \sqcap \mathfrak{C}' = \mathfrak{A}'/_l\mathfrak{B} \sqcap \mathfrak{C}' = (\mathfrak{A}' \cap \mathfrak{C}')/_l(\mathfrak{C}' \cap \mathfrak{B}) = (\mathfrak{A} \cap \mathfrak{C})/_l(\mathfrak{C} \cap \mathfrak{B}),$$
$$\bar{C} \sqcap \mathfrak{A}' = \mathfrak{C}'/_l\mathfrak{D} \sqcap \mathfrak{A}' = (\mathfrak{C}' \cap \mathfrak{A}')/_l(\mathfrak{A}' \cap \mathfrak{D}) = (\mathfrak{C} \cap \mathfrak{A})/_l(\mathfrak{A} \cap \mathfrak{D})$$

and apply the construction described in 4.1 and leading to the coupled coverings $\mathring{A}$, $\mathring{C}$ of the decompositions $\bar{A}$, $\bar{C}$.

The least common covering of $\bar{A} \sqcap \mathfrak{C}'$, $\bar{C} \sqcap \mathfrak{A}'$ is the left decomposition $(\mathfrak{A} \cap \mathfrak{C})/_l(\mathfrak{A} \cap \mathfrak{D})(\mathfrak{C} \cap \mathfrak{B})$ (21.5). The decomposition $\bar{B} = (\mathfrak{A} \cap \mathfrak{C})/_l\mathfrak{U}$ is, with respect to $\mathfrak{A} \cap \mathfrak{C} \supset \mathfrak{U} \supset (\mathfrak{A} \cap \mathfrak{D})(\mathfrak{C} \cap \mathfrak{B})$, a covering of the least common covering of the decompositions $\bar{A} \sqcap \mathfrak{C}'$, $\bar{C} \sqcap \mathfrak{A}'$ (21.3) and therefore a common covering of the latter. In accordance with the mentioned construction, we now define the decomposition $\bar{\bar{A}}$ ($\bar{\bar{C}}$) on $\bar{A}$ ($\bar{C}$) as follows: Each element of $\bar{\bar{A}}$ ($\bar{\bar{C}}$) is the set of all elements of $\bar{A}$ (C) that are incident with the same element of $\bar{B}$. Then the mentioned coupled coverings $\mathring{A}$, $\mathring{C}$ are the coverings of $\bar{A}$, $\bar{C}$, enforced by $\bar{\bar{A}}$, $\bar{\bar{C}}$.

Now let $\bar{\bar{a}} \in \bar{\bar{A}}$ be an arbitrary element. $\bar{\bar{a}}$ is the set of all elements $\bar{a} \in \bar{A}$ incident with an element $\bar{b} \in \bar{B}$. Simultaneously we have $\bar{b} = x\mathfrak{U}$ where $x \in \mathfrak{A} \cap \mathfrak{C}$ is a point of $\mathfrak{A} \cap \mathfrak{C}$. Obviously, there holds $\bar{\bar{a}} = x\mathfrak{U} \sqsubset \bar{A}$ and, moreover (with regard to 20.3.2),

$$\mathring{a} = \mathbf{s}\bar{\bar{a}} = x\mathfrak{U}\mathfrak{B} \in \mathring{A}.$$

Thus we have verified that the elements of the decomposition $\mathring{A}$ are the left cosets, generated by $\mathfrak{U}\mathfrak{B}$, of the points lying in $\mathfrak{A} \cap \mathfrak{C}$. The sum of these cosets is evidently $(\mathfrak{A} \cap \mathfrak{C})\mathfrak{U}\mathfrak{B} = (\mathfrak{A} \cap \mathfrak{C})\mathfrak{B}$. Hence:

$$\mathring{A} = (\mathfrak{A} \cap \mathfrak{C})\mathfrak{B}/_l\mathfrak{U}\mathfrak{B}.$$

Analogously, we obtain $\mathring{C} = (\mathfrak{C} \cap \mathfrak{A})\mathfrak{D}/_l\mathfrak{U}\mathfrak{D}$. It follows that the left decompositions $(\mathfrak{A} \cap \mathfrak{C})\mathfrak{B}/_l\mathfrak{U}\mathfrak{B}$, $(\mathfrak{C} \cap \mathfrak{A})\mathfrak{D}/_l\mathfrak{U}\mathfrak{D}$ are coupled. In accordance with the second equivalence theorem (6.8), they are also equivalent.

Moreover, (by 4.1) we have: $\mathring{A} \sqcap \mathring{C} = \bar{B}$ and therefore (by 2.3):

$$(\mathring{A} \sqcap \mathbf{s}\mathring{C}) \sqcap (\mathring{C} \sqcap \mathbf{s}\mathring{A}) = \bar{B};$$

furthermore, (by 4.1):

$$\mathring{A} \sqcap \mathbf{s}\mathring{C} = \mathring{C} \sqcap \mathbf{s}\mathring{A}.$$

Thus we have arrived at the formulae $\mathring{A} \sqcap \mathbf{s}\mathring{C} = \mathring{C} \sqcap \mathbf{s}\mathring{A} = \bar{B}$ or

$$\begin{aligned} &\big((\mathfrak{A} \cap \mathfrak{C})\mathfrak{B} \cap (\mathfrak{C} \cap \mathfrak{A})\mathfrak{D}\big)/_l\big((\mathfrak{C} \cap \mathfrak{A})\mathfrak{D} \cap \mathfrak{U}\mathfrak{B}\big) \\ &= \big((\mathfrak{C} \cap \mathfrak{A})\mathfrak{D} \cap (\mathfrak{A} \cap \mathfrak{C})\mathfrak{B}\big)/_l\big((\mathfrak{A} \cap \mathfrak{C})\mathfrak{B} \cap \mathfrak{U}\mathfrak{D}\big) = (\mathfrak{A} \cap \mathfrak{C})/_l\mathfrak{U} \end{aligned}$$

from which (1) immediately follows.

Remark. Under the same assumption there, naturally, holds an analogous statement about the right decompositions and so, in particular,

$$(\mathfrak{A} \cap \mathfrak{C})\mathfrak{B}/_r\mathfrak{U}\mathfrak{B} \simeq (\mathfrak{C} \cap \mathfrak{A})\mathfrak{D}/_r\mathfrak{U}\mathfrak{D}.$$

Especially for $\mathfrak{U} = (\mathfrak{A} \cap \mathfrak{D})(\mathfrak{C} \cap \mathfrak{B})$ we have *the general four-group theorem*:

Let $\mathfrak{A} \supset \mathfrak{B}$, $\mathfrak{C} \supset \mathfrak{D}$ be arbitrary subgroups of $\mathfrak{G}$. Suppose that the subgroup $\mathfrak{A} \cap \mathfrak{D}$ is interchangeable with $\mathfrak{C} \cap \mathfrak{B}$, the subgroups $\mathfrak{A} \cap \mathfrak{C}$, $\mathfrak{A} \cap \mathfrak{D}$ are interchangeable with $\mathfrak{B}$ and $\mathfrak{C} \cap \mathfrak{A}$, $\mathfrak{C} \cap \mathfrak{B}$ with $\mathfrak{D}$. Then the left decompositions

$$(\mathfrak{A} \cap \mathfrak{C})\mathfrak{B}/_l(\mathfrak{A} \cap \mathfrak{D})\mathfrak{B}, \quad (\mathfrak{C} \cap \mathfrak{A})\mathfrak{D}/_l(\mathfrak{C} \cap \mathfrak{B})\mathfrak{D}$$

are coupled and therefore equivalent so that

$$(\mathfrak{A} \cap \mathfrak{C})\mathfrak{B}/_l(\mathfrak{A} \cap \mathfrak{D})\mathfrak{B} \simeq (\mathfrak{C} \cap \mathfrak{A})\mathfrak{D}/_l(\mathfrak{C} \cap \mathfrak{B})\mathfrak{D}.$$

Moreover, there holds:

$$(\mathfrak{C} \cap \mathfrak{A})\mathfrak{B} \cap (\mathfrak{C} \cap \mathfrak{B})\mathfrak{D} = (\mathfrak{A} \cap \mathfrak{D})(\mathfrak{C} \cap \mathfrak{B})$$
$$= (\mathfrak{A} \cap \mathfrak{C})\mathfrak{D} \cap (\mathfrak{A} \cap \mathfrak{D})\mathfrak{B}.$$

An analogous statement applies to the right decompositions and so we have, in particular,

$$(\mathfrak{A} \cap \mathfrak{C})\mathfrak{B}/_r(\mathfrak{A} \cap \mathfrak{D})\mathfrak{B} \simeq (\mathfrak{C} \cap \mathfrak{A})\mathfrak{D}/_r(\mathfrak{C} \cap \mathfrak{B})\mathfrak{D}.$$

23.3. Adjoint left decompositions

Let again $\mathfrak{A} \supset \mathfrak{B}$, $\mathfrak{C} \supset \mathfrak{D}$ be arbitrary subgroups of $\mathfrak{G}$ and $A \supset B$, $C \supset D$ their fields.

Our object now is to find out the circumstances under which the left decompositions $\mathfrak{A}/_l\mathfrak{B}$, $\mathfrak{C}/_l\mathfrak{D}$ are adjoint with respect to B, D.

The question is answered by the following theorem:

The left decompositions $\mathfrak{A}/_l\mathfrak{B}$, $\mathfrak{C}/_l\mathfrak{D}$ are adjoint with respect to B, D if and only if the subgroups $\mathfrak{A} \cap \mathfrak{D}$, $\mathfrak{C} \cap \mathfrak{B}$ are interchangeable. Then there holds:

$$(\mathfrak{A} \cap \mathfrak{D})\mathfrak{B} \cap \mathfrak{C} = (\mathfrak{C} \cap \mathfrak{B})\mathfrak{D} \cap \mathfrak{A} = (\mathfrak{A} \cap \mathfrak{D})(\mathfrak{C} \cap \mathfrak{B}). \tag{1}$$

Proof. By 2.6.5 we have

$$D \sqsubset \mathfrak{A}/_l\mathfrak{B} \sqcap C = (D \sqsubset \mathfrak{A}/_l\mathfrak{B}) \sqcap C = D \sqsubset (\mathfrak{A}/_l\mathfrak{B} \sqcap \mathfrak{C}),$$
$$B \sqsubset \mathfrak{C}/_l\mathfrak{D} \sqcap A = (B \sqsubset \mathfrak{C}/_l\mathfrak{D}) \sqcap A = B \sqsubset (\mathfrak{C}/_l\mathfrak{D} \sqcap \mathfrak{A}).$$

Consequently, with regard to 21.2.1, there holds:

$$\mathbf{s}(D \sqsubset \mathfrak{A}/_l\mathfrak{B} \sqcap C) = \mathbf{s}(D \sqsubset \mathfrak{A}/_l\mathfrak{B}) \cap C = \mathbf{s}\big(D \sqsubset (\mathfrak{A} \cap \mathfrak{C})/_l(\mathfrak{C} \cap \mathfrak{B})\big),$$
$$\mathbf{s}(B \sqsubset \mathfrak{C}/_l\mathfrak{D} \sqcap A) = \mathbf{s}(B \sqsubset \mathfrak{C}/_l\mathfrak{D}) \cap A = \mathbf{s}\big(B \sqsubset (\mathfrak{C} \cap \mathfrak{A})/_l(\mathfrak{A} \cap \mathfrak{D})\big)$$

which (by 20.3.2) may be expressed in the form:

$$\begin{aligned}\mathbf{s}(D \sqsubset \mathfrak{A}/_l\mathfrak{B} \sqcap C) &= (A \cap D)B \cap C = (A \cap D)(C \cap B),\\ \mathbf{s}(B \sqsubset \mathfrak{C}/_l\mathfrak{D} \sqcap A) &= (C \cap B)D \cap A = (C \cap B)(A \cap D).\end{aligned} \tag{2}$$

a) Suppose $\mathfrak{A}/_l\mathfrak{B}$, $\mathfrak{C}/_l\mathfrak{D}$ are adjoint with respect to B, D. Then we have, by (2),

$$(A \cap D)(C \cap B) = (C \cap B)(A \cap D).$$

We see that $\mathfrak{A} \cap \mathfrak{D}$, $\mathfrak{C} \cap \mathfrak{B}$ are interchangeable. Consequently, the product $(\mathfrak{A} \cap \mathfrak{D})(\mathfrak{C} \cap \mathfrak{B})$ is a subgroup of $\mathfrak{G}$. Moreover, from (2) we conclude that the field of this subgroup coincides with either of the sets $(A \cap D)B \cap C$, $(C \cap B)D \cap A$, a fact expressed by the formulae (1). Note that neither $\mathfrak{A} \cap \mathfrak{D}, \mathfrak{B}$ nor $\mathfrak{C} \cap \mathfrak{B}, \mathfrak{D}$ are necessarily interchangeable.

b) Suppose the subgroups $\mathfrak{A} \cap \mathfrak{D}$, $\mathfrak{C} \cap \mathfrak{B}$ are interchangeable. Then, by (2), there holds $\mathbf{s}(D \sqsubset \mathfrak{A}/_l\mathfrak{B} \sqcap C) = \mathbf{s}(B \sqsubset \mathfrak{C}/_l\mathfrak{D} \sqcap A)$ and we observe that the decompositions $\mathfrak{A}/_l\mathfrak{B}$, $\mathfrak{C}/_l\mathfrak{D}$ are adjoint with respect to B, D. This accomplishes the proof.

Analogously, for the right decompositions there holds:

The right decompositions $\mathfrak{A}/_r\mathfrak{B}$, $\mathfrak{C}/_r\mathfrak{D}$ are adjoint if and only if the subgroups $\mathfrak{A} \cap \mathfrak{D}$, $\mathfrak{C} \cap \mathfrak{D}$ are interchangeable. In that case:

$$\mathfrak{B}(\mathfrak{A} \cap \mathfrak{D}) \cap \mathfrak{C} = \mathfrak{D}(\mathfrak{C} \cap \mathfrak{B}) \cap \mathfrak{A} = (\mathfrak{A} \cap \mathfrak{D})\,(\mathfrak{C} \cap \mathfrak{B}).$$

23.4. Series of subgroups

In this chapter we shall describe the properties of the series of subgroups on the basis of our theory of series of decompositions, developed in Chapter 10. This new theory will prove extremely useful in connection with invariant subgroups (24.6) considered in the classical theory of groups.

1. *Basic notions.* Let $\mathfrak{A} \supset \mathfrak{B}$ denote arbitrary subgroups of $\mathfrak{G}$. By a *series of subgroups of the group $\mathfrak{G}$, from $\mathfrak{A}$ to $\mathfrak{B}$*, briefly, a *series from $\mathfrak{A}$ to $\mathfrak{B}$*, we mean a finite $\alpha(\geqq 1)$-membered sequence of subgroups $\mathfrak{A}_1, \ldots, \mathfrak{A}_\alpha$ of $\mathfrak{G}$ such that: a) The first and the last member of the sequence is $\mathfrak{A}$ and $\mathfrak{B}$, respectively, i.e., $\mathfrak{A}_1 = \mathfrak{A}$, $\mathfrak{A}_\alpha = \mathfrak{B}$; b) each subsequent member is a subgroup of the subgroup immediately preceding it, thus:

$$(\mathfrak{A} =)\ \mathfrak{A}_1 \supset \mathfrak{A}_2 \supset \cdots \supset \mathfrak{A}_\alpha\ (= \mathfrak{B}).$$

Such a series is briefly denoted by $(\mathfrak{A})$. The subgroups $\mathfrak{A}_1, \ldots, \mathfrak{A}_\alpha$ are the *members of* $(\mathfrak{A})$. $\mathfrak{A}_1$ is the *initial* and $\mathfrak{A}_\alpha$ the *final member of* $(\mathfrak{A})$. By the *length of* $(\mathfrak{A})$ we mean the number of its members.

Each subgroup $\mathfrak{A}$ of $\mathfrak{G}$, for example, is a series of length 1 whose initial as well as final member coincides with $\mathfrak{A}$.

Now let $\big((\mathfrak{A}) =\big)\ \mathfrak{A}_1 \supset \cdots \supset \mathfrak{A}_\alpha$ be a series from $\mathfrak{A}$ to $\mathfrak{B}$.

A member of $(\mathfrak{A})$ is called *essential* if it is either the initial member $\mathfrak{A}_1$ or a proper subgroup (19.4.1) of the member immediately preceding it; otherwise it is said to be *inessential.* If there occurs in $(\mathfrak{A})$ at least one inessential member $\mathfrak{A}_{\gamma+1}$, then we say (since $\mathfrak{A}_{\gamma+1} = \mathfrak{A}_\gamma$) that $(\mathfrak{A})$ is a *series with iteration.* If all the members of $(\mathfrak{A})$ are essential, then $(\mathfrak{A})$ is a *series without iteration.* The number α' of the essential members of $(\mathfrak{A})$ is the *reduced length of* $(\mathfrak{A})$. Obviously, there holds:

$1 \leqq \alpha' \leqq \alpha$, where $\alpha' = \alpha$, characterizes the series without iteration. Analogously as in case of series of decompositions (10.1), $(\mathfrak{A})$ may be reduced by omitting all the inessential members (if there are any) or lengthened by inserting further subgroups. The notion of a *partial series* or a *part of* $(\mathfrak{A})$ does not need any further explanation.

By a *refinement of* $(\mathfrak{A})$ we mean a series of subgroups of $\mathfrak{G}$ containing $(\mathfrak{A})$ as its own part. Every refinement of $(\mathfrak{A})$ has therefore the following form:

$$\mathfrak{A}_{1,1} \supset \cdots \supset \mathfrak{A}_{1,\beta_1-1} \supset \mathfrak{A}_{1,\beta_1} \supset \mathfrak{A}_{2,1} \supset \cdots$$
$$\cdots \supset \mathfrak{A}_{2,\beta_2-1} \supset \mathfrak{A}_{2,\beta_2} \supset \cdots \supset \mathfrak{A}_{\alpha,\beta_\alpha} \supset \mathfrak{A}_{\alpha+1,1} \supset \cdots \supset \mathfrak{A}_{\alpha+1,\beta_{\alpha+1}-1},$$

where $\mathfrak{A}_{\gamma,\beta_\gamma} = \mathfrak{A}_\gamma$, $\gamma = 1, \ldots, \alpha$, and the symbols $\beta_1, \ldots, \beta_{\alpha+1}$ stand for positive integers; if $\beta_\delta = 1$, then the members $\mathfrak{A}_{\delta,1} \supset \cdots \supset \mathfrak{A}_{\delta,\beta_\delta-1}$ are not read.

2. *Associated series of left and right decompositions.* Let

$$\bigl((\mathfrak{A}) =\bigr)\ \mathfrak{A}_1 \supset \cdots \supset \mathfrak{A}_\alpha$$

be a series of subgroups of $\mathfrak{G}$.

Associate, with $(\mathfrak{A})$, the following series of left and right decompositions:

$$\bigl((\mathfrak{G}/_l\mathfrak{A}) =\bigr)\ \mathfrak{G}/_l\mathfrak{A}_1 \geqq \cdots \geqq \mathfrak{G}/_l\mathfrak{A}_\alpha,$$
$$\bigl((\mathfrak{G}/_r\mathfrak{A}) =\bigr)\ \mathfrak{G}/_r\mathfrak{A}_1 \geqq \cdots \geqq \mathfrak{G}/_r\mathfrak{A}_\alpha.$$

Then we speak about *series of left or right decompositions associated with or corresponding to* $(\mathfrak{A})$. It is obvious that the series $(\mathfrak{G}/_l\mathfrak{A})$ or $(\mathfrak{G}/_r\mathfrak{A})$ is obtained by replacing each member $\mathfrak{A}_\gamma$ $(\gamma = 1, \ldots, \alpha)$ of $(\mathfrak{A})$ by $\mathfrak{G}/_l\mathfrak{A}_\gamma$ or $\mathfrak{G}/_r\mathfrak{A}_\gamma$, respectively.

Consider, for example, the series of left decompositions $(\mathfrak{G}/_l\mathfrak{A})$. In the same way we could, of course, consider the series of right decompositions $(\mathfrak{G}/_r\mathfrak{A})$.

First, the statements set out below are evidently correct:

The series $(\mathfrak{A})$ *and* $(\mathfrak{G}/_l\mathfrak{A})$ *have the same length* α.

The series $(\mathfrak{A})$ *and* $(\mathfrak{G}/_l\mathfrak{A})$ *are simultaneously without or with iteration and have the same reduced length* α' $(\leqq \alpha)$.

The series of left decompositions $(\mathfrak{G}/_l\mathring{\mathfrak{A}})$ *associated with an arbitrary refinement* $(\mathring{\mathfrak{A}})$ *of* $(\mathfrak{A})$ *is a refinement of* $(\mathfrak{G}/_l\mathfrak{A})$.

By means of the notion of associated series of left and right decompositions we may study the properties of the series of subgroups on the basis of the theory of the series of decompositions. All we have to do is to apply the considerations relative to the series of decompositions to the series of subgroups. But we must make sure to apply only those properties as are common to both the left and the right decompositions. The importance of this remark will be realized later.

3. *The manifold of local chains.* Consider an arbitrary series $(\mathfrak{A})$ of subgroups of $\mathfrak{G}$:

$$\big((\mathfrak{A}) =\big)\ \mathfrak{A}_1 \supset \cdots \supset \mathfrak{A}_\alpha \qquad (\alpha \geqq 1)$$

and, furthermore, the corresponding series of left and right decompositions on $\mathfrak{G}$:

$$\big((\mathfrak{G}/_l\mathfrak{A}) =\big)\ \mathfrak{G}/_l\mathfrak{A}_1 \geqq \cdots \geqq \mathfrak{G}/_l\mathfrak{A}_\alpha,$$
$$\big((\mathfrak{G}/_r\mathfrak{A}) =\big)\ \mathfrak{G}/_r\mathfrak{A}_1 \geqq \cdots \geqq \mathfrak{G}/_r\mathfrak{A}_\alpha.$$

We know that to each element $\bar{a}$ of $(\mathfrak{G}/_l\mathfrak{A}_\alpha)$ or $(\mathfrak{G}/_r\mathfrak{A}_\alpha)$, respectively, there corresponds a local chain of the series $(\mathfrak{G}/_l\mathfrak{A})$ or $(\mathfrak{G}/_r\mathfrak{A})$ with the base $\bar{a}$. The set of the local chains belonging to the individual elements of $\mathfrak{G}/_l\mathfrak{A}_\alpha$ or $\mathfrak{G}/_r\mathfrak{A}_\alpha$, respectively, is the *left or the right manifold of local chains corresponding to* $(\mathfrak{A})$. Notation: $\tilde{\boldsymbol{A}}_l$, $\tilde{\boldsymbol{A}}_r$.

Our object now is to study the relationship between $\tilde{\boldsymbol{A}}_l$ and $\tilde{\boldsymbol{A}}_r$.

First, let us remark that to every left or right coset $\bar{a}$ with regard to a subgroup of $\mathfrak{G}$ there exists an inverse right or left coset $\bar{a}^{-1}$, respectively; $\bar{a}^{-1}$ consists of all the points inverse of the individual points lying in $\bar{a}$ (20.2.8).

Now consider two mutually inverse cosets $\bar{a} \in \mathfrak{G}/_l\mathfrak{A}_\alpha$, $\bar{a}^{-1} \in \mathfrak{G}/_r\mathfrak{A}_\alpha$ and the corresponding local chains of $(\mathfrak{G}/_l\mathfrak{A})$, $(\mathfrak{G}/_r\mathfrak{A})$ with the bases $\bar{a}$, $\bar{a}^{-1}$:

$$([\bar{K}\bar{a}] =)\ \bar{K}_1\bar{a} \to \cdots \to \bar{K}_\alpha\bar{a},$$
$$([\bar{K}\bar{a}^{-1}] =)\ \bar{K}_1\bar{a}^{-1} \to \cdots \to \bar{K}_\alpha\bar{a}^{-1}.$$

In the above formulae we have denoted the local chains $[\bar{K}\bar{a}]$, $[\bar{K}\bar{a}^{-1}]$ and their members $\bar{K}_\gamma\bar{a}$, $\bar{K}_\gamma\bar{a}^{-1}$ by the same symbol $\bar{K}$ although the local chains or the members of the series $(\mathfrak{G}/_l\mathfrak{A})$, $(\mathfrak{G}/_r\mathfrak{A})$ in question are generally different from one another. This simplification cannot cause any confusion, since the notation of the local chains and their members differs in the symbols of the bases $\bar{a}$, $\bar{a}^{-1}$. A similar simplification will be employed even in the further considerations.

Let $\bar{a}_\gamma$ be an element of the decomposition $\mathfrak{G}/_l\mathfrak{A}_\gamma$ whose subset is $\bar{a}$ $(\gamma = 1, \ldots, \alpha)$. Then the inverse coset $\bar{a}_\gamma^{-1}$ is an element of $\mathfrak{G}/_r\mathfrak{A}_\gamma$ whose subset is $\bar{a}^{-1}$. There evidently holds:

$$\bar{a}_1 \supset \cdots \supset \bar{a}_\alpha\ (= \bar{a}), \qquad \bar{a}_1^{-1} \supset \cdots \supset \bar{a}_\alpha^{-1}\ (= \bar{a}^{-1})$$

and, furthermore,

$$\bar{a}_\gamma = \bar{a}\mathfrak{A}_\gamma, \qquad \bar{K}_\gamma\bar{a} = \bar{a}_\gamma \sqcap \mathfrak{G}/_l\mathfrak{A}_{\gamma+1},$$
$$\bar{a}_\gamma^{-1} = \mathfrak{A}_\gamma\bar{a}^{-1}, \qquad \bar{K}_\gamma\bar{a}^{-1} = \bar{a}_\gamma^{-1} \sqcap \mathfrak{G}/_r\mathfrak{A}_{\gamma+1} \qquad (\mathfrak{A}_{\alpha+1} = \mathfrak{A}_\alpha).$$

Either of the decompositions $\bar{K}_\gamma\bar{a}$, $\bar{K}_\gamma\bar{a}^{-1}$ is mapped, under the extended inversion $\boldsymbol{n}$ of $\mathfrak{G}$, onto the other (21.8.5) and so $\boldsymbol{n}\bar{K}_\gamma\bar{a} = \bar{K}_\gamma\bar{a}^{-1}$, $\boldsymbol{n}\bar{K}_\gamma\bar{a}^{-1} = \bar{K}_\gamma\bar{a}$. With regard to this, any two members $\bar{K}_\gamma\bar{a}$, $\bar{K}_\gamma\bar{a}^{-1}$ with the same index γ $(= 1, \ldots, \alpha)$ are called *mutually inverse*; the same term is employed for the local chains $[\bar{K}\bar{a}]$, $[\bar{K}\bar{a}^{-1}]$. Two mutually inverse members of $[\bar{K}\bar{a}]$ and $[\bar{K}\bar{a}^{-1}]$ are equivalent sets (21.8.5).

It is easy to verify that *the manifolds of local chains, $\tilde{\boldsymbol{A}}_l$, $\tilde{\boldsymbol{A}}_r$, are strongly equivalent.*

Indeed, associating with every local chain $[\bar{K}\bar{a}] \in \tilde{\boldsymbol{A}}_l$ its inverse: $[\bar{K}\bar{a}^{-1}] \in \tilde{\boldsymbol{A}}_r$, we obtain a simple mapping $\boldsymbol{f}$ of the manifold $\tilde{\boldsymbol{A}}_l$ onto $\tilde{\boldsymbol{A}}_r$. The mapping $\boldsymbol{f}$ is a strong equivalence because every two mutually inverse members of $[\bar{K}\bar{a}]$ and $\boldsymbol{f}[\bar{K}\bar{a}] = [\bar{K}\bar{a}^{-1}]$ are equivalent sets.

4. *Pairs of series of subgroups.* Consider a pair of series of subgroups of $\mathfrak{G}$:

$$\big((\mathfrak{A}) =\big)\ \mathfrak{A}_1 \supset \cdots \supset \mathfrak{A}_\alpha \qquad (\alpha \geqq 1),$$
$$\big((\mathfrak{B}) =\big)\ \mathfrak{B}_1 \supset \cdots \supset \mathfrak{B}_\beta \qquad (\beta \geqq 1).$$

To $(\mathfrak{A})$ and $(\mathfrak{B})$ there correspond the following series of left decompositions of $\mathfrak{G}$:

$$\begin{aligned} \big((\mathfrak{G}/_l\mathfrak{A}) =\big)\ \mathfrak{G}/_l\mathfrak{A}_1 \geqq \cdots \geqq \mathfrak{G}/_l\mathfrak{A}_\alpha, \\ \big((\mathfrak{G}/_l\mathfrak{B}) =\big)\ \mathfrak{G}/_l\mathfrak{B}_1 \geqq \cdots \geqq \mathfrak{G}/_l\mathfrak{B}_\beta \end{aligned} \tag{1}$$

and the left manifolds of local chains: $\tilde{\boldsymbol{A}}_l$, $\tilde{\boldsymbol{B}}_l$.

Analogously, to $(\mathfrak{A})$, $(\mathfrak{B})$ there belong the series of right decompositions of $\mathfrak{G}$:

$$\begin{aligned} \big((\mathfrak{G}/_r\mathfrak{A}) =\big)\ \mathfrak{G}/_r\mathfrak{A}_1 \geqq \cdots \geqq \mathfrak{G}/_r\mathfrak{A}_\alpha, \\ \big((\mathfrak{G}/_r\mathfrak{B}) =\big)\ \mathfrak{G}/_r\mathfrak{B}_1 \geqq \cdots \geqq \mathfrak{G}/_r\mathfrak{B}_\beta \end{aligned} \tag{2}$$

and the right manifolds of local chains: $\tilde{\boldsymbol{A}}_r$, $\tilde{\boldsymbol{B}}_r$.

Under these circumstances there applies the theorem:

If the series (1) *or* (2) *are in any of the following four relations, then the series* (2) *or* (1), *respectively, are in the same relation*: *The series* (1) *or* (2), *respectively, are* a) *complementary,* b) *chain-equivalent,* c) *loosely joint or co-basally loosely joint,* d) *joint or co-basally joint.*

Proof. Suppose, for example, that the series (1) are complementary.

In that case each member $\mathfrak{G}/_l\mathfrak{A}_\mu$, of $(\mathfrak{G}/_l\mathfrak{A})$ is complementary to each member $\mathfrak{G}/_l\mathfrak{B}_\nu$ of $(\mathfrak{G}/_l\mathfrak{B})$ (10.8); $\mu = 1, \ldots, \alpha$; $\nu = 1, \ldots, \beta$. Consequently, each member $\mathfrak{A}_\mu$ of $(\mathfrak{A})$ is interchangeable with each member $\mathfrak{B}_\nu$ of $(\mathfrak{B})$ (21.6). Obviously, even each member $\mathfrak{G}/_r\mathfrak{A}_\mu$ of $(\mathfrak{G}/_r\mathfrak{A})$ is complementary to each member $\mathfrak{G}/_r\mathfrak{B}_\nu$ of $(\mathfrak{G}/_r\mathfrak{B})$ (21.6) so that the series (2) are complementary.

Let us now assume that the series (1), for example, are in one of the relations b), c), d). Then the series (1), (2) and therefore even the series $(\mathfrak{A})$, $(\mathfrak{B})$ have the same length $\alpha = \beta$ and in each of the mentioned cases there exists a simple mapping f_l (strong equivalence, equivalence connected with loose coupling, equivalence connected with coupling) of the manifolds $\tilde{A}_l$ onto $\tilde{B}_l$ which may be co-basal. By means of f_l we define a simple mapping f_r of $\tilde{A}_r$ onto $\tilde{B}_r$ by way of associating, with each element $[\bar{K}\bar{a}] \in \tilde{A}_r$, the inverse local chain $[\bar{K}\bar{a}^{-1}] \in \tilde{A}_l$ and, with $[\bar{K}\bar{a}]$, the local chain $f_r[\bar{K}\bar{a}] = [\bar{K}\bar{b}^{-1}] \in \tilde{B}_r$ inverse of $f_l[\bar{K}\bar{a}^{-1}] = [\bar{K}\bar{b}] \in \tilde{B}_l$. If the mapping f_l is co-basal, then $\bar{b} = \bar{a}^{-1}$ and therefore $\bar{b}^{-1} = \bar{a}$; consequently even f_r is co-basal.

Now let $[\bar{K}\bar{a}]$, $f_r[\bar{K}\bar{a}] = [\bar{K}\bar{b}^{-1}]$ be arbitrary elements of the manifolds $\tilde{A}_r$, $\tilde{B}_r$, representing the inverse image and the image under the mapping f_r, respectively. Consider the corresponding inverse local chains $[\bar{K}\bar{a}^{-1}] \in \tilde{A}_l$, $f_l[\bar{K}\bar{a}^{-1}] = [\bar{K}\bar{b}] \in \tilde{B}_l$:

$$([\bar{K}\bar{a}^{-1}] =)\ \bar{K}_1\bar{a}^{-1} \to \cdots \to \bar{K}_\alpha\bar{a}^{-1},$$
$$([\bar{K}\bar{b}] =)\ \bar{K}_1\bar{b} \to \cdots \to \bar{K}_\alpha\bar{b}.$$

Since the series (1) are in one of the relations b), c), d), there exists a permutation $\boldsymbol{p}$ of the set $\{1, \ldots, \alpha\}$ such that every two members $\bar{K}_\gamma\bar{a}^{-1}$, $\bar{K}_\delta\bar{b}$ of the local chains $[\bar{K}\bar{a}^{-1}]$, $[\bar{K}\bar{b}]$ are equivalent or loosely coupled or coupled decompositions in $\mathfrak{G}$; at the same time $\delta = \boldsymbol{p}\gamma$. Let us apply the permutation $\boldsymbol{p}$ to the local chains $[\bar{K}\bar{a}] \in \tilde{A}_r$, $f_r[\bar{K}\bar{a}] = [\bar{K}\bar{b}^{-1}] \in \tilde{B}_r$ by associating, with each member $\bar{K}_\gamma\bar{a}$ of the first local chain, the member $\bar{K}_\delta\bar{b}^{-1}$ of the second. Every pair of such members $\bar{K}_\gamma\bar{a}$, $\bar{K}_\delta\bar{b}^{-1}$ represents decompositions in $\mathfrak{G}$ that are inverse of the equivalent or loosely coupled or coupled decompositions $\bar{K}_\gamma\bar{a}^{-1}$, $\bar{K}_\delta\bar{b}$. Hence even $\bar{K}_\gamma\bar{a}$, $\bar{K}_\delta\bar{b}^{-1}$ are equivalent or loosely coupled or coupled (7.3.4) and the proof is complete.

The symmetry we have just verified in the relations between the series of the left and the right decompositions corresponding to the series $(\mathfrak{A})$, $(\mathfrak{B})$, respectively, leads to the following definition:

The series of subgroups, $(\mathfrak{A})$ and $(\mathfrak{B})$, are called: a) *complementary* or *interchangeable,* b) *chain-equivalent* or *co-basally chain-equivalent,* c) *semi-joint* or *loosely joint,* or *co-basally semi-joint* or *co-basally loosely joint,* d) *joint* or *co-basally joint* if the series of the left decompositions of $\mathfrak{G}$, namely $(\mathfrak{G}/_l\mathfrak{A})$, $(\mathfrak{G}/_l\mathfrak{B})$, and therefore (by the above theorem) even the series of the right decompositions of $\mathfrak{G}$, namely, $(\mathfrak{G}/_r\mathfrak{A})$, $(\mathfrak{G}/_r\mathfrak{B})$ belonging to $(\mathfrak{A})$ and $(\mathfrak{B})$, have the corresponding property.

5. *Complementary series of subgroups.* Consider two complementary series of subgroups of $\mathfrak{G}$:

$$\big((\mathfrak{A}) =\big)\ \mathfrak{A}_1 \supset \cdots \supset \mathfrak{A}_\alpha \qquad (\alpha \geqq 1),$$
$$\big((\mathfrak{B}) =\big)\ \mathfrak{B}_1 \supset \cdots \supset \mathfrak{B}_\beta \qquad (\beta \geqq 1).$$

To these series there belong the corresponding series of the left and the right decompositions of $\mathfrak{G}$, namely, $(\mathfrak{G}/_l\mathfrak{A})$, $(\mathfrak{G}/_l\mathfrak{B})$ and $(\mathfrak{G}/_r\mathfrak{A})$, $(\mathfrak{G}/_r\mathfrak{B})$, respectively, which are more accurately described by the formulae (1), (2).

There holds the following theorem:

The series $(\mathfrak{A})$, $(\mathfrak{B})$ *have co-basally joint refinements* $(\mathfrak{A}_*)$, $(\mathfrak{B}_*)$ *with coinciding initial and final members. The refinements are given by the construction described in part* a) *of the following proof.*

Proof. a) Under the above assumption, every two decompositions $\mathfrak{G}/_l\mathfrak{A}_\gamma$, $\mathfrak{G}/_l\mathfrak{B}_\delta$ $(\gamma = 1, \ldots, \alpha;\ \delta = 1, \ldots, \beta)$ are complementary, hence every two subgroups $\mathfrak{A}_\gamma$, $\mathfrak{B}_\delta$ are interchangeable (21.6). By 22.2.1 even the subgroups $\mathfrak{A}_\gamma$, $\mathfrak{A}_{\gamma-1} \cap \mathfrak{B}_\nu$ or $\mathfrak{B}_\delta$, $\mathfrak{B}_{\delta-1} \cap \mathfrak{A}_\mu$ $(\mathfrak{A}_0 = \mathfrak{B}_0 = \mathfrak{G};\ \mu = 1, \ldots, \alpha;\ \nu = 1, \ldots, \beta)$ are interchangeable and there holds:

$$\begin{aligned} (\mathfrak{A}_{\gamma,\nu} =)\ \mathfrak{A}_\gamma(\mathfrak{A}_{\gamma-1} \cap \mathfrak{B}_\nu) &= \mathfrak{A}_{\gamma-1} \cap \mathfrak{A}_\gamma\mathfrak{B}_\nu, \\ (\mathfrak{B}_{\delta,\mu} =)\ \mathfrak{B}_\delta(\mathfrak{B}_{\delta\;1} \cap \mathfrak{A}_\mu) &= \mathfrak{B}_{\delta-1} \cap \mathfrak{B}_\delta\mathfrak{A}_\mu. \end{aligned} \tag{3}$$

Let us denote:

$$\begin{aligned} &\mathfrak{A}_1\mathfrak{B}_1 = \mathfrak{U}, \quad \mathfrak{A}_\alpha \cap \mathfrak{B}_\beta = \mathfrak{V}, \\ &\mathfrak{A}_0 = \mathfrak{B}_0 = \mathfrak{G}, \quad \mathfrak{A}_{\alpha+1} = \mathfrak{B}_{\beta+1} = \mathfrak{V}. \end{aligned}$$

Then the formulae (3) are true for $\gamma, \mu = 1, \ldots, \alpha + 1;\ \delta, \nu = 1, \ldots, \beta + 1$. From the definition of the subgroups $\mathfrak{A}_{\gamma,\nu}$, $\mathfrak{B}_{\delta,\mu}$ there follows

$$\begin{aligned} &\mathfrak{A}_{\gamma-1} \supset \mathfrak{A}_{\gamma,\nu}, \quad \mathfrak{A}_{\gamma,\beta+1} = \mathfrak{A}_\gamma, \\ &\mathfrak{B}_{\delta-1} \supset \mathfrak{B}_{\delta,\mu}, \quad \mathfrak{B}_{\delta,\alpha+1} = \mathfrak{B}_\delta, \end{aligned}$$

moreover, for $\nu \leqq \beta$, $\mu \leqq \alpha$, we have

$$\mathfrak{A}_{\gamma,\nu} \supset \mathfrak{A}_{\gamma,\nu+1}, \quad \mathfrak{B}_{\delta,\mu} \supset \mathfrak{B}_{\delta,\mu+1}.$$

Thus we arrive at the following series of subgroups from $\mathfrak{A}_{\gamma,1}$ to $\mathfrak{A}_\gamma$ and from $\mathfrak{B}_{\delta,1}$ to $\mathfrak{B}_\delta$:

$$\begin{aligned} &\mathfrak{A}_{\gamma,1} \supset \cdots \supset \mathfrak{A}_{\gamma,\beta+1}, \\ &\mathfrak{B}_{\delta,1} \supset \cdots \supset \mathfrak{B}_{\delta,\alpha+1}. \end{aligned}$$

Consequently, the series of the subgroups of $\mathfrak{G}$ set below are refinements of the series $(\mathfrak{A})$, $(\mathfrak{B})$:

$$\begin{aligned} \big((\mathfrak{A}_*) =\big)\ \mathfrak{U} = \mathfrak{A}_{1,1} \supset \cdots \supset \mathfrak{A}_{1,\beta+1} \supset \mathfrak{A}_{2,1} \supset& \\ \cdots \supset \mathfrak{A}_{2,\beta+1} \supset \cdots \supset \mathfrak{A}_{\alpha+1,1} \supset \cdots \supset \mathfrak{A}_{\alpha+1,\beta+1} &= \mathfrak{V}, \\ \big((\mathfrak{B}_*) =\big)\ \mathfrak{U} = \mathfrak{B}_{1,1} \supset \cdots \supset \mathfrak{B}_{1,\alpha+1} \supset \mathfrak{B}_{2,1} \supset& \\ \cdots \supset \mathfrak{B}_{2,\alpha+1} \supset \cdots \supset \mathfrak{B}_{\beta+1,1} \supset \cdots \supset \mathfrak{B}_{\beta+1,\alpha+1} &= \mathfrak{V}. \end{aligned}$$

We observe that $(\mathfrak{A}_*)$, $(\mathfrak{B}_*)$ have the same length and that their initial and final members coincide:

$$(\mathfrak{U} =)\, \mathfrak{A}_{1,1} = \mathfrak{B}_{1,1}, \qquad \mathfrak{A}_{\alpha+1,\beta+1} = \mathfrak{B}_{\beta+1,\alpha+1}\ (= \mathfrak{B}).$$

The series $(\mathfrak{A}_*)$, $(\mathfrak{B}_*)$ *are the mentioned co-basally joint refinements of the series* $(\mathfrak{A})$, $(\mathfrak{B})$.

b) Let us show that the series of the left decompositions, $(\mathfrak{G}/_l\mathfrak{A}_*)$, $(\mathfrak{G}/_l\mathfrak{B}_*)$, corresponding to $(\mathfrak{A}_*)$, $(\mathfrak{B}_*)$ are co-basally joint. These series are obtained by way of replacing each member $\mathfrak{A}_{\gamma,\nu}$ of $(\mathfrak{A}_*)$ by the left decomposition $\mathfrak{G}/_l\mathfrak{A}_{\gamma,\nu}$ and each member $\mathfrak{B}_{\delta,\mu}$ of $(\mathfrak{B}_*)$ by $\mathfrak{G}/_l\mathfrak{B}_{\delta,\mu}$.

Denote:

$$\bar{A}_\gamma = \mathfrak{G}/_l\mathfrak{A}_\gamma, \quad \bar{B}_\delta = \mathfrak{G}/_l\mathfrak{B}_\delta, \quad \mathring{A}_{\gamma,\nu} = \mathfrak{G}/_l\mathfrak{A}_{\gamma,\nu}, \quad \mathring{B}_{\delta,\mu} = \mathfrak{G}/_l\mathfrak{B}_{\delta,\mu}.$$

Then, on taking account of the formulae (3) and in accordance with 21.4 and 21.5, we have

$$\mathring{A}_{\gamma,\nu} = [\bar{A}_\gamma, (\bar{A}_{\gamma-1}, \bar{B}_\nu)] = (\bar{A}_{\gamma-1}, [\bar{A}_\gamma, \bar{B}_\nu]),$$
$$\mathring{B}_{\delta,\mu} = [\bar{B}_\delta, (\bar{B}_{\delta-1}, \bar{A}_\mu)] = (\bar{B}_{\delta-1}, [\bar{B}_\delta, \bar{A}_\mu]).$$

We see that the series of decompositions, $(\mathfrak{G}/_l\mathfrak{A}_*)$, $(\mathfrak{G}/_l\mathfrak{B}_*)$, corresponding to $(\mathfrak{A}_*)$, $(\mathfrak{B}_*)$ are formed from the complementary series $(\mathfrak{G}/_l\mathfrak{A})$, $(\mathfrak{G}/_l\mathfrak{B})$ by the construction described in 10.7, part a) of the proof. Hence, by 10.8, the series $(\mathfrak{G}/_l\mathfrak{A}_*)$, $(\mathfrak{G}/_l\mathfrak{B}_*)$ are co-basally joint and the proof is complete.

23.5. Exercises

1. Apply the five-group theorem to subgroups of $\mathfrak{Z}$ (18.5.1).
2. Let $\mathfrak{A} \supset \mathfrak{B}$, $\mathfrak{C} \supset \mathfrak{D}$ be subgroups of $\mathfrak{G}$ and $A \supset B$, $C \supset D$ their fields. Suppose the left decompositions $(\bar{A} =)\mathfrak{A}/_l\mathfrak{B}$, $(\bar{C} =)\mathfrak{C}/_l\mathfrak{D}$ in $\mathfrak{G}$ are adjoint with regard to B, D. Realize the construction described in 4.2 and leading to the coupled coverings $\mathring{A}$, $\mathring{C}$ of the decompositions $\bar{A}_1 = C \sqsubset \mathfrak{A}/_l\mathfrak{B}$, $\bar{C}_1 = A \sqsubset \mathfrak{C}/_l\mathfrak{D}$.

24. Invariant (normal) subgroups

24.1. Definition

Let $\mathfrak{A} \supset \mathfrak{B}$ be arbitrary subgroups of $\mathfrak{G}$. If the left and the right coset of every element $a \in \mathfrak{A}$ with regard to $\mathfrak{B}$ coincide, that is to say, if $a\mathfrak{B} = \mathfrak{B}a$, then $\mathfrak{B}$ is said to be *invariant* or *normal in* $\mathfrak{A}$. The left decomposition of $\mathfrak{A}$, generated by $\mathfrak{B}$ is, in that case, the same as the right decomposition; *both decompositions therefore coincide and form a certain decomposition of* $\mathfrak{A}$, called the *decomposition generated by* $\mathfrak{B}$, hence $\mathfrak{A}/_l\mathfrak{B} = \mathfrak{A}/_r\mathfrak{B}$ $(= \mathfrak{A}/\mathfrak{B})$.

In the following study of invariant subgroups lying in the same subgroup $\mathfrak{A}$, we shall restrict our attention to the case $\mathfrak{A} = \mathfrak{G}$. If a subgroup $\mathfrak{B}$ of $\mathfrak{G}$ is invariant in $\mathfrak{G}$, then it is called, for convenience, an *invariant subgroup of* $\mathfrak{G}$.

24.2. Basic properties of invariant subgroups

In $\mathfrak{G}$ there exist at least two (may be coinciding) invariant subgroups: the greatest subgroup, identical with $\mathfrak{G}$ and the least subgroup $\{\underline{1}\}$ consisting of the single element $\underline{1} \in \mathfrak{G}$. They are called the *extreme invariant subgroups of* $\mathfrak{G}$. Groups may also contain subgroups that are not invariant, for example, the subgroup $\mathfrak{A}$ of $\mathfrak{S}_3$ consisting of the two permutations $\underline{1}$, $\boldsymbol{f}$ (notation as in 22.1) is not invariant in $\mathfrak{S}_3$ because, as we have observed, there holds, e.g., $\boldsymbol{a}\mathfrak{A} = \{\boldsymbol{a}, \boldsymbol{c}\}$, $\mathfrak{A}\boldsymbol{a} = \{\boldsymbol{a}, \boldsymbol{d}\}$ and so $\boldsymbol{a}\mathfrak{A} \neq \mathfrak{A}\boldsymbol{a}$.

Let $\mathfrak{A} \supset \mathfrak{B}$ be subgroups of $\mathfrak{G}$. If $\mathfrak{B}$ is invariant in $\mathfrak{G}$, then it naturally has the same property in $\mathfrak{A}$. If, conversely, $\mathfrak{B}$ is invariant in $\mathfrak{A}$, it is not necessarily invariant in $\mathfrak{G}$, since the equality $x\mathfrak{B} = \mathfrak{B}x$ may apply to all the elements x of $\mathfrak{A}$ without applying to all the elements of $\mathfrak{G}$. If, for instance, a subgroup $\mathfrak{A}$ is not invariant in $\mathfrak{G}$, it is invariant in $\mathfrak{A}$ but not in $\mathfrak{G}$.

If a subgroup $\mathfrak{A}$ is invariant in $\mathfrak{G}$, then it is interchangeable with every complex $C \subset \mathfrak{G}$. Indeed, in that case we have $x\mathfrak{A} = \mathfrak{A}x$ for every $x \in \mathfrak{G}$ and therefore even for every $x \in C$. Consequently, $C\mathfrak{A} = \mathfrak{A}C$. We observe, in particular, that any two subgroups $\mathfrak{A}, \mathfrak{C} \subset \mathfrak{G}$ one of which is invariant in $\mathfrak{G}$, are interchangeable.

If, vice versa, some subgroups $\mathfrak{A}, \mathfrak{C} \subset \mathfrak{G}$ are interchangeable, then neither of them is necessarily invariant in $\mathfrak{G}$. That occurs, for example, if $\mathfrak{A}$ is not invariant in $\mathfrak{G}$ and $\mathfrak{A} = \mathfrak{C}$.

Moreover, there applies the following theorem:

If the subgroups $\mathfrak{A}$, $\mathfrak{C}$ *are invariant in* $\mathfrak{G}$, *then even the intersection* $\mathfrak{A} \cap \mathfrak{C}$ *and the product* $\mathfrak{A}\mathfrak{C}$ *are invariant subgroups of* $\mathfrak{G}$.

In fact, if the assumption is satisfied, then there hold, for any $x \in \mathfrak{G}$, the equalities $x\mathfrak{A} = \mathfrak{A}x$, $x\mathfrak{C} = \mathfrak{C}x$. So we have, with respect to 20.2.6 and to an analogous theorem for right cosets,

$$x(\mathfrak{A} \cap \mathfrak{C}) = x\mathfrak{A} \cap x\mathfrak{C} = \mathfrak{A}x \cap \mathfrak{C}x = (\mathfrak{A} \cap \mathfrak{C})x;$$

consequently, $\mathfrak{A} \cap \mathfrak{C}$ is invariant in $\mathfrak{G}$. Furthermore we have, with respect to 12.9.8,

$$x(\mathfrak{A}\mathfrak{C}) = (x\mathfrak{A})\mathfrak{C} = (\mathfrak{A}x)\mathfrak{C} = \mathfrak{A}(x\mathfrak{C}) = \mathfrak{A}(\mathfrak{C}x) = (\mathfrak{A}\mathfrak{C})x;$$

consequently, $\mathfrak{A}\mathfrak{C}$ is invariant in $\mathfrak{G}$.

The information that any two invariant subgroups of $\mathfrak{G}$ are interchangeable and the properties of interchangeable subgroups (22.2) yield a number of results as to invariant subgroups. We shall introduce only the following two theorems:

1. *The Dedekind-Ore theorem. For any three invariant subgroups* $\mathfrak{A} \supset \mathfrak{B}$, $\mathfrak{D}$ *of* $\mathfrak{G}$ *there holds*:

$$\mathfrak{A} \cap \mathfrak{D}\mathfrak{B} = (\mathfrak{A} \cap \mathfrak{D})\mathfrak{B}.$$

2. *The system of all invariant subgroups of* $\mathfrak{G}$ *is closed with regard to the intersections and the products and is, when completed by the multiplications defined by forming intersections and products, a modular lattice with extreme elements.*

24.3. Generating decompositions of groups

1. *First theorem.* Let $\mathfrak{A}$ be a subgroup of $\mathfrak{G}$. As we have seen (21.1) $\mathfrak{A}$ generates a left decomposition $\mathfrak{G}/_l\mathfrak{A}$ and a right decomposition $\mathfrak{G}/_r\mathfrak{A}$ of $\mathfrak{G}$. Let us find out whether, for example, the left decomposition $\mathfrak{G}/_l\mathfrak{A}$ can be generating.

Suppose, first, that $\mathfrak{G}/_l\mathfrak{A}$ is generating and consider two elements $p\mathfrak{A}, q\mathfrak{A} \in \mathfrak{G}/_l\mathfrak{A}$, p, q being arbitrary elements of $\mathfrak{G}$. By the definition of a generating decomposition there exists an element $r\mathfrak{A} \in \mathfrak{G}/_l\mathfrak{A}$ such that:

$$p\mathfrak{A} \,.\, q\mathfrak{A} \subset r\mathfrak{A}.$$

Hence, in particular, $pq\mathfrak{A} = (p\underline{1}) \,.\, q\mathfrak{A} \subset r\mathfrak{A}$, thus $pq\mathfrak{A} \subset r\mathfrak{A}$ and, consequently, $pq = pq \,.\, \underline{1} \in r\mathfrak{A}$ whence, by 20.2.1 and 20.2.4, there follows $r\mathfrak{A} = pq\mathfrak{A}$. So we have, first, $p\mathfrak{A} \,.\, q\mathfrak{A} \subset pq\mathfrak{A}$. Each element of the left coset $pq\mathfrak{A}$ is the product $pq \,.\, x$ of the element pq and some element $x \in \mathfrak{A}$. There obviously holds pqx

$= (p\underline{1})\,(qx) \in p\mathfrak{A}\,.\,q\mathfrak{A}$; hence $p\mathfrak{A}\,.\,q\mathfrak{A} \subset pq\mathfrak{A}$. So we have

$$p\mathfrak{A}\,.\,q\mathfrak{A} = pq\mathfrak{A}, \tag{1}$$

i.e., the product of the left cosets $p\mathfrak{A}$ and $q\mathfrak{A}$ is the left coset $pq\mathfrak{A}$.

The equality (1) yields, in particular, for $q = p^{-1}$ the relations:

$$p\mathfrak{A}p^{-1} = p\mathfrak{A}(p^{-1}\underline{1}) \subset p\mathfrak{A}\,.\,p^{-1}\mathfrak{A} = pp^{-1}\mathfrak{A} = \mathfrak{A}$$

so that $p\mathfrak{A}p^{-1} \subset \mathfrak{A}$. Since p is an arbitrary element of $\mathfrak{G}$, the same holds even for p^{-1} and we have $p^{-1}\mathfrak{A}p \subset \mathfrak{A}$. Consequently,

$$\mathfrak{A} = (pp^{-1})\mathfrak{A}(pp^{-1}) = p(p^{-1}\mathfrak{A}p)p^{-1} \subset p\mathfrak{A}p^{-1},$$

i.e., $p\mathfrak{A}p^{-1} \supset \mathfrak{A}$. Hence

$$p\mathfrak{A}p^{-1} = \mathfrak{A}$$

or, which is the same, $p\mathfrak{A} = \mathfrak{A}p$. Therefore the left coset of each element $p \in \mathfrak{G}$ with regard to $\mathfrak{A}$ is, simultaneously, the right coset of p with regard to $\mathfrak{A}$. We see that $\mathfrak{A}$ is invariant in $\mathfrak{G}$.

Now let us assume, conversely, that the subgroup $\mathfrak{A}$ is invariant in $\mathfrak{G}$. Then, by the definition, there first follows that the left coset $p\mathfrak{A}$ of each element $p \in \mathfrak{G}$ with regard to $\mathfrak{A}$ is, simultaneously, the right coset $\mathfrak{A}p$ of p with regard to $\mathfrak{A}$. Then for any two left cosets $p\mathfrak{A}$, $q\mathfrak{A}$ there holds

$$p\mathfrak{A}\,.\,q\mathfrak{A} = p(\mathfrak{A}q)\mathfrak{A} = p(q\mathfrak{A})\mathfrak{A} = pq(\mathfrak{A}\mathfrak{A}) = pq\mathfrak{A}$$

which yields $p\mathfrak{A}\,.\,q\mathfrak{A} = pq\mathfrak{A}$. Hence, if our assumption is true, the product of $p\mathfrak{A}$ and $q\mathfrak{A}$ is $pq\mathfrak{A}$. Thus we have verified that the decomposition $\mathfrak{G}/_l\mathfrak{A}$ of $\mathfrak{G}$ which is, of course, equal to $\mathfrak{G}/_r\mathfrak{A}$ is generating and we may sum up the above results in the following theorem:

The left as well as the right decomposition of $\mathfrak{G}$ *generated by* $\mathfrak{A}$ *is generating if and only if the subgroup* $\mathfrak{A}$ *is invariant in* $\mathfrak{G}$. *Then the product of any elements* $p\mathfrak{A}$ *and* $q\mathfrak{A}$ *of the decomposition generated by* $\mathfrak{A}$ *is the element* $pq\mathfrak{A}$.

2. *Second theorem.* A remarkable property of the groups consists in that each generating decomposition of a group is generated by some of its invariant subgroups.

Consider a generating decomposition $\overline{G}$ of $\mathfrak{G}$. Since each element of $\mathfrak{G}$ is contained in some element of $\overline{G}$, there exists an element $A \in \overline{G}$ comprising the unit $\underline{1}$ of $\mathfrak{G}$. We shall prove that A is the field of an invariant subgroup $\mathfrak{A}$ of $\mathfrak{G}$ and $\overline{G}$ the decomposition of $\mathfrak{G}$ generated by $\mathfrak{A}$.

To that end, let us first consider that, since $\overline{G}$ is generating, there exists an element $\bar{a} \in \overline{G}$ such that $AA \subset \bar{a}$. As there holds, on the one hand, $\underline{1} = \underline{1}\,.\,\underline{1} \in AA \subset \bar{a}$ and, on the other hand, $\underline{1} \in A$, we have $\bar{a} = A$. Consequently, A is groupoidal.

The corresponding subgroupoid $\mathfrak{A}$ comprises the unit $\underline{1}$ of $\mathfrak{G}$ and, as we shall see, contains with each element a even its inverse a^{-1}.

Assuming $a \in A$, let $\bar{b}$ denote the element of $\bar{G}$ that includes a^{-1}. Since $\underline{1} = aa^{-1} \in A\bar{b}$, the element $\underline{1}$ is contained in the product $A\bar{b}$ and, of course, also in A. As $\bar{G}$ is generating and both subsets $A\bar{b}$ and A comprise the element $\underline{1}$, we have $A\bar{b} \subset A$. Hence: $\underline{1} \,.\, a^{-1} \in A$, i.e., $a^{-1} \in A$ which proves that $\mathfrak{A}$ is a subgroup of $\mathfrak{G}$.

It remains to be shown that $\mathfrak{A}$ is invariant in $\mathfrak{G}$ and that any element $\bar{a} \in \bar{G}$ is the coset of an arbitrary element $a \in \bar{a}$ with regard to $\mathfrak{A}$. Suppose $a \in \mathfrak{G}$ and let $\bar{a}$ denote the element of $\bar{G}$ containing a so that: $a \in \bar{a} \in \bar{G}$. If $x \in \bar{a}$, then $x = \underline{1} \,.\, x \in A\bar{a}$ whence $\bar{a} \subset A\bar{a}$. As $\bar{G}$ is generating and both subsets $A\bar{a}$, $\bar{a}$ comprise the element a, there holds $A\bar{a} \subset \bar{a}$. So we have $A\bar{a} = \bar{a}$ and, analogously, $\bar{a}A = \bar{a}$. Consequently,

$$\bar{a} = A\bar{a} = \bar{a}A. \tag{2}$$

There obviously holds $aA \subset \bar{a}A$. Let us show that there simultaneously holds $\bar{a}A \subset aA$. Let $\bar{b}$ denote the element of $\bar{G}$ comprising a^{-1}. As $\bar{G}$ is generating and both the subsets $\bar{b}\bar{a}$ and A include the element $\underline{1}$, there holds $\bar{b}\bar{a} \subset A$. Thus the product $a^{-1}x$ of a^{-1} and an element $x \in \bar{a}$ is contained in A. Consequently, $x = a(a^{-1}x) \in aA$ and we have $\bar{a} \subset aA$. Hence $\bar{a}A \subset aAA = aA$. So we have $\bar{a}A = aA$. Analogously we arrive at $A\bar{a} = Aa$. From that and from (2) there follows

$$\bar{a} = a\mathfrak{A} = \mathfrak{A}a.$$

From these equalities we, first, see that the subgroup $\mathfrak{A}$ is invariant in $\mathfrak{G}$. Since they hold for every element $a \in \mathfrak{G}$ and the element $\bar{a} \in \bar{G}$ comprising a, they also apply to any $\bar{a} \in \bar{G}$ and $a \in \bar{a}$; every element $\bar{a} \in \bar{G}$ is the coset of an arbitrary element $a \in \bar{a}$ with regard to $\mathfrak{A}$.

Thus we have determined all the generating decompositions of $\mathfrak{G}$:

All generating decompositions of $\mathfrak{G}$ are precisely those decompositions of $\mathfrak{G}$ that are generated by the individual invariant subgroups of $\mathfrak{G}$.

24.4. Properties of the generating decompositions of a group

On $\mathfrak{G}$ there always exist two generating decompositions, namely, the two extreme decompositions $\bar{G}_{\max}$ and $\bar{G}_{\min}$ (14.1) generated by the extreme invariant subgroups $\mathfrak{G}$, $\{\underline{1}\}$ of $\mathfrak{G}$ (24.2).

Let $\bar{A}$, $\bar{B}$ stand for arbitrary generating decompositions on $\mathfrak{G}$. By the above theorem, $\bar{A}$ and $\bar{B}$ are decompositions generated by appropriate subgroups $\mathfrak{A}$ and $\mathfrak{B}$ invariant in $\mathfrak{G}$, respectively. Consequently, $\mathfrak{A}$ and $\mathfrak{B}$ are interchangeable

From the results obtained in 21.3—6 we observe that $\bar{A}, \bar{B}$ have the properties stated below:

The decomposition $\bar{A}$ ($\bar{B}$) *is a covering (refinement) of* $\bar{B}$ ($\bar{A}$) *if and only if the subgroup* $\mathfrak{A}$ *is a supergroup of* $\mathfrak{B}$, *i.e.*, $\mathfrak{A} \supset B$.

The greatest common refinement $(\bar{A}, \bar{B})$ *of* $\bar{A}$ *and* $\bar{B}$ *is generated by the invariant subgroup* $\mathfrak{A} \cap \mathfrak{B}$.

The least common covering $[\bar{A}, \bar{B}]$ *of* $\bar{A}, \bar{B}$ *is generated by the invariant subgroup* $\mathfrak{A}\mathfrak{B}$.

$\bar{A}$ *and* $\bar{B}$ *are complementary.*

Furthermore, there holds:

The system of all generating decompositions of $\mathfrak{G}$ *is, with regard to the operations* (), [], *closed and is, together with the multiplications defined by the latter, a modular lattice with extreme elements. This lattice is isomorphic with the lattice consisting of invariant subgroups of* $\mathfrak{G}$ (24.2).

24.5. Further properties of invariant subgroups

The theorems (24.3) on generating decompositions in groups, together with the study of generating decompositions in groupoids and of decompositions of groups generated by subgroups, lead to fresh information about the properties of invariant subgroups.

1. *Let* $\mathfrak{A} \supset \mathfrak{B}, \mathfrak{C}$ *stand for subgroups of* $\mathfrak{G}$, *the subgroup* $\mathfrak{B}$ *being invariant in* $\mathfrak{A}$. *Then* $\mathfrak{B} \cap \mathfrak{C}$ *is invariant in* $\mathfrak{A} \cap \mathfrak{C}$. *Moreover, the subgroups* $\mathfrak{A} \cap \mathfrak{C}, \mathfrak{B}$ *are interchangeable and* $\mathfrak{B}$ *is invariant in* $(\mathfrak{A} \cap \mathfrak{C})\mathfrak{B}$.

Proof. a) Since $\mathfrak{B}$ is invariant in $\mathfrak{A}$, the decomposition $\mathfrak{A}/_l\mathfrak{B}$ is generating (24.3.1). By 21.2 (1), we have

$$\mathfrak{A}/_l\mathfrak{B} \sqcap \mathfrak{C} = (\mathfrak{A} \cap \mathfrak{C})/_l(\mathfrak{B} \cap \mathfrak{C}).$$

Furthermore, from 14.3.2 we know that the left decomposition in question of $\mathfrak{A} \cap \mathfrak{C}$ with regard to $\mathfrak{B} \cap \mathfrak{C}$ is generating. Consequently, $\mathfrak{B} \cap \mathfrak{C}$ is invariant in $\mathfrak{A} \cap \mathfrak{C}$ (24.3.1).

b) By 19.5.1, $\mathfrak{A} \cap \mathfrak{C}$ is a subgroup of $\mathfrak{A}$. As $\mathfrak{B}$ is invariant in $\mathfrak{A}$, the subgroups $\mathfrak{A} \cap \mathfrak{C}, \mathfrak{B}$ are interchangeable (24.2). In accordance with 21.2 (2), we have

$$\mathfrak{C} \sqsubset \mathfrak{A}/_l\mathfrak{B} = (\mathfrak{C} \cap \mathfrak{A})\mathfrak{B}/_l\mathfrak{B}.$$

Moreover, from 14.3.2 we know that the left decomposition in question of $(\mathfrak{C} \cap \mathfrak{A})\mathfrak{B}$ with regard to $\mathfrak{B}$ is generating. Hence $\mathfrak{B}$ is invariant in $(\mathfrak{A} \cap \mathfrak{C})\mathfrak{B}$ (24.3.1).

In particular, for $\mathfrak{A} = \mathfrak{G}$ we have the following theorem:

If $\mathfrak{B}, \mathfrak{C}$ are subgroups of $\mathfrak{G}$ and $\mathfrak{B}$ is invariant in $\mathfrak{G}$, then $\mathfrak{B} \cap \mathfrak{C}$ is invariant in $\mathfrak{C}$.

2. *Let $\mathfrak{A} \supset \mathfrak{B}, \mathfrak{C} \supset \mathfrak{D}$ be subgroups of $\mathfrak{G}$ while $\mathfrak{B}$ and $\mathfrak{D}$ are invariant in $\mathfrak{A}$ and $\mathfrak{C}$, respectively. Then $\mathfrak{A} \cap \mathfrak{D}$ and $\mathfrak{B} \cap \mathfrak{C}$ are invariant in $\mathfrak{A} \cap \mathfrak{C}$. Let, furthermore, $\mathfrak{U}$ be an invariant subgroup of $\mathfrak{A} \cap \mathfrak{C}$ such that*

$$(\mathfrak{A} \cap \mathfrak{C}) \supset \mathfrak{U} \supset (\mathfrak{A} \cap \mathfrak{D})(\mathfrak{C} \cap \mathfrak{B}). \tag{1}$$

Then $\mathfrak{A} \cap \mathfrak{C}$ and $\mathfrak{U}$ are interchangeable with both $\mathfrak{B}$ and $\mathfrak{D}$ and $\mathfrak{U}\mathfrak{B}$ or $\mathfrak{U}\mathfrak{D}$ is invariant in $(\mathfrak{A} \cap \mathfrak{C})\mathfrak{B}$ or $(\mathfrak{A} \cap \mathfrak{C})\mathfrak{D}$, respectively. Moreover, by 23.2.(1), *there holds:*

$$(\mathfrak{A} \cap \mathfrak{C})\mathfrak{B} \cap \mathfrak{U}\mathfrak{D} = \mathfrak{U} = (\mathfrak{C} \cap \mathfrak{A})\mathfrak{D} \cap \mathfrak{U}\mathfrak{B}.$$

Proof. In accordance with 1, the subgroups $\mathfrak{A} \cap \mathfrak{D}$, $\mathfrak{B} \cap \mathfrak{C}$ are invariant in $\mathfrak{A} \cap \mathfrak{C}$. Since $\mathfrak{A} \cap \mathfrak{C}$ and $\mathfrak{U}$ are subgroups of $\mathfrak{A}$ and $\mathfrak{C}$, respectively, and $\mathfrak{B}$ and $\mathfrak{D}$ are invariant in $\mathfrak{A}$ and $\mathfrak{C}$, respectively, $\mathfrak{A} \cap \mathfrak{C}$ and $\mathfrak{U}$ are interchangeable with both the subgroups $\mathfrak{B}$ and $\mathfrak{D}$.

By 1., $\mathfrak{B}$ is invariant in $\mathfrak{A}' = (\mathfrak{A} \cap \mathfrak{C})\mathfrak{B}$ and $\mathfrak{D}$ in $\mathfrak{C}' = (\mathfrak{C} \cap \mathfrak{A})\mathfrak{D}$. By 24.3.1, the decompositions $\bar{A} = \mathfrak{A}'/_l\mathfrak{B}$, $\bar{C} = \mathfrak{C}'/_l\mathfrak{D}$ are generating and, by 14.3.2, the same applies to the decompositions

$$\bar{A} \sqcap \mathfrak{C}' = (\mathfrak{A} \cap \mathfrak{C})/_l(\mathfrak{B} \cap \mathfrak{C}), \quad \bar{C} \sqcap \mathfrak{A}' = (\mathfrak{A} \cap \mathfrak{C})/_l(\mathfrak{A} \cap \mathfrak{D}).$$

From (1) we conclude that the decomposition $\bar{B} = (\mathfrak{A} \cap \mathfrak{C})/_l\mathfrak{U}$ is a common covering of $\bar{A} \sqcap \mathfrak{C}'$, $\bar{C} \sqcap \mathfrak{A}'$. Since $\mathfrak{U}$ is invariant in $\mathfrak{A} \cap \mathfrak{C}$, $\bar{B}$ is generating. Consequently, the coverings

$$\mathring{A} = (\mathfrak{A} \cap \mathfrak{C})\mathfrak{B}/_l\mathfrak{U}\mathfrak{B}, \qquad \mathring{C} = (\mathfrak{A} \cap \mathfrak{C})\mathfrak{D}/_l\mathfrak{U}\mathfrak{D}$$

of the decompositions $\bar{A}$, $\bar{C}$, enforced by $\bar{B}$, are generating (14.3.3).

On taking account of 24.3.1, we observe that $\mathfrak{U}\mathfrak{B}$ or $\mathfrak{U}\mathfrak{D}$ is invariant in $(\mathfrak{A} \cap \mathfrak{C})\mathfrak{B}$ or $(\mathfrak{A} \cap \mathfrak{C})\mathfrak{D}$, respectively.

In particular $\big($for $\mathfrak{U} = (\mathfrak{A} \cap \mathfrak{D})(\mathfrak{B} \cap \mathfrak{C})\big)$, there holds the following theorem:

Let $\mathfrak{A} \supset \mathfrak{B}$, $\mathfrak{C} \supset \mathfrak{D}$ be subgroups of $\mathfrak{G}$, $\mathfrak{B}$ and $\mathfrak{D}$ invariant in $\mathfrak{A}$ and $\mathfrak{C}$, respectively. Then $\mathfrak{A} \cap \mathfrak{D}$, $\mathfrak{B} \cap \mathfrak{C}$ are invariant in $\mathfrak{A} \cap \mathfrak{C}$. Moreover, $\mathfrak{A} \cap \mathfrak{C}$, $\mathfrak{A} \cap \mathfrak{D}$ are interchangeable with $\mathfrak{B}$ and, similarly, $\mathfrak{A} \cap \mathfrak{C}$, $\mathfrak{B} \cap \mathfrak{C}$ with $\mathfrak{D}$. The subgroup $(\mathfrak{A} \cap \mathfrak{D})\mathfrak{B}$ is invariant in $(\mathfrak{A} \cap \mathfrak{C})\mathfrak{B}$ and the same holds for $(\mathfrak{B} \cap \mathfrak{C})\mathfrak{D}$ in $(\mathfrak{A} \cap \mathfrak{C})\mathfrak{D}$. Furthermore (according to 23.2(2)), *there holds:*

$$(\mathfrak{A} \cap \mathfrak{C})\mathfrak{B} \cap (\mathfrak{B} \cap \mathfrak{C})\mathfrak{D} = (\mathfrak{A} \cap \mathfrak{D})(\mathfrak{B} \cap \mathfrak{C}) = (\mathfrak{A} \cap \mathfrak{C})\mathfrak{D} \cap (\mathfrak{A} \cap \mathfrak{D})\mathfrak{B}.$$

24.6. Series of invariant subgroups

In the classical study of groups, the theory of series of invariant subgroups of $\mathfrak{G}$ is generally based on the assumption that each member of the series, except the first, is an invariant subgroup of the element immediately preceding it. The results are of local character in the sense that they concern only situations in the neighbourhood of the unit of $\mathfrak{G}$. The following study will be restricted, for simplicity, to the special case when each member of the series is an invariant subgroup of $\mathfrak{G}$. On the ground of previous results (23.4), we may immediately proceed to the main part of the theory. Contrary to the classical theory, we shall arrive at results of global character, informative about the situation in the neighbourhood of any point of $\mathfrak{G}$.

Consider two series of subgroups of $\mathfrak{G}$, namely:

$$\big((\mathfrak{A}) =\big) \qquad \mathfrak{A}_1 \supset \cdots \supset \mathfrak{A}_\alpha \qquad (\alpha \geqq 1),$$

$$\big((\mathfrak{B}) =\big) \qquad \mathfrak{B}_1 \supset \cdots \supset \mathfrak{B}_\beta \qquad (\beta \geqq 1)$$

and suppose that all the subgroups in question are invariant in $\mathfrak{G}$.

Then the following theorem is true:

The series $(\mathfrak{A})$, $(\mathfrak{B})$ *have co-basally joint refinements* $(\mathfrak{A}_*)$, $(\mathfrak{B}_*)$ *with coinciding initial and final members, all the subgroups of these refinements being invariant in* $\mathfrak{G}$. $(\mathfrak{A}_*)$, $(\mathfrak{B}_*)$ *are given by the construction described in part* a) *of the proof in* 23.4.5.

Proof. Since the members of the series $(\mathfrak{A})$, $(\mathfrak{B})$ are invariant in $\mathfrak{G}$, the series $(\mathfrak{A})$, $(\mathfrak{B})$ are complementary (23.4.4; 24.4); we can apply to them the construction described in part a) of the proof in 23.4.5. That leads to co-basally joint refinements $(\mathfrak{A}_*)$, $(\mathfrak{B}_*)$ of $(\mathfrak{A})$, $(\mathfrak{B})$; the refinements have the same initial and final members $\mathfrak{U} = \mathfrak{A}_1\mathfrak{B}_1$ and $\mathfrak{V} = \mathfrak{A}_\alpha \cap \mathfrak{B}_\beta$, respectively. In accordance with the construction in question, $(\mathfrak{A}_*)$, $(\mathfrak{B}_*)$ consist of the following subgroups of $\mathfrak{G}$:

$$\mathfrak{A}_{\gamma,\nu} = \mathfrak{A}_\gamma(\mathfrak{A}_{\gamma-1} \cap \mathfrak{B}_\nu) = \mathfrak{A}_{\gamma-1} \cap \mathfrak{A}_\gamma\mathfrak{B}_\nu,$$

$$\mathfrak{B}_{\delta,\mu} = \mathfrak{B}_\delta(\mathfrak{B}_{\delta-1} \cap \mathfrak{A}_\mu) = \mathfrak{B}_{\delta-1} \cap \mathfrak{B}_\delta\mathfrak{A}_\mu$$

$(\gamma, \mu = 1, 2, \ldots, \alpha + 1;\ \delta, \nu = 1, 2, \ldots, \beta + 1;\ \mathfrak{A}_0 = \mathfrak{B}_0 = \mathfrak{G},\ \mathfrak{A}_{\alpha+1} = \mathfrak{B}_{\beta+1} = \mathfrak{V})$. From the results of 24.2 it is obvious that $\mathfrak{A}_{\gamma,\nu}$, $\mathfrak{B}_{\delta,\mu}$ are invariant in $\mathfrak{G}$ and the proof is complete.

24.7. Exercises

1. In the group $\mathfrak{S}_4$ consisting of all permutations of the set $\{a, b, c, d\}$, all the permutations mapping the element d onto itself form a subgroup $\mathfrak{S}_3'$. The permutations which map the elements a, b, c in the same manner as $\boldsymbol{e}, \boldsymbol{a}, \boldsymbol{b}$ in 11.4.2 without changing the element d, form a subgroup of $\mathfrak{S}_4$ which is invariant in $\mathfrak{S}_3'$ but not in $\mathfrak{S}_4$.

2. Let $\mathfrak{A}$ be a subgroup of $\mathfrak{G}$. The set of all elements $p \in \mathfrak{G}$ such that $p\mathfrak{A} = \mathfrak{A}p$ is a subgroup $\mathfrak{N}$ of $\mathfrak{G}$, the so-called *normalizer* of $\mathfrak{A}$. The latter is the greatest supergroup of $\mathfrak{A}$ in which $\mathfrak{A}$ is invariant; that is to say, $\mathfrak{A}$ is invariant in $\mathfrak{N}$ and each subgroup of $\mathfrak{G}$ in which $\mathfrak{A}$ is invariant is a subgroup of $\mathfrak{N}$.
3. The center of $\mathfrak{G}$ is an invariant subgroup of $\mathfrak{G}$.
4. If there exists, in a finite group of order N ($\geqq 2$), a subgroup of order $\frac{1}{2}N$, then the latter is invariant in the former. For example, in the diedric permutation group of order $2n$ ($n \geqq 3$) there is an invariant subgroup of order n consisting of all the elements of the group corresponding to the rotations of the vertices of a regular n-gon about its center (19.7.2).
5. Associating, with every element $p \in \mathfrak{G}$, any element $x^{-1}px \in \mathfrak{G}$ with $x \in \mathfrak{G}$ arbitrary, we obtain a symmetric congruence on $\mathfrak{G}$. The decomposition $\bar{G}$ corresponding to the latter is called the *fundamental decomposition of* $\mathfrak{G}$. The field of each invariant subgroup of $\mathfrak{G}$ is the sum of certain elements of $\bar{G}$. $\bar{G}$ is complementary to every generating decomposition of $\mathfrak{G}$.
6. Let $p \in \mathfrak{G}$ be an arbitrary point and $\mathring{\mathfrak{G}}$ the (p)-group associated with $\mathfrak{G}$ (19.7.11). Consider a subgroup $\mathfrak{A}$ invariant in $\mathfrak{G}$ and the subgroup $\mathring{\mathfrak{A}}$ of $\mathring{\mathfrak{G}}$, lying on the field $p\mathfrak{A} = \mathfrak{A}p$ (20.3.3; 21.8.7). Show that: a) $\mathring{\mathfrak{A}}$ is invariant in $\mathring{\mathfrak{G}}$; b) all generating decompositions of $\mathring{\mathfrak{G}}$ coincide with the generating decompositions of $\mathfrak{G}$.

25. Factor groups

25.1. Definition

Let us now consider a factoroid $\bar{\mathfrak{G}}$ on $\mathfrak{G}$. According to the definition of a factoroid, the field of $\bar{\mathfrak{G}}$ is a generating decomposition of $\mathfrak{G}$ and is therefore generated by a suitable subgroup $\mathfrak{A}$ invariant in $\mathfrak{G}$ (24.3.2). The product $p\mathfrak{A} \,.\, q\mathfrak{A}$ of an element $p\mathfrak{A} \in \bar{\mathfrak{G}}$ and an element $q\mathfrak{A} \in \bar{\mathfrak{G}}$ is, by the definition of multiplication in a factoroid, the element of $\bar{\mathfrak{G}}$ that contains the set $p\mathfrak{A} \,.\, q\mathfrak{A}$. Since the latter coincides, as we know, with $pq\mathfrak{A} \in \bar{\mathfrak{G}}$, the multiplication in $\bar{\mathfrak{G}}$ is given by the following formula:

$$p\mathfrak{A} \circ q\mathfrak{A} = pq\mathfrak{A}. \tag{1}$$

Now we shall show that $\bar{\mathfrak{G}}$ *is a group whose unit is the field of the invariant subgroup* $\mathfrak{A}$ *and the element inverse of an arbitrary element* $a\mathfrak{A}$ *is* $a^{-1}\mathfrak{A}$.

In fact, first, by 15.6.3, $\bar{\mathfrak{G}}$ is associative. Next, by 18.7.5, the field A of the invariant subgroup $\mathfrak{A}$ is the unit of $\bar{\mathfrak{G}}$. Finally we have:

$$p\mathfrak{A} \circ p^{-1}\mathfrak{A} = pp^{-1}\mathfrak{A} = \underline{1}\mathfrak{A} = A$$

and so $p^{-1}\mathfrak{A} \in \bar{\mathfrak{G}}$ is the inverse element of $p\mathfrak{A} \in \bar{\mathfrak{G}}$.

Every factoroid $\overline{\mathfrak{G}}$ on $\mathfrak{G}$ is therefore a group and is uniquely determined by a subgroup $\mathfrak{A}$ invariant in $\mathfrak{G}$; the field of $\overline{\mathfrak{G}}$ is the decomposition of $\mathfrak{G}$ generated by $\mathfrak{A}$. $\overline{\mathfrak{G}}$ is called a *factor group* or a *group of cosets* and is said to be *generated by the invariant subgroup* $\mathfrak{A}$; notation: $\mathfrak{G}/\mathfrak{A}$.

25.2. Factoroids on a group

From the result in 24.3.2 we have the following information about all the possible factoroids on a group $\mathfrak{G}$:

All factoroids on $\mathfrak{G}$ are precisely the factor groups on $\mathfrak{G}$ generated by the individual invariant subgroups of $\mathfrak{G}$.

Note that *the greatest (least) factoroid on $\mathfrak{G}$ is the greatest (least) factor group* $\mathfrak{G}/\mathfrak{G}$ ($\mathfrak{G}/\{\underline{1}\}$); *it is generated by the greatest (least) invariant subgroup of $\mathfrak{G}$, namely, the subgroup $\mathfrak{G}$* ($\{\underline{1}\}$).

25.3. Properties of factor groups

The properties of factor groups follow from the properties of the generating decompositions of groups (24.4).

Let $\mathfrak{G}/\mathfrak{A}$, $\mathfrak{G}/\mathfrak{B}$ be arbitrary factor groups on $\mathfrak{G}$.

$\mathfrak{G}/\mathfrak{A}$ and $\mathfrak{G}/\mathfrak{B}$ are the covering and the refinement of the factor groups $\mathfrak{G}/\mathfrak{B}$ and $\mathfrak{G}/\mathfrak{A}$, respectively, if and only if $\mathfrak{A} \supset \mathfrak{B}$.

The greatest common refinement $(\mathfrak{G}/\mathfrak{A}, \mathfrak{G}/\mathfrak{B})$ of the factor groups $\mathfrak{G}/\mathfrak{A}$, $\mathfrak{G}/\mathfrak{B}$ is the factor group $\mathfrak{G}/(\mathfrak{A} \cap \mathfrak{B})$.

The least common covering $[\mathfrak{G}/\mathfrak{A}, \mathfrak{G}/\mathfrak{B}]$ of the factor groups $\mathfrak{G}/\mathfrak{A}$, $\mathfrak{G}/\mathfrak{B}$ is the factor group $\mathfrak{G}/\mathfrak{A}\mathfrak{B}$.

$\mathfrak{G}/\mathfrak{A}$ and $\mathfrak{G}/\mathfrak{B}$ are complementary.

On every group the system of factor groups is closed with regard to the operations ( ), [ ]. *Together with the multiplications associating with each ordered pair of factor groups either their least common covering or their greatest common refinement, this system is a modular lattice with extreme elements. The latter are the greatest and the least corresponding factor groups.*

Note, in particular, that *the groups belong to the class of groupoids on which every two factoroids are complementary.*

25.4. Factor groups in groups

1. *Intersections and closures.* Let $\mathfrak{A} \supset \mathfrak{B}, \mathfrak{C}$ be subgroups of $\mathfrak{G}$ and $\mathfrak{B}$ invariant in $\mathfrak{A}$. Consider the factoroids $\mathfrak{A}/\mathfrak{B} \sqcap \mathfrak{C}$ and $\mathfrak{C} \sqsubset \mathfrak{A}/\mathfrak{B}$. From 24.5.1 we know that the subgroups $\mathfrak{A} \cap \mathfrak{C}$ and $\mathfrak{B}$ are interchangeable and that the subgroups $\mathfrak{B} \cap \mathfrak{C}$ and $\mathfrak{B}$ are invariant in $\mathfrak{A} \cap \mathfrak{C}$ and $(\mathfrak{A} \cap \mathfrak{C})\mathfrak{B}$, respectively. Moreover, the fields of the factoroids in question are given by the generating decompositions $(\mathfrak{A} \cap \mathfrak{C})/_l(\mathfrak{B} \cap \mathfrak{C})$ and $(\mathfrak{A} \cap \mathfrak{C})\mathfrak{B}/_l\mathfrak{B}$ (21.2.1).

Consequently:

$$\mathfrak{A}/\mathfrak{B} \sqcap \mathfrak{C} = (\mathfrak{A} \cap \mathfrak{C})/(\mathfrak{B} \cap \mathfrak{C}), \quad \mathfrak{C} \sqsubset \mathfrak{A}/\mathfrak{B} = (\mathfrak{C} \cap \mathfrak{A})\mathfrak{B}/\mathfrak{B} \tag{1}$$

from which we conclude that:

The factoroids $\mathfrak{A}/\mathfrak{B} \sqcap \mathfrak{C}$ and $\mathfrak{C} \sqsubset \mathfrak{A}/\mathfrak{B}$ are factor groups given by the formulae (1).

In particular (for $\mathfrak{A} = \mathfrak{G}$), we have the following theorem:

Assuming $\mathfrak{B}, \mathfrak{C}$ to be arbitrary subgroups of $\mathfrak{G}$, $\mathfrak{B}$ invariant in $\mathfrak{G}$, the factoroids $\mathfrak{G}/\mathfrak{B} \sqcap \mathfrak{C}$ and $\mathfrak{C} \sqsubset \mathfrak{G}/\mathfrak{B}$ are factor groups and there holds:

$$\mathfrak{G}/\mathfrak{B} \sqcap \mathfrak{C} = \mathfrak{C}/(\mathfrak{B} \cap \mathfrak{C}), \quad \mathfrak{C} \sqsubset \mathfrak{G}/\mathfrak{B} = \mathfrak{C}\mathfrak{B}/\mathfrak{B}.$$

2. *Special five-group theorem.* Let us return to the situation described in 24.5.2. Consider the factoroids $\mathring{\mathfrak{A}}, \mathring{\mathfrak{C}}$ (15.3.3) which are, as we know, the coverings of the following factor groups, enforced by the factor group $\overline{\mathfrak{B}} = (\mathfrak{A} \cap \mathfrak{C})/\mathfrak{U}$:

$$(\mathfrak{A} \cap \mathfrak{C})\mathfrak{B}/\mathfrak{B} \sqcap (\mathfrak{C} \cap \mathfrak{A})\mathfrak{D} = (\mathfrak{A} \cap \mathfrak{C})/(\mathfrak{C} \cap \mathfrak{B}).$$
$$(\mathfrak{C} \cap \mathfrak{A})\mathfrak{D}/\mathfrak{D} \sqcap (\mathfrak{A} \cap \mathfrak{C})\mathfrak{B} = (\mathfrak{C} \cap \mathfrak{A})/(\mathfrak{A} \cap \mathfrak{D}).$$

The fields of $\mathring{\mathfrak{A}}, \mathring{\mathfrak{C}}$ are the (generating) decompositions

$$(\mathfrak{A} \cap \mathfrak{C})\mathfrak{B}/_l\mathfrak{U}\mathfrak{B} \text{ and } (\mathfrak{C} \cap \mathfrak{A})\,\mathfrak{D}/_l\mathfrak{U}\mathfrak{D}$$

(24.5.2). Consequently:

$$\mathring{\mathfrak{A}} = (\mathfrak{A} \cap \mathfrak{C})\mathfrak{B}/\mathfrak{U}\mathfrak{B}, \quad \mathring{\mathfrak{C}} = (\mathfrak{C} \cap \mathfrak{A})\mathfrak{D}/\mathfrak{U}\mathfrak{D},$$

hence $\mathring{\mathfrak{A}}, \mathring{\mathfrak{C}}$ are factor groups given by these formulae.

From 15.3.3 we know that $\mathring{\mathfrak{A}}, \mathring{\mathfrak{C}}$ are coupled and therefore isomorphic (16.1.2).

Thus we have arrived at *the* so-called *special five-group theorem*:

Let $\mathfrak{A} \supset \mathfrak{B}, \mathfrak{C} \supset \mathfrak{D}$ be subgroups of $\mathfrak{G}$ with $\mathfrak{B}$ and $\mathfrak{D}$ invariant in $\mathfrak{A}$ and $\mathfrak{C}$, respectively. Then $\mathfrak{A} \cap \mathfrak{D}, \mathfrak{C} \cap \mathfrak{B}$ are invariant in $\mathfrak{A} \cap \mathfrak{C}$. Now let $\mathfrak{U}$ be an invariant subgroup of $\mathfrak{A} \cap \mathfrak{C}$ such that

$$\mathfrak{A} \cap \mathfrak{C} \supset \mathfrak{U} \supset (\mathfrak{A} \cap \mathfrak{D})(\mathfrak{C} \cap \mathfrak{B}).$$

Then $\mathfrak{A} \cap \mathfrak{C}$ and $\mathfrak{U}$ are interchangeable with both $\mathfrak{B}$ and $\mathfrak{D}$ and the subgroup $\mathfrak{U}\mathfrak{B}$ or $\mathfrak{U}\mathfrak{D}$ is invariant in $(\mathfrak{A} \cap \mathfrak{C})\mathfrak{B}$ or $(\mathfrak{C} \cap \mathfrak{A})\mathfrak{D}$, respectively. Moreover, the subgroups

$(\mathfrak{A} \cap \mathfrak{C})\mathfrak{B}/\mathfrak{U}\mathfrak{B}$ *and* $(\mathfrak{C} \cap \mathfrak{A})\mathfrak{D}/\mathfrak{U}\mathfrak{D}$ *are coupled, hence isomorphic, so we have:*

$$(\mathfrak{A} \cap \mathfrak{C})\mathfrak{B}/\mathfrak{U}\mathfrak{B} \simeq (\mathfrak{C} \cap \mathfrak{A})\mathfrak{D}/\mathfrak{U}\mathfrak{D}.$$

In particular $\big($for $\mathfrak{U} = (\mathfrak{A} \cap \mathfrak{D})(\mathfrak{C} \cap \mathfrak{B})\big)$, there applies *the four-group theorem* (H. Zassenhaus):

Let $\mathfrak{A} \supset \mathfrak{B}$, $\mathfrak{C} \supset \mathfrak{D}$ *be subgroups of* $\mathfrak{G}$, *with* $\mathfrak{B}$ *invariant in* $\mathfrak{A}$ *and* $\mathfrak{D}$ *in* $\mathfrak{C}$. *Then the subgroups* $\mathfrak{A} \cap \mathfrak{D}$, $\mathfrak{C} \cap \mathfrak{B}$ *are invariant in* $\mathfrak{A} \cap \mathfrak{C}$. *Moreover,* $\mathfrak{A} \cap \mathfrak{C}$ *and* $\mathfrak{A} \cap \mathfrak{D}$ *are interchangeable with* $\mathfrak{B}$ *and* $\mathfrak{C} \cap \mathfrak{A}$, $\mathfrak{C} \cap \mathfrak{B}$ *with* $\mathfrak{D}$. *The subgroup* $(\mathfrak{A} \cap \mathfrak{D})\mathfrak{B}$ *is invariant in* $(\mathfrak{A} \cap \mathfrak{C})\mathfrak{B}$ *and* $(\mathfrak{C} \cap \mathfrak{B})\mathfrak{D}$ *in* $(\mathfrak{C} \cap \mathfrak{A})\mathfrak{D}$. *The factor groups* $(\mathfrak{A} \cap \mathfrak{C})\mathfrak{B}/(\mathfrak{A} \cap \mathfrak{D})\mathfrak{B}$ *and* $(\mathfrak{C} \cap \mathfrak{A})\mathfrak{D}/(\mathfrak{C} \cap \mathfrak{B})\mathfrak{D}$ *are coupled and therefore isomorphic, so we have:*

$$(\mathfrak{A} \cap \mathfrak{C})\mathfrak{B}/(\mathfrak{A} \cap \mathfrak{D})\mathfrak{B} \cong (\mathfrak{C} \cap \mathfrak{A})\mathfrak{D}/(\mathfrak{C} \cap \mathfrak{B})\mathfrak{D}.$$

25.5. Further properties of factor groups

1. *Enforced coverings of factor groups.* Let $\mathfrak{B}$ denote an invariant subgroup of $\mathfrak{G}$ and $\mathfrak{B}_1$ an invariant subgroup of the factor group $\mathfrak{G}/\mathfrak{B}$. Thus the elements of $\mathfrak{B}_1$ are cosets with regard to $\mathfrak{B}$ and one of them is the field B of the invariant subgroup $\mathfrak{B}$. This is true because B is, as we know from 25.1, the unit of the factor group $\mathfrak{G}/\mathfrak{B}$ and is therefore an element of each subgroup of $\mathfrak{G}/\mathfrak{B}$. The sum of all elements of $\mathfrak{B}_1$ is, consequently, a certain supergroup A of B, containing the unit $\underline{1}$ of $\mathfrak{G}$, hence: $\underline{1} \in B \subset A$. The subgroup $\mathfrak{B}_1$ generates, on $\mathfrak{G}/\mathfrak{B}$, a factor group $(\mathfrak{G}/\mathfrak{B})/\mathfrak{B}_1$ and, in accordance with 15.4.1, the latter enforces a certain covering $\overline{\mathfrak{A}}$ of $\mathfrak{G}/\mathfrak{B}$. Note that $\overline{\mathfrak{A}}$ is a factoroid on $\mathfrak{G}$, each of its elements being the sum of all elements of $\mathfrak{G}/\mathfrak{B}$ that are contained in the same element of the factor group $(\mathfrak{G}/\mathfrak{B})/\mathfrak{B}_1$. In particular, the set A is an element of $\overline{\mathfrak{A}}$ and as it contains the unit $\underline{1}$ of $\mathfrak{G}$ it is, by 24.3.2, the field of an invariant subgroup $\mathfrak{A}$ of $\mathfrak{G}$; furthermore, $\overline{\mathfrak{A}}$ is the factor group $\mathfrak{G}/\mathfrak{A}$. Since $\mathfrak{B}$ is invariant in $\mathfrak{G}$, it is also invariant in $\mathfrak{A}$ and it is easy to see that $\mathfrak{B}_1 = \mathfrak{A}/\mathfrak{B}$.

The result:

The covering of the factor group $\mathfrak{G}/\mathfrak{B}$, *enforced by the factor group* $(\mathfrak{G}/\mathfrak{B})/\mathfrak{B}_1$, *is the factor group* $\mathfrak{G}/\mathfrak{A}$; *the field of* $\mathfrak{A}$ *is the sum of all the elements of* $\mathfrak{G}/\mathfrak{B}$ *that are comprised in the invariant subgroup* $\mathfrak{B}_1$ *of* $\mathfrak{G}/\mathfrak{B}$. $\mathfrak{B}_1$ *is the factor group* $\mathfrak{A}/\mathfrak{B}$.

2. *Series of factor groups.* Consider a series of factoroids $(\overline{\mathfrak{A}})$ on $\mathfrak{G}$, namely

$$\big((\overline{\mathfrak{A}}) = \big)\, \overline{\mathfrak{A}}_1 \geqq \cdots \geqq \overline{\mathfrak{A}}_\alpha \ (\alpha \geqq 1).$$

By 25.2, each member $\overline{\mathfrak{A}}_\gamma$ of this series is a factor group $\mathfrak{G}/\mathfrak{A}_\gamma$ of $\mathfrak{G}$, generated by a subgroup $\mathfrak{A}_\gamma$ invariant in $\mathfrak{G}$ $(\gamma = 1, \ldots, \alpha)$. The series $(\overline{\mathfrak{A}})$ therefore consists of the factor groups on $\mathfrak{G}$:

$$((\overline{\mathfrak{A}}) =)\ \mathfrak{G}/\mathfrak{A}_1 \geqq \cdots \geqq \mathfrak{G}/\mathfrak{A}_\alpha.$$

Note that the subgroups $\mathfrak{A}_\gamma$ generate a series $(\mathfrak{A})$ (25.3):

$$((\mathfrak{A}) =)\ \mathfrak{A}_1 \supset \cdots \supset \mathfrak{A}_\alpha.$$

$(\overline{\mathfrak{A}})$ is said to be a *series of factor groups on* $\mathfrak{G}$; notation $(\mathfrak{G}/\mathfrak{A})$.

The theory of series of factor groups on $\mathfrak{G}$ is a special case of the theory of the series of factoroids developed in Chapter 17. The novum of this case consists in the fact that in the theory of the series of factoroids certain situations have to be postulated, whereas in the theory of the series of factor groups they occur automatically. In comparison with the theory of the series of factoroids, this new theory has therefore become simpler and clearer.

Since any two series of factor groups on $\mathfrak{G}$ are complementary (25.3), there holds (17.6; 25.3) the following theorem:

Let

$$((\mathfrak{G}/\mathfrak{A}) =)\ \mathfrak{G}/\mathfrak{A}_1 \geqq \cdots \geqq \mathfrak{G}/\mathfrak{A}_\alpha,$$
$$((\mathfrak{G}/\mathfrak{B}) =)\ \mathfrak{G}/\mathfrak{B}_1 \geqq \cdots \geqq \mathfrak{G}/\mathfrak{B}_\beta$$

be series of factor groups on $\mathfrak{G}$, *of lengths* $\alpha, \beta \geqq 1$, *respectively. The series* $(\mathfrak{G}/\mathfrak{A})$ *and* $(\mathfrak{G}/\mathfrak{B})$ *have co-basally joint refinements* $(\mathfrak{G}/\mathfrak{A}_*)$, $(\mathfrak{G}/\mathfrak{B}_*)$ *with coinciding initial and final members.* $(\mathfrak{G}/\mathfrak{A}_*)$ *and* $(\mathfrak{G}/\mathfrak{B}_*)$ *are given by the construction described in* 17.6. *Their members* $\mathring{\mathfrak{A}}_{\gamma,\nu} = \mathfrak{G}/\mathfrak{A}_{\gamma,\nu}$ *and* $\dot{\mathfrak{B}}_{\delta,\mu} = \mathfrak{G}/\mathfrak{B}_{\delta,\mu}$, *respectively, are factor groups generated by the invariant subgroups*

$$\mathfrak{A}_{\gamma,\nu} = \mathfrak{A}_\gamma(\mathfrak{A}_{\gamma-1} \cap \mathfrak{B}_\nu) \quad (= \mathfrak{A}_{\gamma-1} \cap \mathfrak{A}_\gamma\mathfrak{B}_\nu)$$

and

$$\mathfrak{B}_{\delta,\mu} = \mathfrak{B}_\delta(\mathfrak{B}_{\delta-1} \cap \mathfrak{A}_\mu) \quad (= \mathfrak{B}_{\delta-1} \cap \mathfrak{B}_\delta\mathfrak{A}_\mu),$$

where $\gamma, \mu = 1, 2, \ldots, \alpha + 1$; $\delta, \nu = 1, 2, \ldots, \beta + 1$ *and, furthermore,* $\mathfrak{A}_0 = \mathfrak{B}_0 = \mathfrak{G}$, $\mathfrak{A}_{\alpha+1} = \mathfrak{B}_{\beta+1} = \mathfrak{A}_\alpha \cap \mathfrak{B}_\beta$.

25.6. Exercises

1. The order of a factor group on a finite group of order N is a divisor of N.

2. Consider the complete group $\mathfrak{G}$ of Euclidean motions on a straight line (in a plane); the subgroup of $\mathfrak{G}$, consisting of all Euclidean motions $\boldsymbol{f}[a]$ $(\boldsymbol{f}[\alpha; a, b])$ is invariant in $\mathfrak{G}$ (19.7.1). The corresponding factor group has exactly two elements; one consists of all Euclidean motions $\boldsymbol{f}[a]$ $(\boldsymbol{f}[\alpha; a, b])$, the other of $\boldsymbol{g}[a]$ $(\boldsymbol{g}[\alpha; a, b])$.

3. Let $\mathfrak{A} \supset \mathfrak{B}$, $\mathfrak{C} \supset \mathfrak{D}$ be subgroups of $\mathfrak{G}$ with $\mathfrak{B}$ and $\mathfrak{D}$ invariant in $\mathfrak{A}$ and $\mathfrak{C}$, respectively. Then the factor groups $\mathfrak{A}/\mathfrak{B}$, $\mathfrak{C}/\mathfrak{D}$ are adjoint with regard to the subgroups $\mathfrak{B}$, $\mathfrak{D}$ (15.3.4; 23.3).

4. Every two chains of factor groups in $\mathfrak{G}$, from $\mathfrak{G}$ to $\{\underline{1}\}$, have isomorphic refinements (Jordan-Hölder-Schreier's theorem) (see 16.4.4).

26. Deformations and the isomorphism theorems for groups

26.1. Deformations of groups

Let $\mathfrak{G}$, $\mathfrak{G}^*$ be groupoids and suppose there exists a deformation $\boldsymbol{d}$ of $\mathfrak{G}$ onto $\mathfrak{G}^*$. If one of these groupoids is a group, what can be said about the other?

1. *Deformation of a group onto a groupoid.* There holds the following theorem:

If $\mathfrak{G}$ is a group, then even $\mathfrak{G}^*$ is a group. Moreover, the **d**-*image of the unit of $\mathfrak{G}$ is the unit of $\mathfrak{G}^*$ and to any element $a \in \mathfrak{G}$ there applies* $\boldsymbol{d}a^{-1} = (\boldsymbol{d}a)^{-1}$.

To prove this statement, let us first note that, by 13.6.2, the groupoid $\mathfrak{G}^*$ is associative. Let $\underline{1}^*$ stand for the $\boldsymbol{d}$-image of the unit $\underline{1}$ of $\mathfrak{G}$, hence $\underline{1}^* = \boldsymbol{d}\underline{1}$. By 18.7.4, $\underline{1}^*$ is the unit of $\mathfrak{G}^*$. Let, moreover, a^* be an element of $\mathfrak{G}^*$. Since $\boldsymbol{d}$ is a mapping of $\mathfrak{G}$ *onto* $\mathfrak{G}^*$, there exists at least one element $a \in \mathfrak{G}$ such that $a^* = \boldsymbol{d}a$. The equality $aa^{-1} = \underline{1}$ yields $\boldsymbol{d}(aa^{-1}) = \boldsymbol{d}\underline{1}$, i.e., $a^*\boldsymbol{d}a^{-1} = \underline{1}^*$ and, analogously, from $a^{-1}a = \underline{1}$ we have $\boldsymbol{d}(a^{-1}a) = \boldsymbol{d}\underline{1}$, i.e., $\boldsymbol{d}a^{-1}a^* = \underline{1}^*$. Consequently, $\boldsymbol{d}a^{-1}$ is the inverse of a^*, so we have $\boldsymbol{d}a^{-1} = (\boldsymbol{d}a)^{-1}$, which completes the proof. To sum up: Every deformation maps a group again onto a group and preserves the units as well as the inverse elements in both groups.

Consequently, *if any two groupoids $\mathfrak{G}$, $\mathfrak{G}^*$ are isomorphic and one of them is a group, then the other is also a group.* Because, if $\mathfrak{G}$, $\mathfrak{G}^*$ are isomorphic, then there exists an isomorphism of $\mathfrak{G}$ onto $\mathfrak{G}^*$ and, simultaneously, an (inverse) isomorphism of $\mathfrak{G}^*$ onto $\mathfrak{G}$. Thus each of the groupoids $\mathfrak{G}$, $\mathfrak{G}^*$ is an isomorphic image of the other, and so, if one is a group, then the other is also a group. Every isomorphism, naturally, preserves in both groups the units and the inverse elements as well ast he subgroups and, as we can easily verify, the invariant subgroups.

2. *Deformation of a groupoid onto a group.* Let us now omit any further assumptions as regards the groupoid $\mathfrak{G}$ but suppose that $\mathfrak{G}^*$ is a group. By the first isomorphism theorem for groupoids, $\mathfrak{G}^*$ is isomorphic ($\boldsymbol{i}$) with a suitable factoroid $\bar{\mathfrak{G}}$ on $\mathfrak{G}$. The factoroid $\bar{\mathfrak{G}}$ corresponds to the generating decomposition belonging to the deformation $\boldsymbol{d}$ and under the isomorphism $\boldsymbol{i}$ of $\bar{\mathfrak{G}}$ onto $\mathfrak{G}^*$ each element $\bar{a} \in \bar{\mathfrak{G}}$ is mapped onto that element $a^* \in \mathfrak{G}^*$ which is the $\boldsymbol{d}$-image of the individual elements $a \in \bar{a}$. By the above result, $\bar{\mathfrak{G}}$ is a group because $\mathfrak{G}^*$ is a group. The isomorphism $\boldsymbol{i}$ preserves, in both groups, the units as well as the inverse elements; hence, under the isomorphism $\boldsymbol{i}$, the unit $\bar{\underline{1}}$ of $\bar{\mathfrak{G}}$ is mapped onto the unit $\underline{1}^*$ of $\mathfrak{G}^*$ so that $\boldsymbol{i}\bar{\underline{1}} = \underline{1}^*$ and every two inverse elements $\bar{a}$, $\bar{a}^{-1}$ of $\bar{\mathfrak{G}}$ are mapped onto two inverse elements of $\bar{\mathfrak{G}}^*$, hence $\boldsymbol{i}\bar{a} = a^*$, $\boldsymbol{i}\bar{a}^{-1} = a^{*-1}$. As each $\bar{a} \in \bar{\mathfrak{G}}$ consists of all the $\boldsymbol{d}$-inverse images of the element $a^* \in \mathfrak{G}^*$ for which $\boldsymbol{i}\bar{a} = a^*$, the unit $\bar{1}$ of the group $\bar{\mathfrak{G}}$ consists of all the $\boldsymbol{d}$-inverse images of the element $\underline{1}^*$; analogously, two

inverse elements $\bar{a}$, $\bar{a}^{-1}$ of $\overline{\mathfrak{G}}$ consist of all the $\boldsymbol{d}$-inverse images of two inverse elements a^*, a^{*-1} of $\mathfrak{G}^*$. Consequently, there applies the theorem:

If $\mathfrak{G}^*$ is a group, then the factoroid $\overline{\mathfrak{G}}$ on $\mathfrak{G}$, belonging to the deformation $\boldsymbol{d}$, is a group and is isomorphic with $\mathfrak{G}^*$. The unit of $\overline{\mathfrak{G}}$ is the set of all the $\boldsymbol{d}$-inverse images of the unit of $\mathfrak{G}^*$ and any two inverse elements of $\overline{\mathfrak{G}}$ are sets of all the $\boldsymbol{d}$-inverse images of two inverse elements of $\mathfrak{G}^*$.

Let us introduce a simple example to show that if $\mathfrak{G}^*$ is a group, then $\mathfrak{G}$ not only need not be a group but may be an arbitrary groupoid. In fact, let $\mathfrak{G}^*$ denote the group consisting of a single element $\underline{1}^*$, thus $\underline{1}^*\underline{1}^* = \underline{1}^*$, and $\mathfrak{G}$ be an arbitrary groupoid. We are to show that there exists a deformation of $\mathfrak{G}$ onto $\mathfrak{G}^*$. It is obvious that the mapping associating with each element of $\mathfrak{G}$ the element $\underline{1}^*$ is a deformation of $\mathfrak{G}$ onto $\mathfrak{G}^*$.

26.2. Cayley's theorem and the realization of abstract groups

1. *Left translations.* Let $\mathfrak{G}$ be a group and a an element of $\mathfrak{G}$. Associating with each element $x \in \mathfrak{G}$ the element $ax \in \mathfrak{G}$, we obtain a mapping of $\mathfrak{G}$ into itself. Since the equation $ax = b$, with b denoting an arbitrary element of $\mathfrak{G}$, has a unique solution $x \in \mathfrak{G}$, it is a simple mapping of $\mathfrak{G}$ onto itself, i.e., a permutation of $\mathfrak{G}$. It is called the *left translation determined by the element a* and denoted by ${}_a\boldsymbol{t}$.

The left translation determined by the element $\underline{1}$ is obviously the identical automorphism on $\mathfrak{G}$. If a, b are different elements of $\mathfrak{G}$, then both left translations ${}_a\boldsymbol{t}, {}_b\boldsymbol{t}$ are different because under ${}_a\boldsymbol{t}$ and ${}_b\boldsymbol{t}$ the element $\underline{1}$ is mapped onto a and b, respectively. Composing ${}_a\boldsymbol{t}$ and ${}_b\boldsymbol{t}$, we obviously obtain the left translation determined by ba, hence ${}_b\boldsymbol{t}{}_a\boldsymbol{t} = {}_{ba}\boldsymbol{t}$.

2. *Cayley's theorem.* Let us now consider the groupoid whose field is the set of all left translations determined by the individual elements of $\mathfrak{G}$ and the multiplication defined by the formula ${}_a\boldsymbol{t} \cdot {}_b\boldsymbol{t} = {}_{ab}\boldsymbol{t}$, with ${}_a\boldsymbol{t}$, ${}_b\boldsymbol{t}$ standing for elements of the groupoid. Let us denote it by $\mathfrak{T}_l$. Associating with each element $a \in \mathfrak{G}$ the element ${}_a\boldsymbol{t} \in \mathfrak{T}_l$, we obviously obtain a mapping of $\mathfrak{G}$ onto $\mathfrak{T}_l$; since every two different elements $a, b \in \mathfrak{G}$ are mapped onto two different elements ${}_a\boldsymbol{t}$, ${}_b\boldsymbol{t} \in \mathfrak{T}_l$, the mapping is simple. As the product ab of $a \in \mathfrak{G}$ and $b \subset \mathfrak{G}$ is mapped onto ${}_{ab}\boldsymbol{t} \in \mathfrak{T}_l$, i.e., onto the product ${}_a\boldsymbol{t} \cdot {}_b\boldsymbol{t}$ of the image ${}_a\boldsymbol{t}$ of a and the image ${}_b\boldsymbol{t}$ of b, the mapping is a deformation and therefore an isomorphism of $\mathfrak{G}$ onto $\mathfrak{T}_l$. Consequently, $\mathfrak{T}_l$ is a group and, in fact, a permutation group. Thus we have arrived at Cayley's theorem:

Every group is isomorphic with a suitable permutation group.

The importance of this result lies in the fact that, in studying the common properties of isomorphic groups, one may restrict one's attention to the permutation groups.

3. *Realization of abstract groups.* The above considerations suggest the question whether there exists, given an abstract group $\mathfrak{G}$, a permutation group apt to be deformed on it. Every permutation group of that kind is said *to realize the abstract group* $\mathfrak{G}$, and so we ask whether every abstract group can be realized by permutations.

This question can, with regard to the above results, be answered in the affirmative: every abstract group is isomorphic with the corresponding group of the left translations $\mathfrak{T}_l$; consequently, the group $\mathfrak{T}_l$ realizes $\mathfrak{G}$.

For example, let us realize the abstract group of order 4 whose multiplication table is the second in 19.6.1. The corresponding left translations determined by the individual elements are, by the mentioned table, the following permutations

$$\begin{pmatrix} \underline{1} & a & b & c \\ \underline{1} & a & b & c \end{pmatrix}, \quad \begin{pmatrix} \underline{1} & a & b & c \\ a & \underline{1} & c & b \end{pmatrix}, \quad \begin{pmatrix} \underline{1} & a & b & c \\ b & c & \underline{1} & a \end{pmatrix}, \quad \begin{pmatrix} \underline{1} & a & b & c \\ c & b & a & \underline{1} \end{pmatrix};$$

they generate, together with the multiplication $\boldsymbol{p} \,.\, \boldsymbol{q} = \boldsymbol{pq}$, $\boldsymbol{pq}$ being the composite permutation, a permutation group which realizes the group in question.

4. *Right translations.* Given an element $a \in \mathfrak{G}$ and associating with every element $x \in \mathfrak{G}$ the element $xa \in \mathfrak{G}$, we obtain a permutation of $\mathfrak{G}$, the *right translation* $\boldsymbol{t}_a$ *determined by* a.

To the right translations there apply analogous results as to the left. We leave it to the reader to verify this himself.

26.3. The isomorphism theorems for groups

In 16.1 we have discussed isomorphism theorems for groupoids and now we shall specify them for groups. Let $\mathfrak{G}$, $\mathfrak{G}^*$ be arbitrary groups.

1. *First theorem.* Suppose there exists a deformation $\boldsymbol{d}$ of $\mathfrak{G}$ onto $\mathfrak{G}^*$. As we saw in 16.1.1, the factoroid $\overline{\mathfrak{D}}$ corresponding to $\boldsymbol{d}$ is isomorphic with $\mathfrak{G}^*$. By 25.2, $\overline{\mathfrak{D}}$ is the factor group generated by a subgroup of $\mathfrak{G}$ invariant in $\mathfrak{G}$. The field of the latter is the element of $\overline{\mathfrak{D}}$, containing the unit $\underline{1}$ of $\mathfrak{G}$. Since $\underline{1}$ is a $\boldsymbol{d}$-inverse image of the unit $\underline{1}^*$ of $\mathfrak{G}^*$, it is obvious that the element of $\overline{\mathfrak{D}}$, containing $\underline{1}$, consists of all the $\boldsymbol{d}$-inverse images of $\underline{1}^*$. Consequently, the set of all the $\boldsymbol{d}$-inverse images of the unit of $\mathfrak{G}^*$ is the field of an invariant subgroup $\mathfrak{D}$ of $\mathfrak{G}$ and the factor group $\mathfrak{G}/\mathfrak{D}$ is isomorphic with $\mathfrak{G}^*$.

Now let us assume, conversely, that $\mathfrak{G}^*$ is isomorphic with the factor group $\mathfrak{G}/\mathfrak{D}$ on $\mathfrak{G}$ generated by a subgroup $\mathfrak{D}$ invariant in $\mathfrak{G}$. Then there exists an isomorphism $\boldsymbol{i}$ of $\mathfrak{G}/\mathfrak{D}$ onto $\mathfrak{G}^*$. In accordance with 16.1.1, the mapping $\boldsymbol{d}'$ of $\mathfrak{G}$ onto $\mathfrak{G}/\mathfrak{D}$ such that, for $a \in \mathfrak{G}$, $\boldsymbol{d}'a$ is the element $\bar{a} \in \mathfrak{G}/\mathfrak{D}$ containing a, is a deformation of $\mathfrak{G}$

onto $\mathfrak{G}/\mathfrak{D}$. Consequently, $\boldsymbol{d} = \boldsymbol{id}'$ is a deformation of $\mathfrak{G}$ onto $\mathfrak{G}^*$. By 25.1, the unit of the group $\mathfrak{G}/\mathfrak{D}$ is the field D of $\mathfrak{D}$. Since $\boldsymbol{i}$ maps onto the unit $\underline{1}^*$ of $\mathfrak{G}^*$ precisely the unit of $\mathfrak{G}/\mathfrak{D}$, $\boldsymbol{d}$ maps onto $\underline{1}^*$ exactly those elements of $\mathfrak{G}$ that lie in D. Hence there exists a deformation $\boldsymbol{d}$ of $\mathfrak{G}$ onto $\mathfrak{G}^*$ such that $\mathfrak{D}$ consists of all the $\boldsymbol{d}$-inverse images of the unit of $\mathfrak{G}^*$.

Summing up, we get the first isomorphism theorem for groups:

If there exists a deformation $\mathbf{d}$ *of a group* $\mathfrak{G}$ *onto a group* $\mathfrak{G}^*$*, then the set of all* $\mathbf{d}$*-inverse images of the unit of* $\mathfrak{G}^*$ *is an invariant subgroup* $\mathfrak{D}$ *of* $\mathfrak{G}$ *and the factor group on* $\mathfrak{G}$*, generated by* $\mathfrak{D}$*, is isomorphic with* $\mathfrak{G}^*$*, i.e.,* $\mathfrak{G}/\mathfrak{D} \simeq \mathfrak{G}^*$*. Conversely, if* $\mathfrak{G}^*$ *is isomorphic with the factor group on* $\mathfrak{G}$*, generated by a subgroup* $\mathfrak{D}$ *invariant in* $\mathfrak{G}$*, then there exists a deformation* $\mathbf{d}$ *of* $\mathfrak{G}$ *onto* $\mathfrak{G}^*$ *such that* $\mathfrak{D}$ *consists of all the* $\mathbf{d}$*-inverse images of the unit of* $\mathfrak{G}^*$.

2. *Second theorem:*

Let $\mathfrak{A} \supset \mathfrak{B}, \mathfrak{C} \supset \mathfrak{D}$ *be subgroups of* $\mathfrak{G}$*, with* $\mathfrak{B}$ *and* $\mathfrak{D}$ *invariant in* $\mathfrak{A}$ *and* $\mathfrak{C}$*, respectively. Moreover, let*

$$\mathfrak{A} \cap \mathfrak{D} = \mathfrak{C} \cap \mathfrak{B},$$

$$\mathfrak{A} = (\mathfrak{A} \cap \mathfrak{C})\mathfrak{B}, \quad \mathfrak{C} = (\mathfrak{C} \cap \mathfrak{A})\mathfrak{D}.$$

Then the factor groups $\mathfrak{A}/\mathfrak{B}, \mathfrak{C}/\mathfrak{D}$ *are coupled, hence isomorphic and so* $\mathfrak{A}/\mathfrak{B} \simeq \mathfrak{C}/\mathfrak{D}$*. The mapping of either of the factor groups onto the other, realized by the incidence of the elements, is an isomorphism.*

The proof of this theorem directly follows from the results in 23.1 and 16.1.2.

An important special case concerns the closure and the intersection of an arbitrary subgroup and a factor group in $\mathfrak{G}$.

Let $\mathfrak{A} \supset \mathfrak{B}, \mathfrak{C}$ be subgroups of $\mathfrak{G}$, with $\mathfrak{B}$ invariant in $\mathfrak{A}$. Then, in accordance with 24.5.1, the subgroups $\mathfrak{A} \cap \mathfrak{C}$ and $\mathfrak{B}$ are interchangeable, $\mathfrak{B} \cap \mathfrak{C}$ is invariant in $\mathfrak{A} \cap \mathfrak{C}$ and $\mathfrak{B}$ in $(\mathfrak{A} \cap \mathfrak{C})\mathfrak{B}$. Let us now apply the above theorem to the groups: $\mathfrak{A}' = (\mathfrak{A} \cap \mathfrak{C})\mathfrak{B}$, $\mathfrak{B}' = \mathfrak{B}$, $\mathfrak{C}' = \mathfrak{A} \cap \mathfrak{C}$, $\mathfrak{D}' = \mathfrak{B} \cap \mathfrak{C}$ which, as it is easy to see, satisfy the corresponding conditions. We obtain $(\mathfrak{A} \cap \mathfrak{C})\mathfrak{B}/\mathfrak{B} \simeq (\mathfrak{A} \cap \mathfrak{C})/(\mathfrak{B} \cap \mathfrak{C})$, the isomorphism being realized by the incidence of elements.

Summing up these results, we arrive at the following theorem:

Let $\mathfrak{A} \supset \mathfrak{B}, \mathfrak{C}$ *be subgroups of* $\mathfrak{G}$*, with* $\mathfrak{B}$ *invariant in* $\mathfrak{A}$*. Then* $\mathfrak{A} \cap \mathfrak{C}$ *and* $\mathfrak{B}$ *are interchangeable,* $\mathfrak{B} \cap \mathfrak{C}$ *is invariant in* $\mathfrak{A} \cap \mathfrak{C}$ *and* $\mathfrak{B}$ *in* $(\mathfrak{A} \cap \mathfrak{C})\mathfrak{B}$*. Furthermore, the factor groups* $(\mathfrak{A} \cap \mathfrak{C})\mathfrak{B}/\mathfrak{B}$ *and* $(\mathfrak{A} \cap \mathfrak{C})/(\mathfrak{B} \cap \mathfrak{C})$ *are coupled, hence isomorphic and thus:*

$$(\mathfrak{A} \cap \mathfrak{C})\mathfrak{B}/\mathfrak{B} \simeq (\mathfrak{A} \cap \mathfrak{C})/(\mathfrak{B} \cap \mathfrak{C});$$

the mapping of either of the factor groups onto the other, realized by the incidence of elements, is an isomorphism.

In particular (for $\mathfrak{A} = \mathfrak{G}$), there holds:

Let $\mathfrak{B}, \mathfrak{C}$ be subgroups of $\mathfrak{G}$, with $\mathfrak{B}$ invariant in $\mathfrak{G}$. Then $\mathfrak{B}$ and $\mathfrak{C}$ are interchangeable and the subgroup $\mathfrak{B} \cap \mathfrak{C}$ is invariant in $\mathfrak{C}$. Moreover, the factor groups $\mathfrak{C}\mathfrak{B}/\mathfrak{B}$ and $\mathfrak{C}/(\mathfrak{B} \cap \mathfrak{C})$ are coupled, hence isomorphic and thus

$$\mathfrak{C}\mathfrak{B}/\mathfrak{B} \simeq \mathfrak{C}/(\mathfrak{B} \cap \mathfrak{C});$$

the mapping of either of the factor groups onto the other, realized by the incidence of elements, is an isomorphism.

3. *Third theorem.* As we know (16.1.3), there exists a third isomorphism theorem for groupoids, concerning coverings of a factoroid.

Let $\mathfrak{B}$ denote an invariant subgroup of $\mathfrak{G}$ and $\mathfrak{B}_1$ an invariant subgroup of the factor group $\mathfrak{G}/\mathfrak{B}$. By the third isomorphism theorem for groupoids, the factor group $(\mathfrak{G}/\mathfrak{B})/\mathfrak{B}_1$ is isomorphic with the covering $\bar{\mathfrak{A}}$ of $\mathfrak{G}/\mathfrak{B}$, enforced by $(\mathfrak{G}/\mathfrak{B})/\mathfrak{B}_1$, i.e., $(\mathfrak{G}/\mathfrak{B})/\mathfrak{B}_1 \simeq \bar{\mathfrak{A}}$; the mapping associating, with every element $\bar{\bar{b}} \in (\mathfrak{G}/\mathfrak{B})\mathfrak{B}_1$, the sum $\bar{a} \in \bar{\mathfrak{A}}$ of all the elements $\bar{b} \in \mathfrak{G}/\mathfrak{B}$ lying in $\bar{\bar{b}}$ is an isomorphism. By 25.5.1, the sum of all the elements of $\mathfrak{G}/\mathfrak{B}$ lying in $\mathfrak{B}_1$ is the field of an invariant subgroup $\mathfrak{A}$ of $\mathfrak{G}$ and $\bar{\mathfrak{A}}$ is the factor group $\mathfrak{G}/\mathfrak{A}$. Moreover, we have $\mathfrak{B}_1 = \mathfrak{A}/\mathfrak{B}$.

Hence follows the third isomorphism theorem for groups:

If $\mathfrak{B}$ and $\mathfrak{B}_1$ are invariant subgroups of $\mathfrak{G}$ and $\mathfrak{G}/\mathfrak{B}$, respectively, then the sum of the elements of $\mathfrak{G}/\mathfrak{B}$ that lie in $\mathfrak{B}_1$ is the field of a subgroup $\mathfrak{A}$ invariant in $\mathfrak{G}$ and there holds:

$$(\mathfrak{G}/\mathfrak{B})/(\mathfrak{A}/\mathfrak{B}) \simeq \mathfrak{G}/\mathfrak{A},$$

the isomorphism associating, with every element $\bar{\bar{b}}$ of the factor group on the left-hand side, the sum of all the elements of the factor group $\mathfrak{G}/\mathfrak{B}$ that lie in $\bar{\bar{b}}$.

26.4. Deformations of factor groups

Let us now start from the results concerning deformations of factoroids (16.2) and consider their particular form in case of factor groups.

Let $\boldsymbol{d}$ be a deformation of a group $\mathfrak{G}$ onto a group $\mathfrak{G}^*$ so that we have $\mathfrak{G}^* = \boldsymbol{d}\mathfrak{G}$.

From 26.3.1 we know that the set of all the $\boldsymbol{d}$-inverse images of the unit of $\mathfrak{G}^*$ is an invariant subgroup $\mathfrak{D}$ of $\mathfrak{G}$ and that the factor group $\mathfrak{G}/\mathfrak{D}$ is isomorphic with $\mathfrak{G}^*$.

The deformation $\boldsymbol{d}$ determines the extended mapping $\boldsymbol{d}$ of the system of all subsets of $\mathfrak{G}$ into the system of all subsets of $\mathfrak{G}^*$; the $\boldsymbol{d}$-image of any subset $A \subset \mathfrak{G}$ is the subset $\boldsymbol{d}A \subset \mathfrak{G}^*$ consisting of the $\boldsymbol{d}$-images of the individual elements $a \in A$ (7.1).

Let $\mathfrak{G}/\mathfrak{A}$ be a factor group on $\mathfrak{G}$, generated by an invariant subgroup $\mathfrak{A}$ of $\mathfrak{G}$.

With regard to 25.3, the factor groups $\mathfrak{G}/\mathfrak{A}$, $\mathfrak{G}/\mathfrak{D}$ are complementary. Consequently, $\mathfrak{G}/\mathfrak{A}$ has, under the extended mapping $\boldsymbol{d}$, the image $\boldsymbol{d}(\mathfrak{G}/\mathfrak{A})$; the latter is a factoroid on $\mathfrak{G}^*$ (16.2.1). The partial extended mapping $\boldsymbol{d}$ of $\mathfrak{G}/\mathfrak{A}$ onto the factoroid $\boldsymbol{d}(\mathfrak{G}/\mathfrak{A})$ is a deformation called the *extended deformation* $\boldsymbol{d}$ (16.2.2).

The $\boldsymbol{d}$-image of the field A of $\mathfrak{A}$ contains the unit of $\mathfrak{G}^*$ (26.1.1). Consequently, $\boldsymbol{d}A \in \boldsymbol{d}(\mathfrak{G}/\mathfrak{A})$ is the field of a subgroup $\boldsymbol{d}\mathfrak{A}$ invariant in $\mathfrak{G}^*$ and the factoroid $\boldsymbol{d}(\mathfrak{G}/\mathfrak{A})$ is the factor group generated by the invariant subgroup $\boldsymbol{d}\mathfrak{A}$ (24.3.2), i.e., $\boldsymbol{d}(\mathfrak{G}/\mathfrak{A}) = \boldsymbol{d}\mathfrak{G}/\boldsymbol{d}\mathfrak{A}$.

The least common covering $[\mathfrak{G}/\mathfrak{A}, \mathfrak{G}/\mathfrak{D}]$ of the factor groups $\mathfrak{G}/\mathfrak{A}$, $\mathfrak{G}/\mathfrak{D}$ and the factor group $\boldsymbol{d}\mathfrak{G}/\boldsymbol{d}\mathfrak{A}$ are isomorphic; an isomorphic mapping of the factoroid $[\mathfrak{G}/\mathfrak{A}, \mathfrak{G}/\mathfrak{D}]$ onto $\boldsymbol{d}\mathfrak{G}/\boldsymbol{d}\mathfrak{A}$ is obtained by associating, with every element of the factoroid $[\mathfrak{G}/\mathfrak{A}, \mathfrak{G}/\mathfrak{D}]$, its image under the extended mapping $\boldsymbol{d}$ (16.2.3). The factoroid $[\mathfrak{G}/\mathfrak{A}, \mathfrak{G}/\mathfrak{D}]$ is the factor group $\mathfrak{G}/\mathfrak{A}\mathfrak{D}$ generated by the invariant subgroup $\mathfrak{A}\mathfrak{D}$ (25.3).

The result:

If the group $\mathfrak{G}^*$ is homomorphic ($\boldsymbol{d}$) with the group $\mathfrak{G}$, then the image of every factor group $\mathfrak{G}/\mathfrak{A}$ under the extended mapping $\boldsymbol{d}$ is the factor group $\boldsymbol{d}\mathfrak{G}/\boldsymbol{d}\mathfrak{A}$ and the partial extended mapping $\boldsymbol{d}$ of $\mathfrak{G}/\mathfrak{A}$ onto $\boldsymbol{d}\mathfrak{G}/\boldsymbol{d}\mathfrak{A}$ is a deformation. The factor groups $\mathfrak{G}/\mathfrak{A}\mathfrak{D}$, $\boldsymbol{d}\mathfrak{G}/\boldsymbol{d}\mathfrak{A}$ are isomorphic; an isomorphic mapping of $\mathfrak{G}/\mathfrak{A}\mathfrak{D}$ onto $\boldsymbol{d}\mathfrak{G}/\boldsymbol{d}\mathfrak{A}$ is obtained by associating, with each element of $\mathfrak{G}/\mathfrak{A}\mathfrak{D}$, its image under the extended mapping $\boldsymbol{d}$.

In particular, any factor group which is a covering of $\mathfrak{G}/\mathfrak{D}$ is isomorphic with its image under the extended mapping $\boldsymbol{d}$. An isomorphic mapping is obtained by associating, with each element of the covering, its image under the extended mapping $\boldsymbol{d}$.

26.5. Exercises

1. Realize, by means of permutations, the abstract group of the 4th order whose multiplication table is the first in 19.6.1.

2. Given the multiplication table of a finite group $\mathfrak{G}$, the symbols of the left translations on $\mathfrak{G}$ are obtained by copying, successively, the horizontal heading and writing one line of the table underneath. In a similar way we get, from the vertical heading and the single columns, the symbols of the right translations on $\mathfrak{G}$.

3. A regular octahedron has altogether thirteen axes of symmetry (three of them pass through two opposite vertices, six pass through the centers of two opposite edges and four through the centers of two opposite faces). The rotations of the octahedron about the axes of symmetry which leave the octahedron unaltered form a group of the 24th order, called the *octahedral group* (rotations about the same axis by angles which differ from each other by integer multiples of 360° are considered equal); let us, for the moment, denote the mentioned group by $\mathfrak{O}$. To each rotation which is an element of $\mathfrak{O}$ there corresponds a permutation of the three axes of symmetry passing through two opposite vertices. Associating with each element of $\mathfrak{O}$ the corresponding permutation, we obtain a deformation of $\mathfrak{O}$ onto the symmetric permutation group $\mathfrak{S}_3$. Employing this deformation and taking account of the first and the second isomorphism theorems for groups, prove that $\mathfrak{O}$ contains invariant subgroups of the orders 4 and 12.

27. Cyclic groups

27.1. Definition

A group $\mathfrak{G}$ is called *cyclic* if it contains an element a, called *generator of* $\mathfrak{G}$, such that each element of $\mathfrak{G}$ is a power of a. If $\mathfrak{G}$ is a cyclic group and a its generator, then $\mathfrak{G}$ is denoted by the symbol (a). From the first formula (1) in 19.3 it follows that *every cyclic group is Abelian.*

27.2. The order of a cyclic group

Consider a cyclic group (a). If the powers a^i, a^j of a with any two different exponents i, j are different, then the group (a) has the order 0 because it contains an infinite number of elements

$$\ldots, a^{-2}, a^{-1}, a^0, a^1, a^2, \ldots \tag{1}$$

As each element of (a) is a power of a, the group (a) does not include any other elements but these so that (a) consists of the elements (1). Now suppose that the powers of a with some different exponents i, j are equal and so $a^i = a^j$, $i \neq j$. Hence $a^{-j} \,.\, a^i = a^{-j} \,.\, a^j$, i.e., $a^{i-j} = \underline{1}$. Since one of the numbers $i - j$, $j - i$ is positive and the powers of a with these exponents equal $\underline{1}$, we observe that there exist positive integers x satisfying the equation $a^x = \underline{1}$. One of them is the least; let us denote it n, thus $a^n = \underline{1}$. Now consider the following elements of (a):

$$\underline{1}, a, a^2, \ldots, a^{n-1}. \tag{2}$$

First, it is easy to verify that every two of them are different: in fact, if for any of them there holds $a^i = a^j$, then one of the numbers $i - j$, $j - i$ is a positive integer smaller than n and satisfies the equation $a^x = \underline{1}$; but that contradicts the definition of n. Consequently, the group (a) comprises at least n elements (2) and has therefore the order 0 or $\geqq n$. Moreover, it is easy to show that (a) does not include any other elements, hence its order is n. To that purpose, consider an element a^x of (a). Dividing x by n, we obtain a quotient q and a remainder r whence $x = qn + r$, $0 \leqq r \leqq n - 1$; consequently, a^r is one of the elements (2). The formulae (1) in 19.3 yield

$$a^x = a^{qn+r} = a^{qn} \,.\, a^r = (a^n)^q \,.\, a^r = \underline{1}^q \,.\, a^r = \underline{1} \,.\, a^r = a^r$$

and we have $a^x = a^r$. Thus we have verified that the group (a) consists of the elements (2) and therefore has the order n. Furthermore, the product $a^i \,.\, a^j$ of an element a^i and an element a^j of (a) is the element a^k, k being the remainder of the division of $i + j$ by n because $a^i \,.\, a^j = a^{i+j}$. To sum up, we arrive at the following theorem:

The order n of every cyclic group (a) is either 0, in which case (a) consists of the elements (1), or $n > 0$, and then (a) consists of the elements (2). The product $a^i \,.\, a^j$ of the elements a^i and a^j of (a) is, in the first case, the element a^{i+j} whereas, in the second case, it is a^k, k being the remainder of the division of $i + j$ by n. In the latter, n is the least positive integer such that $a^n = \underline{1}$.

Note that in both cases a^{n-i} is the inverse of a^i.

27.3. Subgroups of cyclic groups

Let us now consider a subgroup $\mathfrak{A}$ of a cyclic group (a). If $\mathfrak{A}$ consists of a single element $\underline{1}$, then it is cyclic and its generator is $\underline{1}$. Suppose that $\mathfrak{A}$ contains besides $\underline{1}$ an element a^i where $i \neq 0$. As $\mathfrak{A}$ comprises with a^i simultaneously its inverse a^{-i} and as one of the numbers $i, -i$ is positive, we see that $\mathfrak{A}$ includes powers of a with positive exponents. One of the latter is the least; let us denote it m, hence $a^m \in \mathfrak{A}$. $\mathfrak{A}$ does not contain any powers of a with positive exponents smaller than m. Let a^x be an arbitrary element of $\mathfrak{A}$. Dividing x by m, we obtain a quotient q and a remainder r, hence $x = qm + r$, $0 \leqq r \leqq m - 1$. In accordance with the formulae (1) in 19.3, there follows: $a^x = a^{qm+r} = a^{qm} \,.\, a^r$. Consequently, a^r is the product of a^{-qm} and a^x. Since a^{-qm} is the inverse of the element $(a^m)^q$ which is, as the qth power of the element $a^m \in \mathfrak{A}$, also included in $\mathfrak{A}$, we see that a^{-qm} is an element of $\mathfrak{A}$. As even a^x is an element of $\mathfrak{A}$, the product $q^{-qm} \,.\, a^x$, namely, the element a^r is included in $\mathfrak{A}$. Consequently, with regard to the inequalities $0 \leqq r \leqq m - 1$ and to the definition of m, there follows $r = 0$. So we have $a^x = (a^m)^q$. Every element of $\mathfrak{A}$ is therefore a power of a^m, hence $\mathfrak{A}$ is cyclic with the generator a^m. Thus we have arrived at the result that *every subgroup of a cyclic group (a) is cyclic.*

Since the cyclic group (a) is Abelian, each of its subgroups is invariant in (a).

27.4. Generators

Do there exist, in the cyclic group (a), any other generators besides a? Let, again, n denote the order of (a) and suppose that some element a^ν of (a) is a generator of (a). Then, in particular, the element a is a power of a^ν, hence $a = a^{\nu q}$, q being an integer. If $n = 0$, then $a = a^{\nu q}$ yields $\nu q = 1$ because, in that case, any two powers of a with different exponents are different; hence $\nu = q = 1$ or $\nu = q = -1$. Consequently, besides a, only a^{-1} can be a generator of (a) and, in fact, each element a^i of (a) is the $-i^{\text{th}}$ power of a^{-1}.

If $n = 0$, then the group (a) has exactly two generators: a, a^{-1}. Note that they are the only two elements of (a) whose exponents are relatively prime to n $(= 0)$.

Let us now consider the case when $n > 0$. The cyclic group (a) consists of the elements $\underline{1}, a, a^2, \ldots, a^{n-1}$. If r is the remainder of the division of νq by n so that $\nu q = nq' + r$ where q' is the quotient and $0 \leqq r \leqq n - 1$, then we have $a^{\nu q} = a^r = a$. Consequently, $r = 1$ because a, a^r belong to the sequence $\underline{1}, a, a^2, \ldots, a^{n-1}$ where any two elements with different exponents are different. So we have $\nu q - nq' = 1$ and therefore ν, n are prime to each other. If, conversely, ν is an integer relatively prime to n, then there exist integers q, q' such that $\nu q - nq' = 1$ and there follows, for every integer i, the relation $i = \nu(qi) - n(q'i)$. Consequently, we have $a^i = (a^\nu)^{qi}$ and so a^ν is a generator of the group (a). If $n > 0$, then the generators of (a) are the powers of a whose exponents are relatively prime to n. We saw that the same applies even if $n = 0$ and can therefore sum up the above results in the following theorem:

The generators of the cyclic group (a) of order $n \geqq 0$ are exactly the powers of a with exponents relatively prime to n.

If $n = 0$, then (a) has precisely two generators whereas, if $n > 0$, then the number of the generators equals the number of the positive integers not greater than n and relatively prime to it.

27.5. Determination of all cyclic groups

1. An important example of a cyclic group of order 0 is the group $\mathfrak{Z}$. Evidently, $\mathfrak{Z} = (1)$. All subgroups of $\mathfrak{Z}$ consist, as we know, of all multiples of a non-negative integer n, hence they are cyclic groups (n). Let $n \geqq 0$ and consider the factor group $\mathfrak{Z}/(n)$. We know that, for $n = 0$, $\mathfrak{Z}/(n)$ consists of the sets $\bar{a}_i = \{i\}$ where $i = \ldots, -2, -1, 0, 1, 2, \ldots$, and, for $n > 0$, it consists of the elements $\bar{a}_0, \ldots, \bar{a}_{n-1}$ where $\bar{a}_j$ denotes the set of all the elements of $\mathfrak{Z}$ that differ from j only by a multiple of n; the factor group $\mathfrak{Z}/(n)$ has, in both cases, the order n. It is easy to show that the factor group $\mathfrak{Z}/(n)$ is cyclic with the generator $\bar{a}_1$. In fact, by the definition of the multiplication in $\mathfrak{Z}/(n)$, any i^{th} power of an element $\bar{a}_k \in \mathfrak{Z}/(n)$ is that element of $\mathfrak{Z}/(n)$ which contains the number ik; hence, in particular, $\bar{a}_j = \bar{a}_1{}^j$, which proves the above assertion. Thus we have simultaneously verified that there exist cyclic groups of an arbitrary order $n \geqq 0$.

Now we shall show that, conversely, every cyclic group is isomorphic with a factor group of $\mathfrak{Z}$. Consider a cyclic group (a). To each element $x \in (a)$ there exists at least one integer ξ such that $a^\xi = x$ and, of course, vice versa, for every integer ξ, a^ξ is an element of (a). Associating with each element $\xi \in \mathfrak{Z}$ the element $a^\xi \in (a)$, we obtain a mapping $\boldsymbol{d}$ of $\mathfrak{Z}$ onto (a). If ξ and η are arbitrary elements of $\mathfrak{Z}$ and $\boldsymbol{d}\xi = x$, $\boldsymbol{d}\eta = y$, then we have $x = a^\xi$, $y = a^\eta$ and therefore $xy = a^\xi a^\eta = a^{\xi+\eta}$, hence $\boldsymbol{d}(\xi + \eta) = xy = \boldsymbol{d}\xi\boldsymbol{d}\eta$. Consequently, the mapping $\boldsymbol{d}$ preserves the multiplications in both groups $\mathfrak{Z}$, (a) and therefore is a homomorphism. We observe, first, that (a) is homomorphic with $\mathfrak{Z}$. By the first isomorphism theorem for groups (26.3.1), the set of all $\boldsymbol{d}$-inverse images of the unit of (a) is an invariant subgroup $\mathfrak{A}$ of $\mathfrak{Z}$ and the factor group on $\mathfrak{Z}$, generated by $\mathfrak{A}$, is isomorphic with (a), i. e., $\mathfrak{Z}/\mathfrak{A} \simeq (a)$. Let n $(\geqq 0)$ be the order of the cyclic group (a). Then even $\mathfrak{Z}/\mathfrak{A}$ has the order n and so $\mathfrak{A}$ consists of all multiples of n. Consequently, the cyclic group (a) of order n is isomorphic with the factor group $\mathfrak{Z}/(n)$ generated by the subgroup (n) of $\mathfrak{Z}$. In particular, every cyclic group of order 0 is isomorphic with $\mathfrak{Z}/(0)$, hence even with $\mathfrak{Z}$.

It is easy to see that any group isomorphic with a cyclic group of order n $(\geqq 0)$ is also cyclic and of order n.

The result:

All cyclic groups of order $n \geqq 0$ are represented by the factor group $\mathfrak{Z}/(n)$ on $\mathfrak{Z}$ in the sense that any cyclic group of order n is isomorphic with $\mathfrak{Z}/(n)$ and, conversely, any group isomorphic with $\mathfrak{Z}/(n)$ is cyclic and of order n.

2. Example. As an example of a cyclic group of order $n > 0$ we may introduce the group consisting of the nth roots of unity with multiplication in the arithmetic sense.

The roots in question are:

$$\varepsilon_0 = 1, \quad \varepsilon_1 = e^{2\pi i/n}, \quad \varepsilon_2 = e^{4\pi i/n}, \ldots, \varepsilon_{n-1} = e^{2(n-1)\pi i/n}$$

and therefore form the cyclic group $(e^{2\pi i/n})$. The points whose coordinates are real and imaginary parts of these roots are the vertices of a regular n-gon. For $n = 6$, for example, we have the vertices of a regular hexagon. The generators of this group of order 6 are $e^{2\pi i/6}$, $e^{10\pi i/6}$.

27.6. Fermat's theorem for groups

The notion of a cyclic group is important even for groups that are not necessarily cyclic. Consider a group $\mathfrak{G}$. Let a be an arbitrary element of $\mathfrak{G}$. The individual powers of a form a cyclic subgroup (a) of $\mathfrak{G}$.

By the *order of the element a* we mean the order of the cyclic subgroup (a). The order n of a is therefore either 0 or the least positive integer x for which $a^x = \underline{1}$; in any case there holds $a^n = \underline{1}$.

Furthermore, it is easy to verify that the order n of each element $a \in \mathfrak{G}$ is a divisor of the order N of $\mathfrak{G}$, i.e., $N = nd$, d integer. For $N = 0$ this statement is obvious. In case of $N > 0$ it is true because the order of any subgroup of $\mathfrak{G}$ is a divisor of the order of $\mathfrak{G}$. From the equality $N = nd$ there follows: $a^N = a^{nd} = (a^n)^d = \underline{1}^d = \underline{1}$. Thus we have arrived at *Fermat's theorem for groups*:

The Nth power of any element of a group of order N is the unit of the group.

27.7. The generating of translations on finite groups by pure cyclic permutations

Let us conclude our study with a remark concerning the generating of, for example, the left translations of a finite group by pure cyclic permutations.

Assume $\mathfrak{G}$ to be a finite group and a an element of $\mathfrak{G}$. As we saw in 26.2.1, the left translation ${}_a\boldsymbol{t}$ of $\mathfrak{G}$ is a permutation of $\mathfrak{G}$ and is therefore generated by a finite number of pure cyclic permutations; that is to say, there exists a decomposition $\overline{G} = \{\bar{a}, \ldots, \overline{m}\}$ of $\mathfrak{G}$ such that each element $\bar{a}, \ldots, \overline{m}$ is invariant under ${}_a\boldsymbol{t}$ and the partial permutations ${}_a\boldsymbol{t}_{\bar{a}}, \ldots, {}_a\boldsymbol{t}_{\overline{m}}$ are pure cyclic permutations of the elements $\bar{a}, \ldots, \overline{m}$. Any element $\bar{x}$ of $\overline{G}$ consists of the elements of the cycle: $x, {}_a\boldsymbol{t}x, ({}_a\boldsymbol{t})^2x, \ldots, ({}_a\boldsymbol{t})^{k-1}x$, with x denoting an arbitrary element of $\bar{x}$ and k being the least positive integer such that $({}_a\boldsymbol{t})^k x = x$. Taking account of the definition of the left translation ${}_a\boldsymbol{t}$, we have

$${}_a\boldsymbol{t}x = ax, \; ({}_a\boldsymbol{t})^2x = a^2x, \ldots, ({}_a\boldsymbol{t})^{k-1}x = a^{k-1}x$$

and from $({}_a\boldsymbol{t})^k x = a^k x = x$ there follows $a^k = \underline{1}$. We observe that the cycle in question is $x, ax, a^2x, \ldots, a^{k-1}x$ and, furthermore, that the set $\{\underline{1}, a, a^2, \ldots, a^{k-1}\}$ is the field of the cyclic subgroup (a) of $\mathfrak{G}$. The element $\bar{x}$ is therefore the right coset of x with respect to (a). Consequently, $\overline{G}$ is the right decomposition of $\mathfrak{G}$ generated by (a).

To sum up:

The cycles of pure cyclic permutations generating a left translation ${}_a\boldsymbol{t}$ of a finite group $\mathfrak{G}$ consist of the same elements as the right cosets with regard to the cyclic subgroup (a) of $\mathfrak{G}$.

27.8. Exercises

1. An element $a \neq \underline{1}$ of a group $\mathfrak{G}$ has the order 2 if and only if it is inverse of itself.
2. In every finite group of an even order there exist elements of the order 2.
3. If an element a of a group $\mathfrak{G}$ is of the order n, then the order of each element of the cyclic subgroup (a) of $\mathfrak{G}$ is a divisor of n.
4. Every group whose order is a prime number is cyclic.
5. The order of each element $\bar{a}$ of any factor group on a finite group $\mathfrak{G}$ is a divisor of the order of each element of $\mathfrak{G}$ contained in $\bar{a}$. If the order of $\bar{a}$ is a power of a prime number p, then there exists in $\bar{a}$ an element a whose order is also a power of p.

BIBLIOGRAPHY

1937

1. DUBREIL, P., et M.-L. DUBREIL-JACOTIN: Propriétés algébriques des relations d'équivalence, C. R. Acad. Sci. Paris **205**, 704—706.
2. DUBREIL, P., et M.-L. DUBREIL-JACOTIN: Propriétés algébriques des relations d' équivalence. Théorèmes de Schreier et de Jordan-Hölder. C. R. Acad. Sci. Paris **305**, 1349—1351.
3. VANDIVER, H. S.: On concept of co-sets in a semigroup. Proc. Nat. Acad. Sci. USA **23**, 552—555.

1938

4. KRASSNER, M.: Une généralisation de la notion de corps, J. Math. pures appl. (9) **17**, 367—385.

1939

5. BORŮVKA, O.: Teorie grupoidů, část první (Gruppoidtheorie, I. Teil). Spisy vyd. Přírodověd. fak. Mas. univ., čís. 275, 17 p.
6. BOURBAKI, N.: Eléments de Mathématique (Livre I, Théorie des Ensembles), Paris.
7. DUBREIL, P., et M.-L. DUBREIL-JACOTIN: Théorie algébrique des relations d' équivalence, J. Math. pures appl. (9) **18**, 63—95.

1940

8. BIRKHOFF, G.: Lattice Theory, 1st ed., New York, V + 155 p.
9. DUBREIL, P., et M.-L. DUBREIL-JACOTIN: Équivalences et opérations, Ann. Univ. Lyon Sect. A (3) **3**, 7—23.

1941

10. BORŮVKA, O.: Über Ketten von Faktoroiden, Math. Ann. **118**, 41—64.
11. DUBREIL, P.: Contribution à la théorie des demi-groupes I, Mém. Acad. Sci. Inst. France **63**, Nr. 3, 1—52.

1942

12. FUNAYAMA, N.: On the congruence relations on a lattice, Proc. Imp. Acad. Tokyo **18**, 530—531.
13. FUNAYAMA, N., and T. NAKAYAMA: On the distributivity of a lattice-congruence, Proc. Imp. Acad. Tokyo **18**, 553—554.
14. ORE, O.: Theory of equivalence relations, Duke Math. J. **9**, 573—627.

1943

15. Borůvka, O.: O rozkladech množin, Rozpravy II. tř. České akademie LIII, **23**, 1–26.
16. Borůvka, O.: Über Zerlegungen von Mengen, Mitteilungen d. Tschech. Akad. Wiss. LIII, **23**, 14 p.
17. Конторович, П.: Группы с базисом расщепления. I, II, Мат. сборник **12** (54), 56–70; **19** (61) (1946), 287–305.
18. Richardson, A. R.: The class-ring in multiplicative systems, Ann. of Math. (2) **44**, 21–39.

1944

19. Borůvka, O.: Úvod do teorie grup, Praha, 80 p.

1945

20. Krishnan, V. S.: The theory of homomorphisms and congruences for partially ordered sets, Proc. Indian Acad. Sci., Sect. A, **22**, 1–19.
21. Krishnan, V. S: Homomorphisms and congruences in general algebra, Math. Student **13**, 1–9.

1946

22. Borůvka, O.: Teorie rozkladů v množině, část I, Spisy vyd. Přírodověd. fak. Mas. univ., čís. 278, 1–37.
23. Dubreil, P.: Algèbre I (Équivalences, Opérations, Groupes, Anneaux, Corps), Paris, X + 305 p.
24. Richardson, A. R.: Congruences in multiplicative systems, Proc. London Math. Soc. (2) **49**, 195–210.
25. Whitman, Ph. M.: Lattice, equivalence relations and subgroups, Bull. Amer. Math. Soc. **52**, 507–522.

1947

26. Châtelet, A.: Algèbre des relations de congruence, Ann. Sci. École Norm. Sup. (3) **64**, 339–368.
27. Châtelet, A.: Algèbre des relations de congruence, Révue Sci. **85**, 579–596.

1948

28. Balachandran, V. K.: Ideals of the distribution lattice, J. Indian Math. Soc. (N. S.) **12**, 49–56.
29. Bates, G. F., and F. Kiokemeister: A note on homomorphic mappings of quasigroups into multiplicative systems, Bull. Amer. Math. Soc. **54**, 1180–1185.
30. Birkhoff, G.: Lattice Theory, Revised Edition, New York, XIII + 283 p.
31. Croisot, R.: Une interprétation des relations d'équivalence dans un ensemble, C. R. Acad. Sci. Paris **226**, 616–617.
32. Croisot, R.: Condition suffisante pour l'égalité des longueurs de deux chaînes de mêmes extrémités dans une structure, Application aux relations d'équivalence et aux sous-groupes, C. R. Acad. Sci. Paris **226**, 767–768.
33. Kiokemeister, F.: A theory of normality for quasigroups, Amer. J. Math. **70**, 99–106.
34. Riquet, J.: Relations binaires, Bull. Soc. Math. France **76**, 114–155.

1949

35. Balachandran, V. K.: The Chinese remainder theorem for the distributive lattices, J. Indian Math. Soc. (N. S.) **13**, 76–80.

36. Foster, A. L.: On the permutational representation of general sets of permutations by partition lattices, Trans. Amer. Math. Soc. **66**, 366–388.
37. Kurepa, D.: The concept of a binary relation, Equivalence relation, Order relation, Bull. Soc. Math. Phys. Serbie **1**, Nr. 3–4, 53–58.
38. Stoll, R. R.: Equivalence relations in algebraic systems, Amer. Math. Monthly **56**, 372–377.
39. Škrášek, J.: Applications des méthodes mathématiques à la théorie des classifications, Spisy vyd. Přírodověd. fak. Mas. univ., čís. 316, 39 p.
40. Tamari, D.: Groupoïdes ordonnés, L'ordre lexicographique pondéré, C. R. Acad. Sci. Paris **228**, 1909–1911.

1950

41. Ballieu, R.: Une relation d'équivalence dans les groupoïdes et son application à une classe de demi-groupes, IIIe Congrès Nat. Sci. Bruxelles **2**, 46–50.
42. Dilworth, R. P.: A decomposition theorem for partially ordered sets, Ann. of Math (2) **51**, 161–166.
43. Dubreil-Jacotin, M.-L.: Quelques propriétés des applications multiformes, C. R. Acad. Sci. Paris **230**, 806–808.
44. Dubreil-Jacotin, M.-L.: Applications multiformes et relations d'équivalence, C. R. Acad. Sci. Paris **230**, 906–908.
45. Dubreil, P.: Relations binaires et applications, C. R. Acad. Sci. Paris **230**, 1028–1030.
46. Dubreil, P.: Comportement des relations binaires dans une application multiforme, C. R. Acad. Sci. Paris **230**, 1242–1243.
47. Dubreil, P.: Sur une classe de relations d'équivalence, Int. Congress of Math. Harvard **1**, 305.
48. Goldie, A. W.: The Jordan-Hölder Theorem for general abstract algebras, Proc. London Math. Soc. (2) **52**, 107–131.
49. Trevisan, G. A.: A proposito delle relazioni di congruenza sui quasi-gruppi, Rend. Sem. Mat. Univ. Padova **19**, 367–370.

1951

50. Dubreil, P.: Contribution à la théorie des demi-groupes II, Univ. Roma Ist. Naz. Alta Mat. Rend. Mat. e Appl. (5) **10**, 183–200.
51. Dubreil-Jacotin, M.-L.: Quelques propriétés des équivalences régulières par rapport à la multiplication et à l'union, dans un treillis à multiplication commutative avec élément unitée, C. R. Acad. Sci. Paris **232**, 287–289.
52. Dubreil-Jacotin, M.-L., et R. Croisot: Sur les congruences dans les ensembles où sont définies plusieurs opérations, C. R. Acad. Sci. Paris **233**, 1162–1164.
53. Marczewski, E.: Sur les congruences et les propriétés positives d'algebrès abstraites, Colloqu. Math. **2**, 220–228.
54. Stoll, R. R.: Homomorphisms of a semigroup onto a group, Amer. J. Math. **73**, 475–481.
55. Šik, F.: Sur les décompositions créatrices sur les quasi-groupes, Spisy vyd. Přírodověd. fak. Mas. univ., čís. 329, 169–186.
56. Teissier, M.: Sur les équivalences dans les demi-groupes, C.R.Acad. Sci. Paris **232**, 1987–1989.

1952

57. Borůvka, O.: Uvod do teorie group, 2. vyd., Praha, 154 p.
58. Dubreil-Jacotin, M.-L., et R. Croisot: Équivalences régulières dans un ensemble ordonné, Bull. Soc. Math. France **80**, 11–35.
59. Goldie, A.W.: The scope of the Jordan-Hölder theorem in abstract algebra, Proc. London Math. Soc. (3) **3**, 349–368.

60. Hashimoto, J.: Ideal theory for lattices, Math. Japonicae **2**, 149—186.
61. Мальцев, А. И.: Симметрические группоиды, Мат. сборник **31** (73), 136—151.
62. Schmidt, J.: Über die Rolle der transfiniten Schlußweisen in einer Allgemeinen Idealtheorie, Math. Nachr. **7**, 165—182.
63. Thurston, H. A.: Certain congruences on quasigroups, Proc. Amer. Math. Soc. **3**, 10—12.
64. Thurston, H. A.: Noncommuting quasigroup congruences, Proc. Amer. Math. Soc. **3**, 363—366.
65. Thurston, H. A.: Equivalences and mappings, Proc. London Math. Soc. (3) **2**, 175—182.

1953

66. Croisot, R.: Demi-groupes inversifs et demi-groupes réunions de demi-groupes simples, Ann. Sci. Ecole Norm. Sup. (3) **70**, 361—379.
67. Dubreil-Jacotin, M.-L., L. Lesieur et R. Croisot: Leçons sur la théorie des treillis des structures algébriques ordonnées et des treillis géométriques, Paris, VIII + 385 p.
68. Dubreil, P.: Contribution à la théorie des demi-groupes III, Bull. Soc. Math. France **81**, 289—306.
69. Jonsson, Bjarni: On the representation of lattices, Math. Scand. **1**, 193—206.
70. Мальцев, А. И.: Мультипликативные сравнения матриц, Доклады Акад. Наук СССР **90**, 333—335.
71. Sade, A.: Contribution à la théorie des quasi-groupes, diviseurs singuliers, C. R. Acad. Sci. Paris **237**, 372—374.
72. Thierrin, G.: Sur la caractérisation des équivalences régulières dans les demi-groupes, Acad. Roy. Belgique Bull. Cl. Sc. (5) **39**, 942—947.
73. Thierrin, G.: Sur quelques classes de demi-groupes, C. R. Acad. Sci. Paris **236**, 33—35.
74. Thierrin, G.: Sur quelques équivalences dans les demi-groupes, C .R. Acad. Sci. Paris **236**, 565—567.
75. Thierrin, G.: Quelques propriétés des équivalences réversibles généralisées dans un demi-groupe D, C. R. Acad. Sci. Paris **236**, 1399—1401.
76. Thierrin, G.: Sur une équivalence en relation avec l'équivalence réversible généraliseé, C. R. Acad. Sci. Paris **236**, 1723—1725.
77. Trevisan, G.: Construzione di quasigruppi con relazioni di congruenza non permutabili, Rend. Sem. Mat. Univ. Padova **22**, 11—22.
78. Воробьев, Н. Н.: О конгруенциях алгебр, Доклады Акад. Наук СССР **93**, 607 to 608.
79. Wang, S. Ch.: Notes on the permutability of congruence relations, Acta Math. Sinica **3**, 133—141.

1954

80. Croisot, R.: Automorphismes intérieurs d'un semi-groupe, Bull. Soc. Math. France **83**, 161—194.
81. Dubreil, P.: Les relations d'équivalence et leur principales applications. Les conférences du Palais de la Découverte, Série A, Nr. 194, Univ. de Paris, Paris, 22 p.
82. Dubreil, P.: Algèbre I (Équivalences, Opérations, Groupes, Anneaux, Corps), 2ème éd., Paris, 467 p.
83. Fujiwara, T.: On the structure of algebraic systems, Proc. Japan Acad. **30**, 74—79.
84. Jakubík, J., a M. Kolibiar: О некоторых свойствах пар структур, Чехослов. мат. журнал **4** (79), 1—27.
85. Jakubík, J.: Системы отношений конгруентности в структурах, Чехослов. мат. журнал **4** (79), 248—273.
86. Jakubík, J.: О отношениях конгруентности на абстрактных алгебрах, Чехослов. мат. журнал **4** (79), 314—317.

87. Kolibiar, M.: Poznámka k representácii sväzu pomocou rozkladov množiny, Mat. fys. čas. Slov. akad. vied **4**, 79—80.
88. Pierce, R. S.: Homomorphisms of semi-groups, Ann. of Math. **59**, 287—291.
89. Мальцев, А. И.: К общей теории алгебраических систем, Мат. сборник **35** (77), 3—20.
90. Numakura, K.: A note on the structure of commutative semi-groups, Proc. Japan Acad. **20**, 262—265.
91. Šik, F.: Über Charakterisierung kommutativer Zerlegungen, Spisy Přírodověd. fak. Mas. univ., čís. 354, 6 p.
92. Šik, F.: Über abgeschlossene Kongruenzen auf Quasigruppen, Spisy Přírodověd. fak. Mas. univ., čís. 354, 10 p.
93. Tamura, T., and N. Kimura: On decompositions of a commutative semi-group, Kôdai Math. Sem. Rep. **6**, 109—112.
94. Thierrin, G.: Sur la caractérisation des groupes par leurs équivalences régulières, C. R. Acad. Sci. Paris **238**, 1954—1956.
95. Thierrin, G.: Sur la caractérisation des groupes par leurs équivalences simplifiables, C. R. Acad. Sci. Paris **238**, 2046—2048.
96. Thurston, H. A.: Congruences on a distributive lattice, Proc. Edinburgh Math. Soc. (2) **10**, 76—77.

1955

97. Fujiwara, T.: Remarks on the Jordan-Hölder-Schreier theorem, Proc. Japan. Acad. **31**, 137—140.
98. Jakubík, J.: Relácie kongruentnosti a slabá projektivnost vo sväzoch, Čas. pěst. mat. **80**, 206—216.
99. Thierrin, G.: Contribution à la théorie des équivalences dans les demi-groupes (Thèse), Paris.
100. Thierrin, G.: Demi-groupes inversés et rectangulaires, Acad. Roy. Belg. Bull. Cl. Sci. (5) **41**, 83—92.
101. Thierrin, G.: Contribution à la théorie des équivalences dans les demi-groupes, Bull. Soc. Math. France **83**, 103—159.
102. Yamada, M.: On the greatest semi-lattice decomposition of a semi-group, Kôdai Math. Sem. Rep. **7**, 59—62.

1956

103. Cowell, W. R.: Concerning a class of permutable congruence relations on loops, Proc. Amer. Math. Soc. **7**, 583—588.
104. Dilworth, R. P.: Homomorphisms of distributive lattices, Bull. Amer. Math. Soc. **62**, 550.
105. Edmondson, D. E.: Modular lattices, Bull. Amer. Math. Soc. **62**, 349.
106. Jakubík, J.: O existenčných algebrách. Čas. pěst. mat. **81**, 43—54.
107. Kolibiar, M.: O kongruenciách na distributívnych sväzoch, Acta Sci. Math. Com. **1**, 247—253.
108. Nicolescu, M.: The notion of equivalence and its importance in mathematics, Gaz. Mat. Fiz. (A) **8**, 337—345.
109. Schwarz, Š.: O pologrupách splňujúcich zoslabené pravidlá krátenia, Mat.-fyz. čas. Slov. Akad. vied **6**, 149—158.
110. Thierrin, G.: Sur quelques décompositions des groupoïdes, C. R. Acad. Sci. Paris **242**, 596—598.
111. Thierrin, G.: Sur la théorie des demi-groupes, Comment. math. Helvet. **30**, 211—223.

1957

112. Benedicty, M.: Alcune applicazioni della nozione di insiemi quoziente, Archimede **9**, 1–5.
113. Croisot, R.: Equivalences principales bilatères définies dans un demi-groupe, J. Math. pures appl. (9) **36**, 373–417.
114. Dwinger, Ph.: Complete homomorphisms of complete lattices. Bull. Amer. Math. Soc. **63**, 266.
115. Dwinger, Ph.: Complete homomorphisms of complete lattices, Indagationes Math. **19**, 412–420.
116. Hashimoto, J.: Direct, subdirect decompositions and congruence relations, Osaka Math. J. **9**, 87–112.
117. Ono, K.: On some properties of binary relations, Nagoya Math. J. **12**, 160–161.
118. Schneider, M.: Bemerkungen über Kongruenzrelationen in Quasigruppen, Nachr. der Österreich. Math. Ges., Nr. 47/48, 26–27.
119. Steiner, H. G.: Einführung in die Relationentheorie, Math.-Phys. Semesterber. **5**, 261–271.

1958

120. Brunovský, P.: O zevšeobecnených algebraických systémoch, Acta Fak. Sci. mat. Univ. Comm. **3**, 41–45.
121. Grätzer, G., and E. T. Schmidt: Ideals and congruence relations in lattices, Acta Math. Acad. Sci. Hung. **9**, 137–175.
122. Grätzer, G., and E. T. Schmidt: Two notes on lattice congruences, Ann. Univ. Sci. Budapest, Sect. Math., **1**, 83–87.
123. Grätzer, G., and E. T. Schmidt: On ideal theory for lattices, Acta Sci. Math. Szeged **19**, 82–92.
124. Jakubík, J.: O zamenitelných kongruenciách ve svätoch, Mat.-fyz. čas. Slov. Akad. vied. 8, 155–162.
125. Thurston, H. A.: Derived operations and congruences. Proc. Math. Soc. **8**, 127–134.
126. Wagner, K.: Verbandstheoretische Charakterisierung der Cantorschen Äquivalenzrelation, Math. Ann. **134**, 295–297.

1959

127. Preston, G. B.: Congruences on Brandt semi-groups, Math. Ann. **139**, 91–94.

Books

1. Александров, П. С.: Введение в теорию групп, Москва 1951. (*English translation*: Indroduction to the theory of groups, London and Glasgow 1959; *German translation*: Einführung in die Gruppentheorie, 8. Aufl., Berlin 1973.)
2. Borůvka, O.: Úvod do teorie grup, 2. vyd., Praha 1952.
3. Borůvka, O.: Grundlagen der Gruppoid- und Gruppentheorie, Berlin 1960. (*Czech version*: Základy teorie grupoidů a grup, Praha 1962.)
4. Bruck, R. H.: A survey of binary systems, Berlin 1958.
5. Burnside, W.: Theory of groups of finite order, 2nd ed., Cambridge 1911, New York 1955.
6. Carmichael, R. D.: Introduction to the theory of groups of finite order, New York 1956.
7. Dubreil, P.: Algèbre I, 2ème éd., Paris 1954.
8. Fuchs, L.: Abelian groups, 3rd ed. Budapest 1966.

9. Hall Jr. M.: The theory of groups, New York 1959.
10. Jacobson, N.: Lectures in abstract algebra I, II, Toronto—New York—London 1951, 1953.
11. Kaplansky, I.: Infinite abelian groups, Ann Arbor 1954.
12. Курош, А. Г.: Теория групп, 3-ое изд., Москва 1967. (*German translation*: Gruppentheorie. I, II, Berlin 1970, 1972.)
13. Курош, А. Г.: Теория групп, 2-ое перераб. изд., Москва 1953. (*English translation*: The theory of groups I, New York 1955.)
14. Ledermann, W.: Introduction to the theory of finite groups, Edinbourgh—London—New York 1949.
15. Ляпин, Е. С.: Полугруппы, Москва 1960.
16. Miller, G. A., H. F. Blichfeldt and L. E. Dickson: Theory and applications of finite groups, New York—London 1916.
17. Scorza, G.: Gruppi astratti, Roma 1942.
18. Specht, W.: Gruppentheorie, Berlin—Göttingen—Heidelberg 1956.
19. Speiser, A.: Theorie der Gruppen von endlicher Ordnung, 3. Aufl., Berlin 1937.
20. Шмидт, О. Ю.: Абстрактная теория групп, Москва 1933.
21. van der Waerden, B. L.: Algebra I, II, 8. bzw. 5. Aufl., Berlin—Heidelberg—New York 1971 bzw. 1967.
22. Zappa, G.: Gruppi, corpi, equazioni, 2a, ed., Napoli 1954.
23. Zassenhaus, H.: Lehrbuch der Gruppentheorie, I, Leipzig—Berlin 1937. (*English translation*: The theory of groups, New York 1949.)

INDEX

Printed in the German Democratic Republic